Novel Approaches in Metronomic Chemotherapy for Breast Cancer Treatment

This book provides the latest research and developments in the field of metronomic chemotherapy for breast cancer. It presents the principles and mechanisms of metronomic chemotherapy, preclinical and clinical studies, and the latest developments in drug delivery systems and nanoformulations. The clinical pharmacology of metronomic chemotherapy, including pharmacokinetics, pharmacogenetics, pharmacoeconomics, and adverse drug reactions, are also examined.

Key Features:

- Introduces metronomics therapy in the neoadjuvant and adjuvant treatment of breast cancer
- Explores the potential of metronomics in terms of personalized chemotherapy
- Presents pharmacological bases of metronomic chemotherapy
- Covers the latest developments in drug delivery systems, nanotechnology, and nanoformulations
- Discusses antiangiogenic effects and the impact of metronomics on immunity

This book is useful for students, researchers, oncologists, pharmacologists, and healthcare experts interested in understanding the clinical potential of metronomic chemotherapy in breast cancer.

Novel Approaches in Metronomic Chemotherapy for Breast Cancer Treatment

Edited by
Manzoor Ahmad Mir
University of Kashmir, Srinagar, India

CRC Press
Taylor & Francis Group
Boca Raton London New York

CRC Press is an imprint of the
Taylor & Francis Group, an **informa** business

Cover image: Shutterstock

First edition published 2024
by CRC Press
2385 NW Executive Center Drive, Suite 320, Boca Raton FL 33431

and by CRC Press
4 Park Square, Milton Park, Abingdon, Oxon, OX14 4RN

CRC Press is an imprint of Taylor & Francis Group, LLC

ISBN: 9781032620459 (hbk)
ISBN: 9781032632650 (pbk)
ISBN: 9781032632674 (ebk)

DOI: 10.1201/9781032632674

Typeset in Times
by KnowledgeWorks Global Ltd.

This book is dedicated to my family

Ayzel Manzoor Mir

Aariz Manzoor Mir

Sumaira Manzoor Mir

Contents

Foreword

I am pleased to provide the foreword to Dr. Manzoor Ahmad Mir's valuable book, *Novel Approaches in Metronomic Chemotherapy for Breast Cancer Treatment.*

Breast cancer is a major public health concern worldwide, affecting millions of women every year. According to the World Health Organization (WHO), worldwide, cancer is currently the prominent source of death, with a clear-cut tendency toward increasing prevalence and a growing social burden over the next 30 years. According to an estimate, by 2040 more than 29 million people are expected to be diagnosed with some type of cancer. Major problems in breast cancer treatment are limited effectiveness, side effects, and increasing treatment resistance, which emphasize the need for novel therapeutic approaches to deal with this terrible illness. Although the standard chemotherapy regimens have improved survival rates, they have many drawbacks such as toxicity and development of drug resistance. Metronomic chemotherapy, a novel approach, has emerged as an alternative treatment option for breast cancer patients. It involves the administration of low doses of chemotherapeutic agents at regular intervals without extended drug-free periods.

Thus the main aim of *Novel Approaches in Metronomic Chemotherapy for Breast Cancer* is to provide an overview of the latest research and developments in the field of metronomic chemotherapy for breast cancer. The book also explores various novel approaches to metronomic chemotherapy, including the use of drug combinations and the development of new drugs. The scope of the book includes a comprehensive analysis of the current state of the art in metronomic chemotherapy for breast cancer and a discussion of the challenges and opportunities in this field. The book also aims to provide insights into the future directions of research in this area, including new drug combinations, biomarkers for patient selection, and personalized approaches to treatment.

Prof. Raid Saleem Albaradie
College of Applied Medical Sciences
University of Majmaah
Kingdom of Saudi Arabia

Preface

INTRODUCTION TO THE BOOK

Chemoresistance and toxic side effects are significant obstacles in conventional chemotherapy. Conventional chemotherapy drugs have various dosing schedules, toxicity profiles, and antitumor effects from molecularly targeted treatments. While traditional chemotherapy has several advantages, it also has many disadvantages, such as the requirement for high doses, poor absorption, side effects, minimal therapeutic indices, non-specific targeting, and the formation of multiple drug resistance (MDR). Even modern drug delivery techniques have unfavorable therapeutic effects that restrict the dosage and reduce the potency of anticancer pharmaceutics. Therefore, improving cancer treatment requires identifying new therapeutic targets as well as new treatment modalities. To overcome the shortcomings of conventional chemotherapy, novel treatment techniques called "metronomic chemotherapy" (MCT) have been developed. The benefits of this treatment strategy include minimal side effects and a low risk of developing acquired drug resistance. This strategy targets activated endothelial cells in cancers by often providing traditional chemotherapeutic medications at extremely low dosages. It is a brand-new and encouraging approach for treating solid and hematologic cancers and entails the repeated administration of chemotherapeutic pharmaceutics at low doses that maintain long-lasting, active plasma levels and produce favorable tolerability in addition to its direct impact on tumor cells. The main methods of its activity are inhibiting tumor angiogenesis and modifying the immunological response of the host, directly affecting tumor cells, their progenitors, and surrounding stromal cells. Due to these factors, MCT can be viewed as a multitarget therapy in and of itself. The key advantages metronomic chemotherapy has over conventional medicines are its reduced cost, easier use, and improved toxicity profile. Growing data supports the use of MCT in personalized medicine, beginning with unfit elderly people and also for palliative care. This book provides a brief review of metronomic chemotherapy as a better substitute for conventional chemotherapy.

Here in this book, we provide a comprehensive overview of the current knowledge and recent advances in MCT for breast cancer treatment. It covers a wide range of topics, including the principles and mechanisms of MCT, preclinical and clinical studies, and the latest developments in drug delivery systems and nanoformulations. Further, we discuss the lessons learned from the preclinical and clinal trials experienced with low-dose MCT for the treatment of solid tumors.

Target Audience

The book is designed in such a way so as to target a huge number of audiences, including the following:

- Graduate and postgraduate students of applied medical fields such as nursing, oncology, biotechnology, biochemistry, immunology, medical laboratories, biomedical lab technology, physiotherapy, etc.
- Postgraduate and graduate students pursuing normal graduations in life sciences with one of the subjects as cancer biology and medicine.
- Researchers and medicos in the field of cancer biology and molecular medicine.
- Health professionals related to pharmacy, drug design, and oncology.
- Paramedical staff involved in oncology and chemotherapy.
- Staff working in pharma companies related to oncology and cancer research.

Importance and Highlights of the Book

- The book is most important in terms of the significance of MCT used in breast cancer.
- It will be the first book to highlight metronomics in breast cancer.
- The book will highlight up-to-date information on various parameters of metronomics in breast cancer. Last but not least, it also throws light on the preclinical and clinical responses of breast cancer patients to metronomics and has illustrations, flowcharts, and diagrams highlighting the role of metronomics in breast cancer.
- It will also highlight the metronomics in the neoadjuvant and adjuvant treatment of breast cancer.
- It will discuss metronomics in terms of personalized chemotherapy.
- It will also shed light on nanoformulation-based metronomic chemotherapy.

Acknowledgements

I would like to take this opportunity to pen down my appreciation for a great number of people whose contribution in numerous ways immensely helped me to move toward my destination. It gives me immense pleasure to express my sense of gratitude and respect to my mentor, Dr. Javed Naim Agrewala, whose constant support, utmost patience, invaluable advice, and guidance rekindled my interest in science from time to time. A special vote of thanks goes to Prof. Neolofar Khan for her efficient guidance to work hard with sincerity, dedication, and devotion both in professional and personal life. The staff, scholars (especially Miss Shazia, Miss Nusrat, Miss Hina Qayoom, Mr Burhan Ul Haq, and Miss Aisha Shabir), and students at the Department of Bioresources deserve a special mention for all the help rendered at the hour of need. I am also thankful to my soulmate, Miss Sumaira Manzoor, and my parents, Mohammad Abdullah Mir, Mohd Ashraf Malik, Zona Begum, and Posha Begum for their constant support, love, and care throughout the whole journey. I would like to mention here the affection and love showered by my angels, Ayzel Manzoor and Aariz Manzoor, throughout the writing process of the book. Finally, the author would like to thank and acknowledge the Jammu and Kashmir Science Technology and Innovation Council (JKST&IC) for providing financial support vide Project No. JKST&IC/SRE/885-87, dated 27-10-2021 sanctioned to Dr Manzoor Ahmad Mir (Principal Investigator), and this book as part of the same project entitled "Targeting Drug-Resistant Cancer Stem Cells by Combinational Delivery of Paclitaxel and Quercetin Zerumbone in Breast Cancer."

About the Editor

Dr. Manzoor Ahmad Mir is a gold medalist and holds his master's degree in life sciences from HNBG Central University and PhD in immunology (life sciences) from Jawaharlal Nehru University, New Delhi. He has qualified the CSIR-JRF-NET exam thrice and worked as JRF at CSIR-IMTECH from 2004–2006 and as SRF from 2006–2009. He started his teaching job as assistant professor from 2009 at the J&K higher education department and worked as nodal officer and coordinator of Kupwara University Campus for 2011–2013. Dr. Manzoor started working as assistant professor in the department of bioresources and has been heading the department since 2019. He has 15 years of outstanding teaching and research experience. He was recently named among the top 2% globally recognized Scientists of the World for the year 2023 published by Stanford University. He has published more than 60 highly impactful scientific research papers in reputed peer reviewed international journals with a cumulative impact factor of 368 and more than 2000 citations. He has also published 15 international books with Web of Science publishers like Elsevier, Springer-Nature, Academic Press, NOVA Biomedical and OMICS international. Additionally he has 56 book chapters to his credit with the publishers of international repute. Dr. Manzoor has developed more than 200 MOOC modules of 4 quadrant in immunology, endocrinology and human genetics for SWAYAM MHRD Govt. of India for undergraduate students. He has been awarded Summer Research Fellowship by Indian National Science Academies for year 2019, Teacher Associate for Research Excellence award by the SERB-DST Govt. of India in 2020. Dr. Manzoor has attended many international conferences at USA, UK, Kuwait, China, UAE, Scotland and Saudi Arabia. His basic research interests include molecular immunology and cancer biology especially drug repurposing for breast cancer. Dr. Manzoor is on the editorial board of prestigious journals and is a member of many scientific organizations and societies including the American Association of Cancer Research, Fellow of Royal Society of Biology, American Society of Oncology, European Medical Oncology Society, International Immunology Association, Indian Association of Immunology, CHASCON, Indian National Science Association, and Immuno-oncology society of India.

List of Contributors

Sania Bashir
Department of Bioresources
School of Biological Sciences
University of Kashmir
Srinagar, India

Mary Chatterjee
Department of Biotechnology
University Institute of Engineering and Technology
Punjab University
Chandigarh, India

Shristi Handa
Department of Biotechnology
University Institute of Engineering and Technology
Punjab University
Chandigarh, India

Kirti Hooda
Centre for Nanoscience and Nanotechnology
Punjab University
Chandigarh, India

Asma Jan
Department of Bioresources
School of Biological Sciences
University of Kashmir
Srinagar, India

Nusrat Jan
Department of Bioresources
School of Biological Sciences
University of Kashmir
Srinagar, India

Tabish Javed
Department of Bioresources
School of Biological Sciences
University of Kashmir
Srinagar, India

Fatima Manhas
Department of Bioresources
School of Biological Sciences
University of Kashmir
Srinagar, India

Manzoor Ahmad Mir
Department of Bioresources
School of Biological Sciences
University of Kashmir
Srinagar, India

Hina Qayoom
Department of Bioresources
School of Biological Sciences
University of Kashmir
Srinagar, India

Nasreen Rashid
Department of Bioresources
School of Biological Sciences
University of Kashmir
Srinagar, India

Aisha Shabir
Department of Bioresources
School of Biological Sciences
University of Kashmir
Srinagar, India

Shazia Sofi
Department of Bioresources
School of Biological Sciences
University of Kashmir
Srinagar, India

Burhan Ul Haq
Department of Bioresources
School of Biological Sciences
University of Kashmir
Srinagar, India

1 Introduction to Metronomic Chemotherapy

Manzoor Ahmad Mir, Nusrat Jan, and Shazia Sofi

1.1 INTRODUCTION

Cancer is a group of complicated illnesses, and millions of individuals are affected by some type of cancer globally. Following heart disease as the second biggest cause of mortality in recent years, cancer has emerged as one of the most significant health concerns (1, 2). According to an estimate, more than 29 million individuals are expected to receive a cancer diagnosis in 2040, making cancer one of the major causes of mortality globally. Breast cancer (BC), which has a high mortality rate, is the most common kind of carcinoma among females globally. This high death rate is mostly due to the emergence of medication resistance, one of the biggest obstacles to be overcome in the treatment of BC. As a result, research has been concentrated on discovering new therapeutical tools, particularly those that enable personalized therapy based on the features of patients. While ensuring the quality of life (QOL) for cancer patients has been a priority for the scientific community, researchers are also mindful of the rising costs. There are some drawbacks to developing new pharmaceuticals since it is a lengthy, expensive process that involves clinical studies that frequently fail in the early stages (3). Uncontrolled cell proliferation that results in tumor development is a hallmark of cancer (4). It can happen due to somatic mutations in cells' upstream signaling pathways or genetic abnormalities in any gene that codes for cell cycle proteins. Even if more traditional cancer treatment options are available, millions of patients still die from cancer each year (5). Many conventional therapies are unsuccessful against many malignant tumors because of the metastasis, recurrence, heterogeneity, and resistance of cancer cells to chemotherapy and radiation (6). Another claimed explanation for the therapy's failure is that tumor cells can escape the immune response. Conventional cancer therapy approaches have many shortcomings that are frequently insufficient for achieving positive results. Chemotherapy is thought to be the therapeutic option employed the most frequently for various cancer therapies. However, it has been documented in a number of publications that standard chemotherapeutic drugs have a non-specific distribution constraint that results in low bioavailability, quick blood clearance, and relatively poor solubility in body fluids (7, 8). Any cancer therapy agent's ultimate objective is to destroy tumor cells in a targeted manner while minimizing damage to the surrounding healthy cells and tissue, which is still a goal that has not been attained.

A novel treatment approach known as "metronomic chemotherapy" (MCT) has been developed to address the limitations of traditional chemotherapy. This therapeutic strategy addresses the persistent, evenly spaced administration of several chemotherapy medicines in low strengths without longer rest intervals. The uniqueness of this treatment approach also lies in a change in the cell target, now concentrating on tumor-endothelial cells, in addition to its success against tumors with exceptionally low toxicity. A shift in the creation of treatment recommendations against cancer will be made possible by the skills and expertise gained in the preclinical MCT settings, together with the improved experience gained in clinical practice (9, 10).

1.2 METRONOMIC CHEMOTHERAPY

MCT involves administering low doses of chemotherapeutics continuously, resulting in sustained and active plasma levels of medication and improved tolerability. It is a novel and promising therapeutic strategy for treating solid and blood cancers. In addition to its direct impact on tumor cells,

DOI: 10.1201/9781032632674-1

MCT also affects the cell microenvironment by preventing tumor angiogenesis or enhancing the response of the immune system. MCT for these reasons is a candidate for multitargeted treatment in and of itself (11). In terms of delivery schedules, toxicity profiles, and ultimately the antitumor effect, molecular targeted drugs are distinct from conventional chemotherapeutic drugs (12). In the 2000s, Kerbel and Hanahan originally used the term "MCT" to describe the frequent administration of traditional chemotherapeutic medications at low dosages with no extended drug-free pauses. The main mechanism of action does not directly inhibit tumor growth. In addition to stimulating the host's response to the immune system against tumors and acting on stromal tissue, MCT mostly affects the tumor cells indirectly, particularly in their microenvironment by preventing tumor angiogenesis (13).

Although MCT was initially limited to palliative settings for a considerable amount of time since it was mistakenly believed to be devoid of anticancer activity, preliminary results on MCT in a variety of malignancy types, primarily BC and lung cancer, showed fascinating and encouraging outcomes. Oral medications are the ones that are most frequently utilized in a metronomic regimen since they can be taken for longer periods than conventional chemotherapy. The usage context, the kind of cancer, and patient preferences can all influence how different metronomic medication doses and regimens operate. The most extensively researched medications for MCT in this context include cyclophosphamide (CTX), methotrexate (MTX), capecitabine (CAPE), and oral vinorelbine (VNR) for BC; CPB for gastrointestinal malignancies; and oral VNR for non–small cell lung cancer (NSCLC). With the introduction of new medications, particularly immune checkpoint inhibitors, and a better understanding of the specific modes of action of MCT, it is hoped that these regimens will be employed more frequently in clinical settings (11).

MCT is a method of administering cytotoxic drugs that is distinct from traditional chemotherapy regimens. Conventional therapy relies on the administration of therapy given at a maximum dosage along with chemotherapy regimens, whereas MCT entails administering chemotherapeutic drugs continuously or often at low doses that are noticeably below the maximum tolerated dose (MTD), without lengthy pauses in between doses. MCT was once thought to have angiogenesis inhibition as its mode of action. However, it is now generally acknowledged that MCT has a variety of modes of action, including antiangiogenic, antiproliferative, and immunomodulatory properties (Figure 1.1). This alternative method of treatment may raise

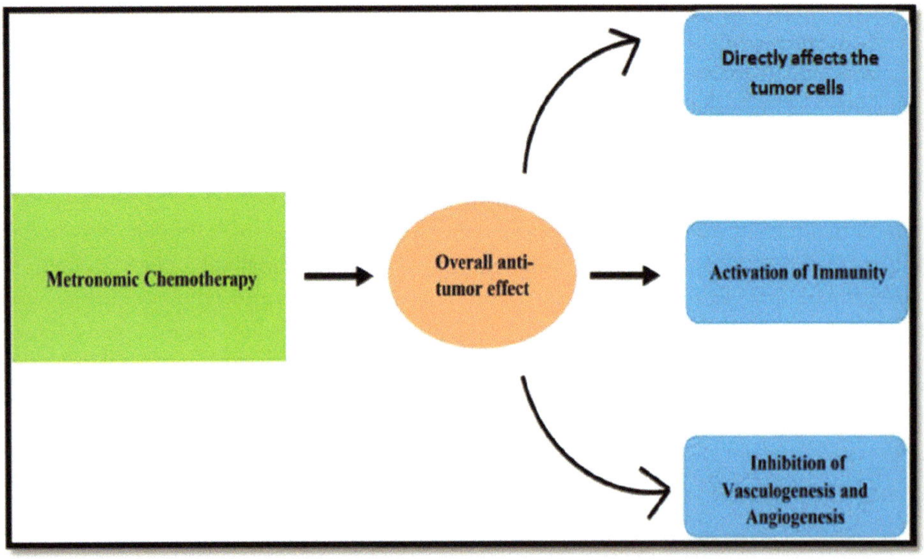

FIGURE 1.1 MCT mechanism of action.

the therapeutic index of medications since it provides a better balance between activity and side effects of the treatment, allowing for a longer course of therapy and enhancing survival. Because MCT requires regular medication delivery, oral medicines are a more practical choice for individuals (14).

MCT is a treatment with many targets, as it affects cancer cells and their surrounding tissue both directly and indirectly. In addition to inducing tumor dormancy, it can decrease tumor angiogenesis and boost the capability of the immune system to fight against cancer (15). MCT principally exhibits its anticancer properties by delaying tumor angiogenesis. CTX and taxanes are two anticancer medicines with considerable in vitro potential against angiogenesis that have been shown to benefit from metronomic drug administration. Additionally, it has been demonstrated after in vivo that MCT possesses an antiangiogenic mode of action that selectively prevents the proliferation of stimulated endothelial cells and inhibits the migration of endothelial cells. The expression of thrombospondin-1, the innermost barrier of angiogenesis, is increased, and the numbers and viability of bone marrow–derived endothelial progenitor cells continue to drop with time (16).

The growth and control of cancer are significantly influenced by both inherent and acquired immune systems. Neutropenia and lymphopenia are the most frequent side effects of chemotherapy when it comes to the immune system. However, extensive research has demonstrated that only a few anticancer medications, including taxanes, anthracyclines, and CPM, have immune-triggering properties (17). This is because metronomic therapies are somewhat compatible with the effect on modulatory T cells. Additionally, T cells, which are CD+, CD25+, Foxp3+ lymphocytes, can also suppress the immune system's response to an antigen in two ways: By secreting cytokines or by interacting with other cells. Inhibiting the activity of both tumor-unspecific effector cells such as natural killer cells (NK) and NK T cells and tumor-specific effector cells (CD8+ cytotoxic T lymphocytes and CD4+ T helper cells) allows T cells to delay the response of the immune system to the cancer. T cells have also been found in high amounts for various human malignancies, which correspond to the growth of tumors and a poor response to therapy (18). To boost the immune system's response to antigens linked with the tumor, it is therefore necessary to reduce or specifically halt the function of T cells. Additionally, the effect of low-dose CTX on T cells has been demonstrated in numerous experimental and clinical trials. It reduces the percentage of T cells, impairs their functionality, and speeds up lymphocyte and memory T-cell growth (19). Additionally, this treatment promotes the maturation of dendritic cells (DCs), which stimulates the immune system. According to Tanaka and colleagues, MCT schedules of various chemotherapeutic agents, including vinblastine, paclitaxel, and etoposide, might stimulate maturation of DCs at nontoxic doses (20).

Tumor dormancy can result from a cell cycle disruption or from a dynamically balanced scenario where the start of apoptosis balances the growth of the cell. During the early stages of tumor growth and following the completion of anticancer therapy during the remission period, tumors might enter a dormant state where they can resume their progression from lingering residual illness (21). Three key mechanisms through which MCT encourages tumor dormancy are immune surveillance, angiogenesis suppression, and programmed cell death of malignant cells (22) (Figure 1.2).

1.3 DRUGS USED IN MCT

The typical MCT treatment often consists of a variety of medications from diverse classes having properties such as stimulation of the immune system, inhibition of angiogenesis, and apoptotic effects. Frequent and recurrent low-dose treatment of various anticancer medicines like CTX, MTX, etoposide, vinblastine, and paclitaxel causes cytotoxicity in both CECs and circulating endothelial progenitors (CEPs) but has no impact on leukocytes and nonendothelial cells (23). Endothelial cells are genetically more stable than tumor cells, which makes them more resilient to the emergence of treatment resistance (24). In some circumstances, switching to a metronomic schedule has been shown to significantly suppress tumor development when standard dosages of chemotherapy medicines are no longer effective (Figure 1.3).

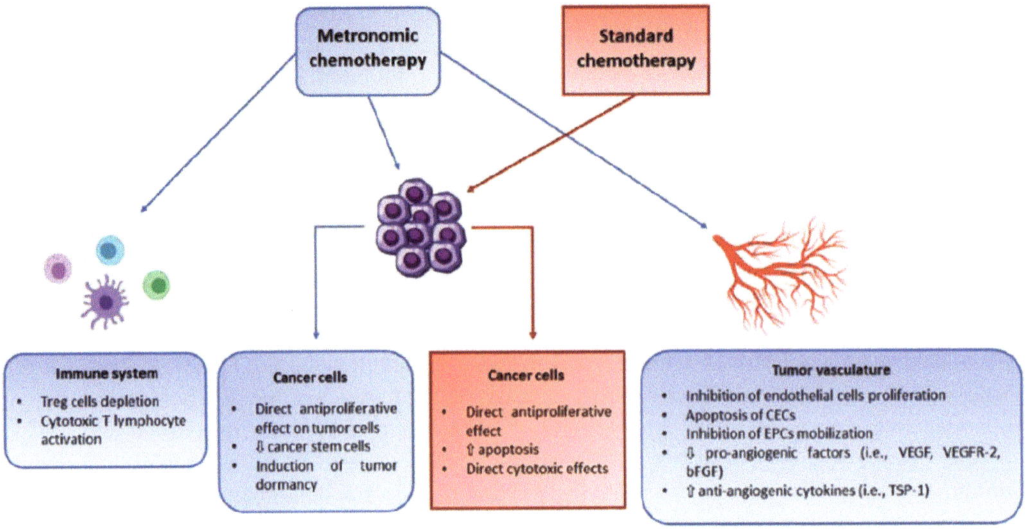

FIGURE 1.2 Main differences between the mechanisms of action of metronomic and conventional chemotherapy. Basic fibroblast growth factor (bFGF), endothelial progenitor cells (EPCs), circulating endothelial cells (CECs), thrombospondin-1 (TSP-1), regulatory T cells (Treg), and vascular endothelial growth factor (VEGF) are some of the other terms used.

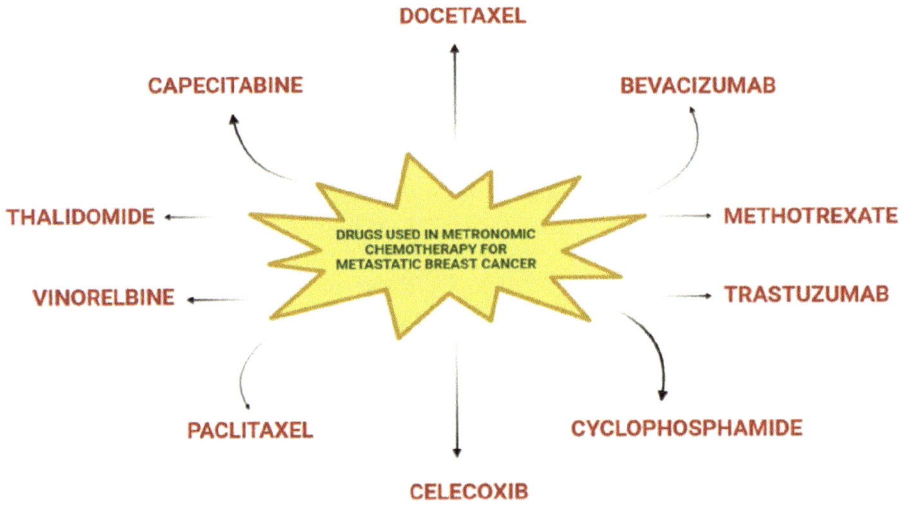

FIGURE 1.3 Drugs used in metronomic chemotherapy.

1.3.1 Metronomic Chemotherapy in Adults

Adult individuals with BC have been the subject of the majority of clinical studies on metronomic treatment (25, 26). Numerous researchers have utilized different MCT schedules for individuals with advanced and recurring ovarian cancer, advanced multiple myeloma, hormone-resistant prostate cancer, non-Hodgkin lymphoma, and other cancers (27). Table 1.1 lists a few metronomic schedules that have been thoroughly investigated for adults. The majority of these studies revealed a moderate response rate to MCT as well as overall clinical benefit (Figure 1.4).

TABLE 1.1
Metronomic Schedules Investigated for Adults

Patient Profile	Metronomic Chemotherapy Regimen	Reference
MBC earlier treated with traditional chemotherapeutics	• CTX was given orally every day • MTX was given orally two times a week	(25)
Without treatment or earlier treated BC with traditional chemotherapeutics	• CTX was given orally every day • MTX was given orally two times a week or CTX was given orally every day • MTX was given orally two times a week • TD was given orally every day	(28)
HER2+ MBC treated earlier with trastuzumab and conventional chemotherapeutics	• CTX was given orally every day • MTX was given orally two times a week • TZM was given every 3 weeks	(26)
Recurrent ovarian cancer treated earlier with conventional chemotherapeutics	• CTX was given orally every day • BZM was given every 2 weeks	(27)
Hormone-refractory prostate cancer treated earlier with androgen deprivation	• CTX was given orally every day • DXM was given orally every day or • DXM was given orally every day • Celecoxib was given orally two times a day	(29)
Aggressive relapsed or refractory non-Hodgkin lymphoma	• CTX was given orally every day • Celecoxib was given orally two times a day	(30)
Progressive multiple myeloma treated earlier with conventional chemotherapeutics	• CTX was given orally every day • Prednisone was given orally every day	(31)
Metastatic or locally advanced neuroendocrine cancer	• 5-Fluorouracil (IV daily) • Long-acting release octreotide (monthly)	(32)

Abbreviations: BC, breast cancer; BZM, bevacizumab; CTX, cyclophosphamide; DXM, dexamethasone; MBC, metastatic breast cancer; MTX, methotrexate; TD, thalidomide; TZM, trastuzumab.

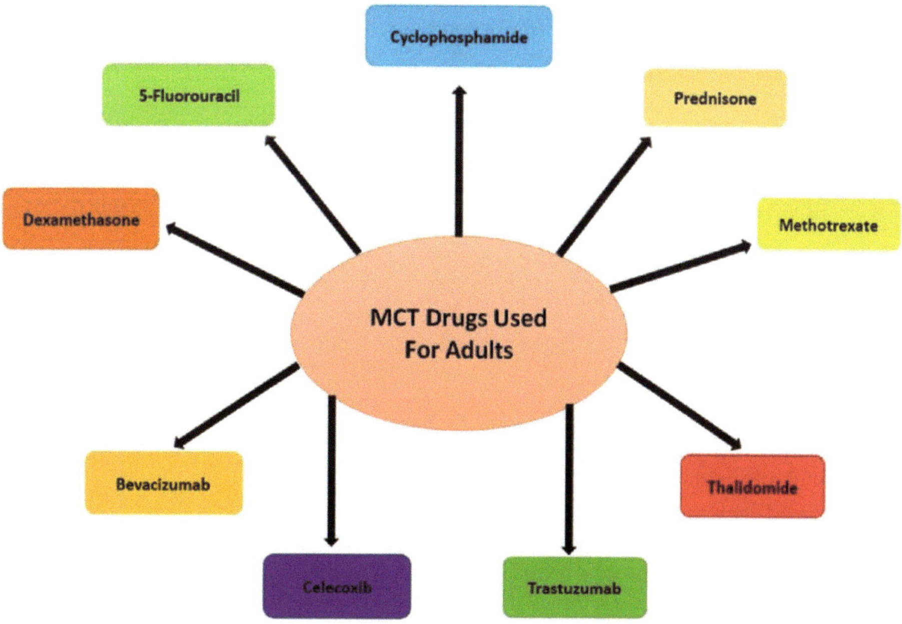

FIGURE 1.4 Metronomic schedules investigated for adults.

TABLE 1.2
MCT in Pediatric Patients

Patient Profile	Metronomic Chemotherapy Regimen	Reference
Refractory, relapsing, or "high risk of relapse" cancers of various types	• ETP was given orally every day for 3 weeks, alternating with TZM given orally every day for 6 weeks • Celecoxib was given orally every day • RA given orally every day for 2 weeks	(34)
Refractory or relapsing cancers of various types	• ETP was given orally every day for 3 weeks alternating with CTX given orally every day for 3 weeks • Thalidomide was given orally every day • Celecoxib was given orally every day	(35)
Refractory relapsing cancers of various types	• VIN (IV three times a week) • Celecoxib was given orally every day or CTX was given orally every day • Celecoxib was given orally every day	(36)

Abbreviations: CTX, cyclophosphamide; ETP, etoposide; RA, retinoic acid; TZM, temozolomide; VIN, vinblastine.

1.3.2 Metronomic Chemotherapy in Pediatric Patients

By decreasing the burden of side effects, MCT has the potential to increase the survival and QOL of pediatric cancer individuals. Although there have only been a small number of clinical studies with MCT in children, some of the outcomes have been quite encouraging (Table 1.2). Sterba and colleagues showed the benefit of metronomic TZM combined with radiation in kids with medulloblastoma in a research study that was published in 2002. Later, they carried out a pilot research study in kids with relapsed solid tumors using a four-drug MCT schedule (33). In 2006, Sterba and colleagues examined metronomic etoposide and metronomic temozolomide using a combination of oral maintenance bio-differentiating and antiangiogenic treatment approaches (34). In a different trial, thalidomide and celecoxib were used as part of MCT regimen that alternated between etoposide and CPM to treat and assess twenty children with recurrent or progressing malignancy of different tumors (35) (Figure 1.5).

1.4 TOXICITY OF MCT

MCT is generally well tolerated, according to the majority of clinical research. Further, MCT may also be an affordable therapeutic option, as shown by the results of many clinical trials (35). There were either few or no effects of high-grade toxicity. The most frequent harmful side effects were low-grade tiredness, mild nausea and/or vomiting, mild to moderate anemia, neutropenia, leukopenia, and lymphopenia (37). MCT delivers a considerable clinical benefit overall, including a better QOL with little harm. Data, however, are currently insufficient to draw firm conclusions about the tolerance of various drug combinations.

Several possible hazards and issues need to be taken into account while taking MCT for a prolonged duration, particularly in kids. Prolonged MCT may lead to antineoplastic drug toxicity over time, which may result in secondary illnesses. A cumulative dosage of etoposide or temozolomide, for instance, can result in secondary leukemia (38, 39). Even though tumor neoangiogenesis differs from physiologic angiogenesis, young children undergoing MCT require extra care because of their rapid growth and development. Recent research suggests that MCT and subdural hematoma in a child with medulloblastoma may be related (40). Therefore, any future research must carefully follow children undergoing MCT treatment.

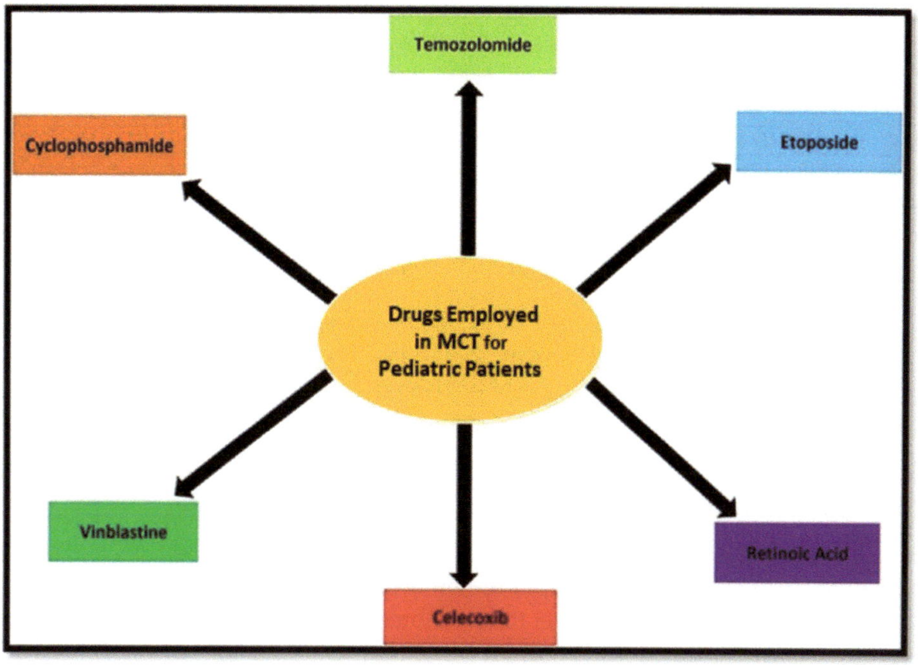

FIGURE 1.5 MCT in pediatric patients.

1.5 OPTIMAL BIOLOGICAL DOSE (OBD) AND BIOMARKERS

A potential method of treating cancer is low-dose MCT, confirming benefits including good action against resistant tumors, minimal toxicity, enhanced effectiveness, and the capacity to mix chemotherapeutic drugs. The empiricism involved in determining the ideal "low dose" or OBD and in accessing the therapeutic effectiveness throughout therapy, however, is a significant drawback of this form of regimen, since antiangiogenesis is regarded to be the main mechanism for this type of chemotherapy. Targeting both proliferating endothelial cells in a tumor's expanding vasculature and circulating vascular endothelial growth factor (VEGFR) 2+ cells originating from bone marrow, including CEPs, is assumed to be the primary method for MCT. To determine the OBD of various metronomic schedules, Yuval Shaked and colleagues have looked into some cellular pharmacodynamic biomarkers. These include:

- Prior findings demonstrating significant and long-lasting decreases in circulating VEGFR 2+ CEPs due to extended administration of low-dose MCT every day.
- Preclinical evidence of using these cells' concentrations as a substitute blood-based indicator of angiogenesis and the effectiveness of certain antiangiogenic medications, including the OBD.
- CEPs potentially serving as an indicator of targeted antiangiogenic medication effectiveness.

Additionally, they discovered a link between each OBD and the greatest decline in VEGFR 2+ CEPs in viable peripheral blood. According to these findings, CEPs may serve as a stand-in biomarker to identify the OBD of MCT schedules and can lower the amount of empiricism (41).

1.6 RESISTANCE

The emergence of resistance against the treatment schedule used is one of the major challenges in chemotherapy. Given that endothelial cells were predicted to have a stable genetic makeup, an antiangiogenic treatment schedule targeting these cells is unlikely to result in acquired drug resistance. Several experimental and clinical investigations that reveal the formation of escape mechanisms against antiangiogenic therapy regimens support the concept that two mechanisms of unconventional resistance, evasive resistance and intrinsic or preexisting tolerance, might arise. Evasive resistance is a strategy for avoiding a particular antiangiogenic action.

Resistance that is both evasive and intrinsic may arise in response to various treatments for various tumor types. MCT-resistant cells go through a complicated molecular evolution. The development of a resistant phenotype is influenced by changes in a wide variety of variables and processes. "Reduced vascular dependence" is a significant category of resistance to antiangiogenic therapy. This resistance mechanism is in action because the resistant tumor develops in the absence of new tumor vasculature, including the presence of hypoxia and nutritional restriction. Blood arteries are damaged, and coagulation occurs during the antiangiogenic treatment. The cells' anticoagulation abilities may thus be a part of the complicated mechanism of resistance. Therefore, MCT can potentially result in resistance, necessitating an evaluation of its occurrence and significance (42, 43).

1.7 PHARMACOGENETICS

Pharmacogenetic information about MCT is currently insufficient. The optimal metronomic medication may be chosen using gene expression profiling and in-depth gene expression analysis of the resistant cancers. The gene expression of resistant tumors varies from that of nonresistant tumors, and genome-wide microarray studies are needed to study the molecular basis of the in vivo resistance mechanism. According to a few conducted research studies, resistance-related genes express differently in vivo than they do in vitro, pointing to the role of microenvironmental variables in the development of the observed in vivo resistance (43).

In individuals with advanced castration-resistant prostate malignancy undergoing treatment with metronomic celecoxib, dexamethasone, and CTX, Orlandi and colleagues examined the relationship between the variations in VEGF-A sequence and prostate-specific antigen advancement, progression-free survival and overall survival (44). Shaughnessy and colleagues evaluated the pharmacogenomics of bortezomib (45). Mercurio and colleagues measured the expression of multiple genes (*ICAM1*, *CRK*, *CD36*, and *IQGAP1*) in pilocytic astrocytomas and glioblastomas in a different investigation using reverse transcriptase quantitative polymerase chain reaction. These genes were all shown to be considerably overexpressed in hypothalamic-chiasmatic pilocytic astrocytomas compared to cerebellar pilocytic astrocytomas, whereas *CRK* and *ICAM1* were found to be significantly overexpressed in pilocytic astrocytomas compared to glioblastomas. The treatment of refractory multifocal pilocytic astrocytoma with fluvastatin/celecoxib for 18 months was effective because it targets these overexpressed genes. This novel approach shows promise in the management of refractory pilocytic astrocytoma (46). Kubisch and colleagues studied the effects of metronomic antiangiogenic therapy on xenografted prostate cancer tumors in mice with severe combined immunodeficiency using CTX. When comparing resistant to in vivo passaged tumor samples, they conducted an extensive gene expression analysis of the resistant tumors in vivo to investigate the complex molecular changes occurring during resistance formation. They found a wide range of differentially expressed genes, including neurotensin, annexin A3, and plasminogen activator tissue. Such research can investigate novel MCT molecular targets and shed light on the mechanisms behind resistance (43).

1.8 CLINICAL IMPLICATIONS AND FUTURE PERSPECTIVES

Conventional chemotherapy schedules have many significant limitations, and the metronomic administration of cancer chemotherapy medicines shows great potential in addressing these issues. Among these is the development of drug resistance, suppressing the antitumor immunity, toxicity, and low QOL while undergoing treatment. Additionally, MCT costs less for the patient than targeted treatments while still being effective (47). Unfortunately, MCT may occasionally be less effective for advanced cancer due to the delay between an antitumor action and a noticeable decrease in tumor size. For instance, even a small amount of advancement in the treatment of brainstem gliomas might be fatal to the patient, necessitating more aggressive action such as fast tumor mass reduction by surgery or radiation. Similarly, lymphoblastic leukemia treatment regimens call for a time of intense induction, then a period of light dosage consolidation, and then 2–3 years of lower-dosage, higher-frequency maintenance therapy (48). Any attempts to exclude the maintenance therapy result in worse outcomes, and this technique provides survival rates of 90–95% (49–51). The treatment of a few kids with high-dose chemotherapy up-front and metronomic maintenance therapy was reported to have shown positive outcomes by Choi and colleagues (52). Eight of the ten patients in that research, including six with metastatic illness, had stable clinical and radiological conditions 20 months after their diagnoses. Additionally, encouraging outcomes were noted when metronomic treatment was used for kids with medulloblastoma, with over 65% survival rates after 24 months, necessitating additional research (50). A rapid antitumor response that results in a return of the illness, although the long-standing aim of conventional chemotherapy, does not enhance patient outcomes. While the antineoplastic responses to MCT may be delayed, for example, because of the time needed to destroy tumor blood vessels or activate an antineoplastic immune response, the antitumor response is more likely to be sustained (24, 53), thanks to the decreased selection of resistant tumor cell clones and the suppression of antineoplastic immunity with a lower likelihood of disease relapse.

Many of the drugs currently used in conventional chemotherapy schedules for BC, such as VNR, CTX, MTX, and fluoropyrimidines, have been investigated in the context of metronomic schedules, frequently in combination with other drugs like hormones, targeted drugs (such as thalidomide [TZM] or bevacizumab [BZM]), or vaccines (54, 55). The different studies conducted give data on the clinical effectiveness of MCT in metastatic breast cancer (MBC), all of which have distinct designs (56). A special type of chemotherapy drug called oral vinorelbine targets microtubules. Inhibition of angiogenesis, suppression of CEPs, and inhibition of the HIF-1 pathway are just a few of the particular actions that these drugs can do (41, 57).

These qualities make vinorelbine one of the most promising drugs to be examined in MCT schedules, combined with the ability for oral administration and proven effectiveness in other solid tumors such as lung, BC, and prostate. In comparison to intravenous bolus administration, oral chemotherapy has advantages such as prolonged plasma drug concentration or an expanded therapeutic window, sustained plasma drug concentration below the MTD, fewer side effects, an improvement in patient QOL, and lower medical expenses (58–61). However, to determine the best use of MCT and to pinpoint the patients most likely to benefit from this approach, more research is required (56).

MCT will need to be used more widely as a primary up-front treatment modality, which will necessitate better means to assess therapeutic efficacy, such as the discovery of biomarkers that may be employed to assess therapeutic success. Cancer antigens CA 15-3 and CA 19-9, prostate-specific antigen, platelet biomarkers (62–66), serum VEGF and other angiogenic cytokines (67), plasma levels of platelet-derived growth factor (PDGF)-BB (68), thrombospondin-1 expression (69), VEGF single-nucleotide polymorphisms (SNPs) (44), CD133 gene expression (70), and various immune response genes (98) are examples of existing biomarkers. The ratios of angiogenesis regulators are also being used to track tumor response (71).

To compare the relevance of these indicators to the outdated paradigm of gauging the efficacy of MTD chemotherapy by the degree of tumor shrinking, more research is still required. Although early evidence suggests this may be the case, it is still unknown whether MCT will eventually be more successful than MTD-based regimens in the treatment of metastatic illness (8, 25, 28, 72–74). MCT will be easier to apply in the clinic with the development and greater usage of computer models to help with patient-specific optimal doses and timing regimens (75). The previously mentioned set of experimental and clinical data, along with the theoretical considerations, indicate MCT as the optimum treatment option.

The results of experimental and clinical studies have established MCT as a unique treatment approach to prevent many different types of cancers. Even after the completion of clinical assessment for more than 10 years, optimizing metronomic anticancer therapy remains difficult. Therefore, soon, cancer research should concentrate on identifying the most effective therapeutic agents to use for the treatment of tumors before deciding whether to apply each medication alone or in combination, as well as how to interpret the scheduling of drug delivery. In addition, innovative medical oncology strategies are being developed to combine MCT with conventional cancer treatment, targeted therapy, and/or radiation. Such techniques can increase a lot of possibilities for powerful combinations. In the meanwhile, a thorough analysis of tumor-endothelial cells' pharmacoproteomic and pharmacogenetic properties is now necessary to determine how susceptible they are to MCT and to choose the most advantageous medication combination to apply (15). On the other hand, even in the most developed nations, the rising cost of cancer treatment is a significant social and political issue. The majority of cancer occurrences and deaths occur in low- or middle-income nations (LMINs).

The approach to many current therapeutic options is completely out of the ordinary in LMINs, where, as mentioned, the majority of cancer occurrences and fatalities occur. Due to their expense, fatality, and complex structure and technology needed, the standards of treatment in industrialized countries are beyond what LMINs can afford. According to earlier research that has been published, metronomics may be an additional positive method for the enhancement of cancer management in LMINs. Due to its acceptable toxicity profile and reasonable cost, this treatment method can be more advantageous even in a maintenance scenario when continued administration is required (76).

Additionally, metronomic therapeutic techniques were appropriately explored for replacing, enhancing, or supplementing conventional regimens in some tumors about two decades ago. Until recently, phase I and phase II experimental research has dominated published research. Phase III trials are required to begin to study the impact of MCT on cancer care. It is possible to show that, for the most part, those data are reasonably matching conventional regimens, but they do not hypothesize the clinical results. As opposed to dismissing metronomic treatment, continual assessments and their gradual integration into our most recent practice are currently the more acceptable methods (76, 77).

1.9 CONCLUSION

MCT has traditionally only been used in the palliative setting; however, given the importance of preclinical and clinical data, it is time to expand its application to other cancer types, including those with early-onset BC and colon cancer, specific populations of NSCLC patients, and pediatric cancers. Patients in resource-limited nations may benefit greatly from oral regimens that can be delivered at home and have a highly favorable toxicity profile.

For further reading, some textbooks mentioned here also give a further understanding of MCT:

- *Progress in Anti-cancer Chemotherapy* by Gabriel N. Hortobagyi and David Khayat.
- *Triple-Negative Breast Cancer: Biomarkers, Emerging Therapeutic Strategies and Clinical Challenges* by Marion Montgomery.
- *Current Cancer Treatment: Novel Beyond Conventional Approaches* by Oner Ozdemir.
- *Breast Cancer: Current and Alternative Therapeutic Modalities* by Esra Gunduz and Mehmet Gunduz.

The following links for video lectures may also be helpful:

https://youtu.be/TAqO0Mq19JQ
https://youtu.be/6wEnYfvoBqk
https://youtu.be/A9fRRCUlMms
https://youtu.be/zIWvmCsKr34

For more insight about the topic, we would suggest detailed findings from the books of (Mir MA, 2022) https://doi.org/10.1016/C2021-0-02565-7, https://doi.org/10.1016/C2022-0-00074-X, and (Mir MA, 2021) https://doi.org/10.52305/WXJL6770, (Mir MA, 2015) https://doi.org/10.1016/C2014-0-02898-5 and from the cancer.net website on the following links:

https://www.cancer.net/cancer-types/breast-cancer/types-treatment
https://discovery.ucl.ac.uk/id/eprint/1472740/
https://www.jmedsciences.com/doi/JMEDS/pdf/10.5005/jp-journals-10045-00138

For diagrammatic illustrations and descriptive tables, see (Lazzeroni, 2012): http://www.eurekaselect.com/article/49928

For video links on the overall status of cancer, its various types, and current new treatment possible options available, see: https://www.sciencedirect.com/science/article/pii/S205970292032278X

REFERENCES

1. Siegel RL, Miller KD, Fedewa SA, Ahnen DJ, Meester RGS, Barzi A, et al. Colorectal cancer statistics, 2017. CA: A Cancer Journal for Clinicians. 2017;67(3):177–93.
2. Yang X, Cao W, Wang X, Zhang X, Zhang W, Li Z, et al. Down-regulation of 14-3-3zeta reduces proliferation and increases apoptosis in human glioblastoma. Cancer Gene Therapy. 2020;27(6):399–411.
3. Rodrigues R, Duarte D, Vale N. Drug repurposing in cancer therapy: Influence of patient's genetic background in breast cancer treatment. International Journal of Molecular Sciences. 2022;23(8):4280.
4. Shen Z, Song J, Yung BC, Zhou Z, Wu A, Chen X. Emerging strategies of cancer therapy based on ferroptosis. Advanced Materials. 2018;30(12):1704007.
5. Torre LA, Bray F, Siegel RL, Ferlay J, Lortet-Tieulent J, Jemal A. Global cancer statistics, 2012. CA: A Cancer Journal for Clinicians. 2015;65(2):87–108.
6. Sun Y. Translational horizons in the tumor microenvironment: Harnessing breakthroughs and targeting cures. Medicinal Research Reviews. 2015;35(2):408–36. doi: 10.1002/med.21338
7. Cho K, Wang XU, Nie S, Chen Z, Shin DM. Therapeutic nanoparticles for drug delivery in cancer. Clinical Cancer Research. 2008;14(5):1310–6.
8. Manso L, Valdiviezo N, Sepúlveda J, Ciruelos E, Mendiola C, Ghanem I, et al. Safety and efficacy of metronomic non-pegylated liposomal encapsulated doxorubicin in heavily pretreated advanced breast cancer patients. Clinical and Translational Oncology. 2013;15:467–71.
9. Scharovsky OG, Mainetti LE, Rozados VR. Metronomic chemotherapy: Changing the paradigm that more is better. Current Oncology. 2009;16(2):7–15.
10. Liu J, He M, Wang Z, Li Q, Xu B. Current research status of metronomic chemotherapy in combination treatment of breast cancer. Oncology Research and Treatment. 2022;45(11):681–92.
11. Cazzaniga ME, Cordani N, Capici S, Cogliati V, Riva F, Cerrito MG. Metronomic chemotherapy. Cancers. 2021;13(9):2236.
12. Kerbel RS, Kamen BA. The anti-angiogenic basis of metronomic chemotherapy. Nature Reviews Cancer. 2004;4(6):423–36.
13. Hanahan D, Bergers G, Bergsland E. Less is more, regularly: Metronomic dosing of cytotoxic drugs can target tumor angiogenesis in mice. The Journal of Clinical Investigation. 2000;105(8):1045–7.
14. Cazzaniga ME, Biganzoli L, Cortesi L, De Placido S, Donadio M, Fabi A, et al. Treating advanced breast cancer with metronomic chemotherapy: What is known, what is new and what is the future?: Taylor & Francis; 2019;12:2989–97. doi: 10.2147/OTT.S189163
15. Maiti R. Metronomic chemotherapy. Journal of Pharmacology and Pharmacotherapeutics. 2014;5(3):186–92.
16. Bocci G, Nicolaou KC, Kerbel RS. Protracted low-dose effects on human endothelial cell proliferation and survival in vitro reveal a selective antiangiogenic window for various chemotherapeutic drugs. Cancer Research. 2002;62(23):6938–43.
17. Zitvogel L, Apetoh L, Ghiringhelli F, Kroemer G. Immunological aspects of cancer chemotherapy. Nature Reviews Immunology. 2008;8(1):59–73.

18. Kono K, Kawaida H, Takahashi A, Sugai H, Mimura K, Miyagawa N, et al. CD4 (+) CD25 high regulatory T cells increase with tumor stage in patients with gastric and esophageal cancers. Cancer Immunology, Immunotherapy. 2006;55:1064–71.
19. Lutsiak MEC, Semnani RT, De Pascalis R, Kashmiri SVS, Schlom J, Sabzevari H. Inhibition of CD4+ 25+ T regulatory cell function implicated in enhanced immune response by low-dose cyclophosphamide. Blood. 2005;105(7):2862–8.
20. Ding M, Zhang Y, Li J, Pu K. Bioenzyme-based nanomedicines for enhanced cancer therapy. Nano Convergence. 2022;9(1):7.
21. Bergers G, Benjamin LE. Tumorigenesis and the angiogenic switch. Nature Reviews Cancer. 2003;3(6):401–10.
22. Bahl A, Bakhshi S. Metronomic chemotherapy in progressive pediatric malignancies: Old drugs in new package. The Indian Journal of Pediatrics. 2012;79:1617–22.
23. Browder T, Butterfield CE, Kräling BM, Shi B, Marshall B, O'Reilly MS, et al. Antiangiogenic scheduling of chemotherapy improves efficacy against experimental drug-resistant cancer. Cancer Research. 2000;60(7):1878–86.
24. Klement G, Baruchel S, Rak J, Man S, Clark K, Hicklin DJ, et al. Continuous low-dose therapy with vinblastine and VEGF receptor-2 antibody induces sustained tumor regression without overt toxicity. The Journal of Clinical Investigation. 2000;105(8):R15–R24.
25. Colleoni M, Rocca A, Sandri MT, Zorzino L, Masci G, Nole F, et al. Low-dose oral methotrexate and cyclophosphamide in metastatic breast cancer: antitumor activity and correlation with vascular endothelial growth factor levels. Annals of Oncology. 2002;13(1):73–80.
26. Orlando L, Cardillo A, Ghisini R, Rocca A, Balduzzi A, Torrisi R, et al. Trastuzumab in combination with metronomic cyclophosphamide and methotrexate in patients with HER-2 positive metastatic breast cancer. BMC Cancer. 2006;6(1):1–8.
27. Garcia AA, Hirte H, Fleming G, Yang D, Tsao-Wei DD, Roman L, et al. Phase II clinical trial of bevacizumab and low-dose metronomic oral cyclophosphamide in recurrent ovarian cancer: A trial of the California, Chicago, and Princess Margaret Hospital phase II consortia. Journal of Clinical Oncology. 2008;26(1):76–82.
28. Colleoni M, Orlando L, Sanna G, Rocca A, Maisonneuve P, Peruzzotti G, et al. Metronomic low-dose oral cyclophosphamide and methotrexate plus or minus thalidomide in metastatic breast cancer: Antitumor activity and biological effects. Annals of Oncology. 2006;17(2):232–8.
29. Fontana A, Galli L, Fioravanti A, Orlandi P, Galli C, Landi L, et al. Clinical and pharmacodynamic evaluation of metronomic cyclophosphamide, celecoxib, and dexamethasone in advanced hormone-refractory prostate cancer. Clinical Cancer Research. 2009;15(15):4954–62.
30. Buckstein R, Kerbel RS, Shaked Y, Nayar R, Foden C, Turner R, et al. High-dose celecoxib and metronomic "Low-dose" cyclophosphamide is an effective and safe therapy in patients with relapsed and refractory aggressive histology non–Hodgkin's lymphoma. Clinical Cancer Research. 2006;12(17):5190–8.
31. de Weerdt O, van de Donk NWCJ, Veth G, Bloem AC, Hagenbeek A, Lokhorst HM. Continuous low-dose cyclophosphamide–prednisone is effective and well tolerated in patients with advanced multiple myeloma. The Netherlands Journal of Medicine. 2001;59(2):50–6.
32. Brizzi MP, Berruti A, Ferrero A, Milanesi E, Volante M, Castiglione F, et al. Continuous 5-fluorouracil infusion plus long acting octreotide in advanced well-differentiated neuroendocrine carcinomas. A phase II trial of the Piemonte oncology network. BMC Cancer. 2009;9(1):1–8.
33. Sterba J, Pavelka Z, Slampa P. Concomitant radiotherapy and metronomic temozolomide in pediatric high-risk brain tumors. Neoplasma. 2002;49(2):117–20.
34. Sterba J, Valik D, Mudry P, Kepak T, Pavelka Z, Bajciova V, et al. Combined biodifferentiating and antiangiogenic oral metronomic therapy is feasible and effective in relapsed solid tumors in children: Single-center pilot study. Oncology Research and Treatment. 2006;29(7):308–13.
35. Kieran MW, Turner CD, Rubin JB, Chi SN, Zimmerman MA, Chordas C, et al. A feasibility trial of antiangiogenic (metronomic) chemotherapy in pediatric patients with recurrent or progressive cancer. Journal of Pediatric Hematology/Oncology. 2005;27(11):573–81.
36. Stempak D, Gammon J, Halton J, Moghrabi A, Koren G, Baruchel S. A pilot pharmacokinetic and antiangiogenic biomarker study of celecoxib and low-dose metronomic vinblastine or cyclophosphamide in pediatric recurrent solid tumors. Journal of Pediatric Hematology/Oncology. 2006;28(11):720–8.
37. Mross K, Steinbild S. Metronomic anti-cancer therapy–an ongoing treatment option for advanced cancer patients. Journal of Cancer Therapeutics and Research. 2012;1(1):32.

38. Le Deley MC, Vassal G, Taïbi A, Shamsaldin A, Leblanc T, Hartmann O. High cumulative rate of secondary leukemia after continuous etoposide treatment for solid tumors in children and young adults. Pediatric Blood & Cancer. 2005;45(1):25–31.
39. De Vita S, De Matteis S, Laurenti L, Chiusolo P, Reddiconto G, Fiorini A, et al. Secondary Ph+ acute lymphoblastic leukemia after temozolomide. Annals of Hematology. 2005;84:760–2.
40. Rome A, André N, Scavarda D, Gentet JC, Maues De Paula A, Padovani L, et al. Metronomic chemotherapy-induced bilateral subdural hematoma in a child with meningeal carcinomatosis. Pediatric Blood & Cancer. 2009;53(2):246–7.
41. Shaked Y, Emmenegger U, Man S, Cervi D, Bertolini F, Ben-David Y, et al. Optimal biologic dose of metronomic chemotherapy regimens is associated with maximum antiangiogenic activity. Blood. 2005;106(9):3058–61.
42. Bergers G, Hanahan D. Modes of resistance to anti-angiogenic therapy. Nature Reviews Cancer. 2008;8(8):592–603.
43. Kubisch R, Meissner L, Krebs S, Blum H, Günther M, Roidl A, et al. A comprehensive gene expression analysis of resistance formation upon metronomic cyclophosphamide therapy. Translational Oncology. 2013;6(1):1–IN3.
44. Orlandi P, Fontana A, Fioravanti A, Di Desidero T, Galli L, Derosa L, et al. VEGF-A polymorphisms predict progression-free survival among advanced castration-resistant prostate cancer patients treated with metronomic cyclophosphamide. British Journal of Cancer. 2013;109(4):957–64.
45. Shaughnessy Jr JD, Qu P, Usmani S, Heuck CJ, Zhang Q, Zhou Y, et al. Pharmacogenomics of bortezomib test-dosing identifies hyperexpression of proteasome genes, especially PSMD4, as novel high-risk feature in myeloma treated with Total Therapy 3. Blood, The Journal of the American Society of Hematology. 2011;118(13):3512–24.
46. Mercurio S, Padovani L, Colin C, Carré M, Tchoghandjian A, Scavarda D, et al. Evidence for new targets and synergistic effect of metronomic celecoxib/fluvastatin combination in pilocytic astrocytoma. Acta Neuropathologica Communications. 2013;1:1–14.
47. Klement GL, Kamen BA. Nontoxic, fiscally responsible, future of oncology: Could it be beginning in the Third World? Journal of Pediatric Hematology/Oncology. 2011;33(1):1.
48. Toyoda Y, Manabe A, Tsuchida M, Hanada R, Ikuta K, Okimoto Y, et al. Six months of maintenance chemotherapy after intensified treatment for acute lymphoblastic leukemia of childhood. Journal of Clinical Oncology. 2000;18(7):1508–16.
49. Dekker AW, Van't Veer MB, Sizoo W, Haak HL, Van Der Lelie J, Ossenkoppele G, et al. Intensive postremission chemotherapy without maintenance therapy in adults with acute lymphoblastic leukemia. Dutch Hemato-Oncology Research Group. Journal of Clinical Oncology. 1997;15(2):476–82.
50. Peyrl A, Chocholous M, Kieran MW, Azizi AA, Prucker C, Czech T, et al. Antiangiogenic metronomic therapy for children with recurrent embryonal brain tumors. Pediatric Blood & Cancer. 2012;59(3):511–7.
51. Wernli M, Tichelli A, Fliedner Vv, Re GBd, Chaputs B, Fey MF, et al. Intensive induction/consolidation therapy without maintenance in adult acute lymphoblastic leukaemia: A pilot assessment. British Journal of Haematology. 1994;87(1):39–43.
52. Choi LMR, Rood B, Kamani N, La Fond D, Packer RJ, Santi MR, et al. Feasibility of metronomic maintenance chemotherapy following high-dose chemotherapy for malignant central nervous system tumors. Pediatric Blood & Cancer. 2008;50(5):970–5.
53. Pasquier E, Kavallaris M, André N. Metronomic chemotherapy: New rationale for new directions. Nature reviews Clinical Oncology. 2010;7(8):455–65.
54. Romiti A, Cox MC, Sarcina I, Di Rocco R, D'Antonio C, Barucca V, et al. Metronomic chemotherapy for cancer treatment: A decade of clinical studies. Cancer Chemotherapy and Pharmacology. 2013;72:13–33.
55. Wildiers H, Tryfonidis K, Dal Lago L, Vuylsteke P, Curigliano G, Waters S, et al. Pertuzumab and trastuzumab with or without metronomic chemotherapy for older patients with HER2-positive metastatic breast cancer (EORTC 75111-10114): an open-label, randomised, phase 2 trial from the Elderly Task Force/Breast Cancer Group. The Lancet Oncology. 2018;19(3):323–36.
56. Cazzaniga ME, Camerini A, Addeo R, Nolè F, Munzone E, Collovà E, et al. Metronomic oral vinorelbine in advanced breast cancer and non-small-cell lung cancer: Current status and future development. Future Oncology. 2016;12(3):373–87.
57. Pasquier E, Andre N, Braguer D. Targeting microtubules to inhibit angiogenesis and disrupt tumour vasculature: Implications for cancer treatment. Current Cancer Drug Targets. 2007;7(6):566–81.
58. Thanki K, Gangwal RP, Sangamwar AT, Jain S. Oral delivery of anticancer drugs: Challenges and opportunities. Journal of Controlled Release. 2013;170(1):15–40.

59. Mahmud F, Chung SW, Alam F, Choi JU, Kim SW, Kim I-S, et al. Metronomic chemotherapy using orally active carboplatin/deoxycholate complex to maintain drug concentration within a tolerable range for effective cancer management. Journal of Controlled Release. 2017;249:42–52.
60. Mahmud F, Jeon O-C, Alam F, Maharjan R, Choi JU, Park J, et al. Oral pemetrexed facilitates low-dose metronomic therapy and enhances antitumor efficacy in lung cancer. Journal of Controlled Release. 2018;284:160–70.
61. Moes J, Koolen S, Huitema A, Schellens J, Beijnen J, Nuijen B. Development of an oral solid dispersion formulation for use in low-dose metronomic chemotherapy of paclitaxel. European Journal of Pharmaceutics and Biopharmaceutics. 2013;83(1):87–94.
62. Cervi D, Yip T-T, Bhattacharya N, Podust VN, Peterson J, Abou-Slaybi A, et al. Platelet-associated PF-4 as a biomarker of early tumor growth. Blood, The Journal of the American Society of Hematology. 2008;111(3):1201–7.
63. Lakka Klement G, Yip T-T, Cassiola F, Kikuchi L, Cervi D, Podust V, et al. Platelets actively sequester angiogenesis regulators. Blood, The Journal of the American Society of Hematology. 2009;113(12):2835–42.
64. Mayer EL, Isakoff SJ, Klement G, Downing SR, Chen WY, Hannagan K, et al. Combination antiangiogenic therapy in advanced breast cancer: A phase 1 trial of vandetanib, a VEGFR inhibitor, and metronomic chemotherapy, with correlative platelet proteomics. Breast Cancer Research and Treatment. 2012;136:169–78.
65. Peterson JE, Zurakowski D, Italiano JE, Michel LV, Connors S, Oenick M, et al. VEGF, PF4 and PDGF are elevated in platelets of colorectal cancer patients. Angiogenesis. 2012;15:265–73.
66. Peterson JE, Zurakowski D, Italiano Jr JE, Michel LV, Fox L, Klement GL, et al. Normal ranges of angiogenesis regulatory proteins in human platelets. American Journal of Hematology. 2010;85(7):487–93.
67. Tang J-H, Zhao J-H, Lu J-W, Yan F, Qin J-W, Xu B. Circulating levels of angiogenic cytokines in advanced breast cancer patients with system chemotherapy and their potential value in monitoring disease course. Journal of Cancer Research and Clinical Oncology. 2011;137:55–63.
68. Treiber G, Wex T, Malfertheiner P. Impact of different anticancer regimens on biomarkers of angiogenesis in patients with advanced hepatocellular cancer. Journal of Cancer Research and Clinical Oncology. 2009;135:271–81.
69. Allegrini G, Falcone A, Fioravanti A, Barletta MT, Orlandi P, Loupakis F, et al. A pharmacokinetic and pharmacodynamic study on metronomic irinotecan in metastatic colorectal cancer patients. British Journal of Cancer. 2008;98(8):1312–9.
70. Allegrini G, Di Desidero T, Barletta MT, Fioravanti A, Orlandi P, Canu B, et al. Clinical, pharmacokinetic and pharmacodynamic evaluations of metronomic UFT and cyclophosphamide plus celecoxib in patients with advanced refractory gastrointestinal cancers. Angiogenesis. 2012;15:275–86.
71. Perroud HA, Rico MJ, Alasino CM, Pezzotto SM, Rozados VR, Scharovsky OG. Association between baseline VEGF/sVEGFR-2 and VEGF/TSP-1 ratios and response to metronomic chemotherapy using cyclophosphamide and celecoxib in patients with advanced breast cancer. Indian Journal of Cancer. 2013;50(2):115–21.
72. Montagna E, Cancello G, Bagnardi V, Pastrello D, Dellapasqua S, Perri G, et al. Metronomic chemotherapy combined with bevacizumab and erlotinib in patients with metastatic HER2-negative breast cancer: Clinical and biological activity. Clinical Breast Cancer. 2012;12(3):207–14.
73. Munoz R, Man S, Shaked Y, Lee CR, Wong J, Francia G, et al. Highly efficacious nontoxic preclinical treatment for advanced metastatic breast cancer using combination oral UFT-cyclophosphamide metronomic chemotherapy. Cancer Research. 2006;66(7):3386–91.
74. Munoz R, Shaked Y, Bertolini F, Emmenegger U, Man S, Kerbel RS. Anti-angiogenic treatment of breast cancer using metronomic low-dose chemotherapy. The Breast. 2005;14(6):466–79.
75. Barbolosi D, Ciccolini J, Meille C, Elharrar X, Faivre C, Lacarelle B, et al. Metronomics chemotherapy: Time for computational decision support. Cancer Chemotherapy and Pharmacology. 2014;74:647–52.
76. Malik PS, Raina V, André N. Metronomics as maintenance treatment in oncology: Time for chemo-switch. Frontiers in Oncology. 2014;4:76.
77. Stanway S, Lodge M, Sullivan R, Diprose K, Young AM, Crisp N, et al. The UK's contribution to cancer control in low-income and middle-income countries. The Lancet Oncology. 2021;22(9):e410–e8.

2 Metronomic Chemotherapy

A Multitargeted Therapy

Manzoor Ahmad Mir, Aisha Shabir, and Burhan Ul Haq

2.1 INTRODUCTION

According to the GLOBOCAN Reports 2020, it has been estimated that around 19.3 million new cancer cases and about 10 million cancer deaths occurred in 2020. It was also reported that among the newly diagnosed cancers, breast cancer in females has surpassed lung cancer, making it the most frequently detected cancer, with around 2.3 million new cases (11.7%), followed by lung cancer (11.4%), colorectal cancer (10%), and prostate cancer (7.3%). The number of cancer cases worldwide is predicted to reach 28.4 million in 2040, which will be around 47% from 2020. This increase will be more pronounced in transitioning (64–95%) than in transitioned (32–56%) countries due to demographic changes, though it may worsen by rising risk factors brought on by globalization and an expanding economy [1].

The advancements made in the fields of molecular biology and tumor biology have helped a lot in treating various types of cancers to a great extent. The treatment nowadays is focused on two major targets, i.e., use of personalized medicines to subgroups of tumor subtypes harboring certain genetic aberrations and targeting elements of the tumor microenvironment, including immunological and anti-cancer systems [2]. Even though cancer research has grown so far, still treatment processes are sometimes unsuccessful in treating certain malignancies [3]. Therefore, certain types of chemotherapies were developed concerning the treatment of various types of cancers.

Chemotherapy has been used widely in cancer therapeutics. It is typically employed to combat cancer cells that have metastasized from the original tumor, given its systemic nature [4]. Chemotherapy is now used in four primarily distinct contexts:

a. Chemotherapy is used as induction treatment to reduce the size of the tumor as much as feasible and achieve therapeutic benefits ranging from alleviation of symptoms to curative effects depending on the kind of cancer.
b. Adjuvant treatment is used to eliminate any cancer cells that may have escaped the initial tumor and might later expand to produce metastasis. Chemotherapy is given after using surgery/radiation. The main objective of this treatment is to increase the number of patients who are cured.
c. The preoperative therapy in the main goal is to boost the likelihood of achieving local tumor control and in which chemotherapy is administered for small and inoperable cancers.
d. Non-adjuvant treatment, which is administered before the surgery, aims to improve the long-term tumor management in cases when the tumor is operable by having an impact on microscopic pathology. To maximize the likelihood of cancer cell elimination and control at that location, chemotherapy can also be administered as instillations, often as an addition to general therapy.

2.2 TRADITIONAL CHEMOTHERAPY

Conventional chemotherapy is defined by the maximum tolerated dose (MTD) strategy; it refers to administering the maximum amount of medicine that a patient can tolerate. [5]. This approach has

DOI: 10.1201/9781032632674-2

Traditional Chemotherapy	Metronomic Chemotherapy
1. Follows the maximum tolerated dose approach.	1. Doses are given in much lower amounts.
2. Doses are given at a specific time interval with breaks of 2–4 weeks between each dose.	2. Continuous dosage without long breaks.
3. Plasma concentration either rises or falls.	3. Plasma concentration is often sustained.
4. Increased chances of toxicity and side effects.	4. Comparatively lower toxicity.
5. High cost.	5. Very cost-effective.
6. Delivered in hospital settings.	6. Delivered in cost settings.
7. Time-defined protocols.	7. Prolonged use.

FIGURE 2.1 A tabular representation of the comparison between traditional chemotherapy and metronomic chemotherapy.

shown encouraging outcomes in terms of heightened effectiveness and decreased treatment duration. However, this is accompanied by a disadvantage concerning the extended intervals between the therapy sessions [6].

The lengthy (3–4 weeks) intervals between chemotherapy treatments cause cancer cells to gradually recover and acquire resistance, which is then followed by other side effects like hair loss, vomiting, nausea, bone marrow suppression, organ damage, and peripheral neuropathy [7]. The MTD strategy is therefore concerned with the maximum amount of dosage that is medically feasible to treat the patients.

The enduring impact of the MTD approach has come under more and more severe examination over time. For the most advanced solid tumors like metastatic breast cancers (MBCs), lung, prostate, and colorectal cancer, clinical implementation of this strategy in the late 1980s and 1990s did not result in higher cure rates. Extensive usage of colony-stimulating factors to increase the dosage intensity also didn't show any significant increase in the effectiveness of these toxic chemotherapy regimens [8]. Last but not least, the MTD approach's most severe implementation was the usage of high-dosage chemotherapy given along with an individual's stem cell support for solid tumor therapy, but the ineffectiveness of the approach on conditions like MBC is currently being explored by randomized clinical trials [9].

Although conventional chemotherapy gave positive results to some extent, due to associated side effects, researchers looked forward to some alternatives, and this is what led to the development of metronomic chemotherapy (MCT) [10]. The introduction of novel molecularly targeted medicines and the development of MCT present a recent scientific justification for rejecting the MTD paradigm (Figure 2.1).

2.3 METRONOMIC CHEMOTHERAPY

Due to the side effects and other drawbacks of the MTD approach, MCT came into being as a significant alternative to MTD. Kerbel and Hanahan in two distinct articles coined the term metronomic chemotherapy in the 2000s. MCT is therefore a method of administering cytotoxic drugs that are distinct from traditional chemotherapy regimens.

In contrast with MTD, MCT drugs are continuously administered at low doses without long gaps, and because MCT requires regular medication delivery, oral medicines are a more practical choice for patients [11–13]. Hence, unlike conventional chemotherapy, MCT incorporates the

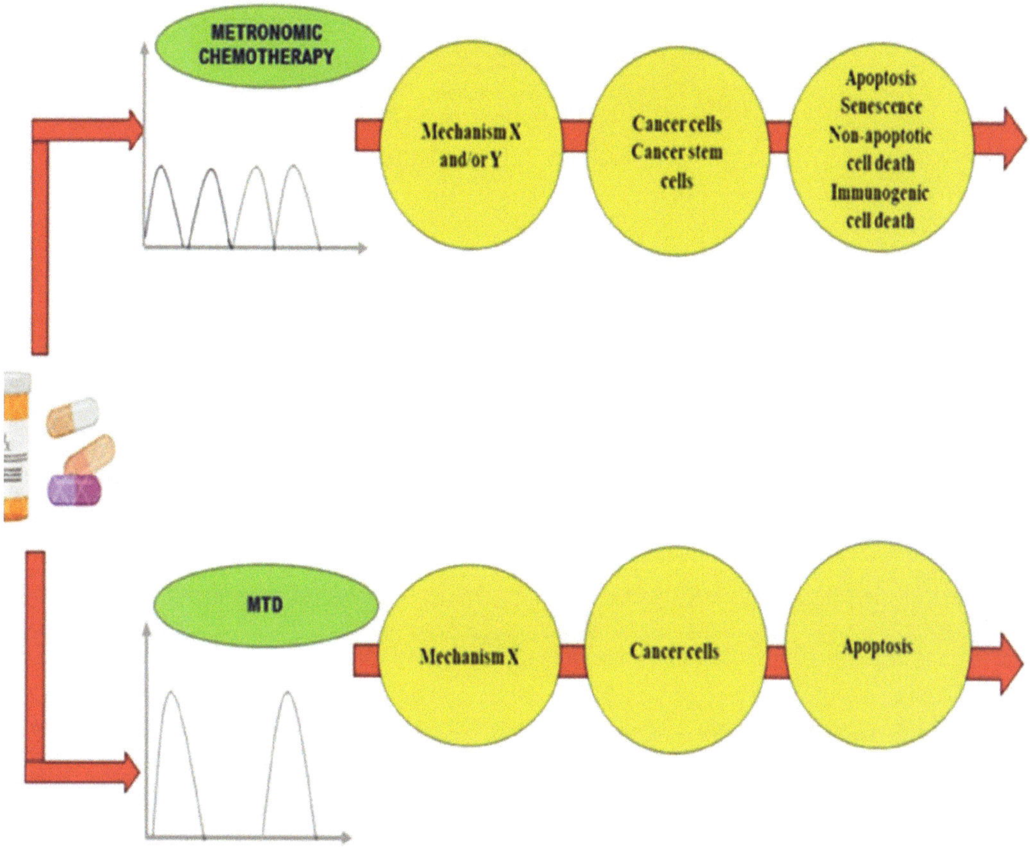

FIGURE 2.2 Differences in doses and schedules between MCT and maximal tolerated dose (MTD) chemotherapy led to different impacts on tumor compartments. They also lead to different mechanisms of action on anticancer cells and cancer stem cells (CSCs), which in turn lead to potentially different types of cell death.

use of less dangerous medication and more useful drug administration modalities, i.e., oral [14] (Figure 2.2). MCT's fundamental tenets include continuous drug delivery, modest drug dosage, immune system manipulation, and utilization of certain anti-angiogenic actions of chemotherapeutic medicines [15]. Chemotherapy that is administered orally has advantages over IV bolus administration, encompassing a heightened therapeutic range, prolonged plasma drug levels beneath MTD, reduced adverse effects, enhanced quality of life, and decreased medical cost [16–18]. Kerbel and Kamen hypothesized that a synergistic effect may result from combining MCT with conventional MTD therapy [19]. Before the beginning of the low-dose MCT, MTD would shrink the tumor; this would prevent endothelial cell rebound and have an anti-angiogenic impact [20].

2.4 MECHANISM OF ACTION OF METRONOMIC CHEMOTHERAPY

Earlier it was thought that MCT's mode of action was the suppression of angiogenesis, but it is now well-acknowledged that MCT has several modes of action, including anti-angiogenic, anti-proliferative, and immune-modulatory properties [11, 21, 22]. Therefore, MCT provides a significant balance between the action and side effects of therapy. It may also increase the therapeutic index of medication and enable longer enhanced treatment, which may ultimately increase the life span of patients [23] (Figure 2.3).

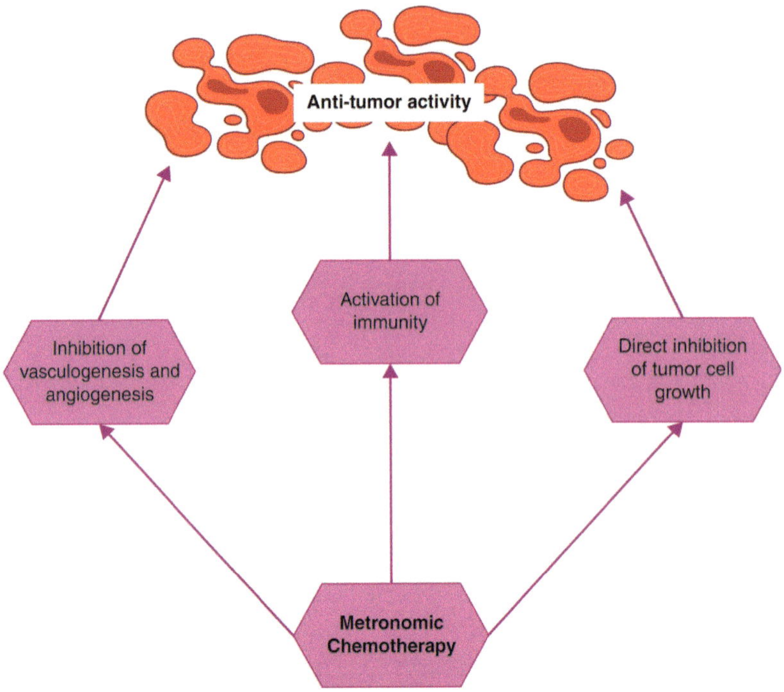

FIGURE 2.3 Mechanism of action of metronomic chemotherapy. The effectiveness of metronomic chemotherapy stems from its ability to hinder tumor cell growth. The critical role of MCT lies in inhibiting vasculogenesis and angiogenesis, achieved by directly curbing endothelial cell proliferation, increasing the levels of natural angiogenic inhibitors like thrombospondin (TSP)-1 and hypoxia-inducible factor (HIF)-1α.

Angiogenesis is a complicated process in which certain bodily biomolecules play a role. It results in the production of new blood vessels from pre-existing ones and further promotes the growth of cancer cells and metastasis [24]. The anti-angiogenic drugs used to treat cancer prevent angiogenesis, which inhibits the expansion of tumors [25]. According to clinical studies, angiogenesis has an important role in the formation and likely spread of breast cancer [26]. Low-dose chemotherapy has been shown to increase the immune response against cancerous cells by decreasing the number and inhibitory activity of regulatory T cells (Treg), which are involved in immunological tolerance and have the capacity to stop the function of additional immune cells, which include those involved in the immune response against tumors [27] (Figure 2.4).

2.5 MCT AS A MULTITARGETED THERAPY

Multitargeted therapies are designed to inhibit several pathways and proteins involved in tumor growth and progression. This can potentially increase the effectiveness of treatment and reduce the likelihood of drug resistance, as targeting multiple pathways simultaneously makes it more difficult for cancer cells to adapt [28]. Multitargeted therapies can be developed using a variety of strategies such as designing molecules that have affinities for multiple targets or combining multiple drugs with different specificities [29].

Multitargeted therapies include some types of targeted cancer therapies that inhibit both angiogenesis (formation of new blood vessels) and specific oncogenic signaling pathways [30]. Multitargeted therapies involve targeting both cancer and the tumor microenvironment, minimizing toxicities to enable combinations. Numerous criteria for an ideal multitargeted treatment seem to be addressed by MCT [31] (Table 2.1).

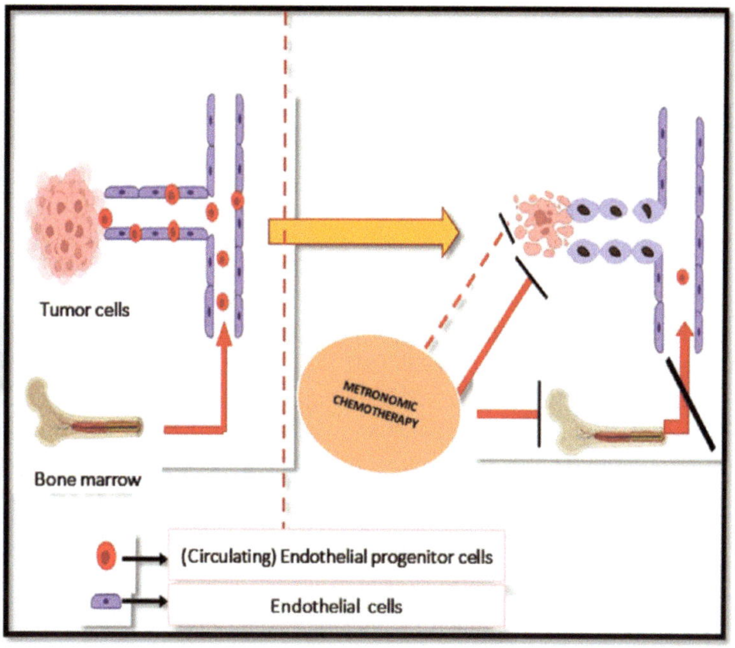

FIGURE 2.4 Effects of metronomic chemotherapy on blood vessels. The migration and survivability of bone marrow–derived circulating endothelial progenitor cells (CEPs), which can aid in tumor neo-angiogenesis, are inhibited by metronomic chemotherapy. Additionally, the prolonged usage of low-dose drugs like cyclophosphamide (CPA) may potentially have immuno-modulatory effects.

Finding the appropriate target molecules to inhibit to provide the optimum anti-tumor effects is the initial objective of targeted therapy. According to conventional wisdom, tumor cells should be favored for the best targets to allow for selectivity, and medications that target a single molecule should be favored over less targeted therapies that can interfere with several targets. One of the challenges in targeted therapy is identifying the most suitable target molecules to inhibit optimal

TABLE 2.1
Advantages of Multitargeted Therapy

Advantages	Explanation
Enhanced efficacy	Cancer involves interconnected pathways. Multitargeted therapies can address these complexities by targeting multiple points of intervention, leading to increased effectiveness in controlling disease progression.
Synergistic effects	Combining drugs that target different pathways can lead to synergistic effects, where the combined actions of drugs are more potent than the sum of their individual actions. This enhances therapeutic outcomes and minimizes the chances of drug resistance.
Reduced drug resistance	Resistance to therapy can emerge when a disease evolves to bypass the effects of a single-target drug. Multitargeted therapies make it harder for the disease to develop resistance, as multiple pathways must be simultaneously altered for therapy to lose efficacy.
Flexible adaptability	Diseases can exhibit considerable heterogeneity among patients. Multitargeted therapies can be tailored to an individual patient's specific molecular profiles, allowing for a more personalized treatment approach
Reduced treatment time	The combination of multiple targeted agents can result in faster and more comprehensive disease control, potentially leading to shorter treatment duration and quicker patient recovery.

anti-tumor effects. MCT's multifaceted actions, including its impact on angiogenesis, proliferation, and immune response, demonstrate its potential as an effective multitargeted treatment. This approach maximizes therapeutic outcomes by simultaneously addressing multiple vulnerabilities of cancer cells [32].

A wider therapeutic window can be provided by multitargeted medicines, enabling successful treatment even in circumstances when a single target therapy might not be practical because of toxicity worries [33]. Combinations of already existing medicines that were first created for several uses can be used in multitargeted therapy. Utilizing the current understanding of medication safety and pharmacokinetics, this method can hasten the creation of novel medicines [34].

Single-targeted therapies frequently result in compensatory mechanisms being triggered by diseases, which gradually reduces their efficacy. Multitargeted therapies can block these pathways, preserving a longer-lasting therapeutic impact [35].

Recently it was proposed that low-dose MCT may be used in combination with conventional chemotherapy in a procedure known as 'chemo-switching,' in which short rounds of traditional chemotherapy are followed by long-term maintenance chemotherapy, which is then supplemented with targeted treatment [36, 37].

2.6 MCT IN BREAST CANCER

MCT has shown promising results in the treatment of breast cancer. Several drugs that are used in conventional chemotherapy schedules for breast cancer have now been studied in the context of MCT, frequently in combination with some hormones and other targeted drugs like trastuzumab, bevacizumab, vinorelbine, cyclophosphamide methotrexate, and fluoropyrimidines [38, 39]. Numerous studies have revealed the therapeutic effectiveness of MCT in treating MBC [40]. Microtubule targeting agents like oral vinorelbine are a distinct family of chemotherapeutic compounds, These substances have specialized effects, including prevention of angiogenesis, suppression of endothelial progenitor cells, and blocking the HIF-I pathway [41]. These qualities make vinorelbine one of the promising drugs to be examined in MCT schedules, combined with the ability for oral administration and proven effectiveness in other solid tumors such as breast, lungs, and prostate [42]. The study conducted by Fedele et al. in phase II trials demonstrated that 58 patients with MBC who had received intensive pre-treatment had shown the effectiveness of capecitabine when given on a metronomic schedule. The dosage of the drug was 1500 mg per day, and the clinical benefit rate was around 62%, with the median duration of response around 7 months. Hematologic toxicity was rare (around 5%), and the regimen was safe [43]. Moreover the phase II trials of pegylated liposomal-doxorubicin has undergone metronomic dose testing in MBC with 52 patients who underwent a metronomic schedule of a dose at 20 mg/m^2 every 2 weeks. The clinical benefit rate was 45% with minimal toxicities. According to scientists, this medication may have benefits over traditional medicines have benefits over traditional anthracyclines, including improved quality of life [44] (Table 2.2).

2.7 MCT IN PROSTATE CANCER

Prostate cancer is usually formed from the epithelial cells located in lobules and ducts of the prostate gland. The development and differentiation of these epithelial cells are strictly controlled by androgens, particularly testosterone. Due to this, androgen deprivation is effective in slowing the development of well-differentiated prostate tumors, and since most prostate tumors are not well-differentiated, endocrine therapy has a low and brief response rate. Chemotherapy does not effectively treat prostate cancer. Before the development of docetaxel, mitoxantrone was the only cytotoxic medication in use that had any effect on prostate cancer [50].

For metastatic castrate-resistant prostate cancer (mCRPC) patients who have exhausted all other therapeutic options or are ineligible for traditional chemotherapy or lack access to standard mCRPC treatment such as in low- or middle-income nations, MCT provides a feasible therapeutic alternative [51].

TABLE 2.2
Metronomic Chemotherapy Trials in Breast Cancer

Metronomic Chemotherapy Protocol	Patient Population	n	Response Rate	Reference
Cyclophosphamide, methotrexate, thalidomide	Both pretreated and not-treated MBC	171	ORR: 20.9% CBR: 41.5%	[45]
Cyclophosphamide, methotrexate	previously treated MBC	61	ORR: 52%	[46]
Capecitabine	Both pretreated and not-treated MBC	58	ORR: 62%	[43]
Pegylated liposomal-doxorubicin	Both pretreated and not-treated MBC	52	ORR: 57%	[44]
Bevacizumab, cyclophosphamide, capecitabine	Pretreated MBC	46	ORR: 48% CBR: 68%	[47]
Vinorelbine	Elderly MBC	34	ORR: 38%	[48]
Vinorelbine, capecitabine	Pretreated MBC	31	ORR: 55%	[49]

Abbreviations: CBR, clinical benefit rate; mBC, metastatic breast cancer; n, number of patients; ORR, overall response rate.

A trial was conducted with 17 patients with CRPC, and patients were given a weekly oral medication of ketoconazole in combination with cyclophosphamide or etoposide in combination with estramustine (KEES combination) given on alternate weeks (Table 2.3). It was seen that around 87% of patients had shown reduced prostate-specific antigen (PSA). Therapy had few side effects, including thrombocytopenia and anemia. Therefore, it was concluded that KEES might be a potential choice for CRPC patients, causing a significant drop in the level of PSA with few side effects [52].

2.8 MCT IN LUNG CANCER

Lung cancer is one of the leading causes of cancer-related deaths globally, which accounted for around 22–23% of all cancer-related fatalities in 2020 [56]. Numerous clinical trials have been conducted over the past 10 years to examine MCT in first- and second-line treatment as well as to

TABLE 2.3
Metronomic Chemotherapy Trials in Prostate Cancer

Metronomic Chemotherapy Protocol	Patient Population	n	Response Rate	Reference
Cyclophosphamide, methotrexate, celecoxib	CRPC after taxane	67	ORR: 5%	[53]
Cyclophosphamide, methotrexate	CRPC after taxane	58	PSA 50% Reduction 30% ORR: 42%	[54]
Ketoconazole, doxorubicin/ vinblastine, estramustine regimen	CRPC	45	PSA 50% Reduction: 58%	[55]
Ketoconazole, cyclophosphamide, etoposide, estramustine	CRPC	17	PSA 50% Reduction: 87%	[52]

Abbreviations: CRPC, castrate-resistant prostate cancer; n, number of patients; ORR, overall response rate; PSA, prostate-specific antigen.

TABLE 2.4
Metronomic Chemotherapy Trials in Lung Cancer

Metronomic Chemotherapy Protocol	Patient Population	n	Response Rate	References
Cisplatin, etoposide, bevacizumab → bevacizumab, erlotinib	III–IV stage lung cancer	45	ORR: 65%	[59]
Vinorelbine	Pre-treated lung cancer	21	ORR: 29%	[58]

Abbreviations: n, number of patients; ORR, overall response rate.

look at the continuation of therapy for metastatic lung cancer. Comparing targeted therapies to conventional chemotherapies has shown greater response rates, longer progression-free survival rates, and longer overall survival [57].

According to a study that conducted a clinical phase II trial of oral vinorelbine at 40 mg/m^2 on 21 patients, which was administered on days 1, 3, and 5 every week continuously in previously treated lung cancer patients, it was seen that the disease control was around 29% and the median progression-free survival was 4.3 months. Hematologic toxicity included neutropenia [58] (Table 2.4).

Another phase II trial was conducted in advanced lung cancer patients, in which around 45 patients were administered cisplatin (30 mg/m^2 on the first to third days) along with oral etoposide (30 mg, days 1–15) and bevacizumab (5 mg/kg, day 3) every 3 weeks, and the patients who responded well were then treated with bevacizumab in combination with erlotinib until progression occurred. Side effects included hair loss and grade I and II hematological and mucosal toxicity [59].

2.9 MCT IN COLON CANCER

One of the most prevalent malignancies and leading causes of mortality globally is colorectal cancer [56]. Multiple lines of systemic therapy have increased the overall survival of patients with metastatic colorectal cancer; nonetheless, the illness still has a dismal prognosis despite advancements in chemotherapy and biological medicines. Since 2004 phase III trials have authorized anti-angiogenic therapy for the treatment of metastatic colorectal cancer, and hence MCT based on decreasing angiogenesis may be a useful therapeutic approach [60]. Many clinical trials were conducted for testing MCT; one such trial included a schedule of a combination of oxaliplatin, capecitabine, and bevacizumab on 75 patients with metastatic colorectal cancer. The overall response rate was 50.1%, the clinical benefit rate was around 86.7%, and the median progression-free survival was 12.3 months. Grade 4 toxicity was not observed in any patient (Table 2.5). Another phase I study was conducted

TABLE 2.5
Metronomic Chemotherapy Trials in Colorectal Cancer

Metronomic Chemotherapy Protocols	Patient Population	n	Response Rate	References
Oxaliplatin, capecitabine, bevacizumab	Elderly metastatic colorectal cancer	75	ORR: 50.01% CBR: 86.7%	[61]
Cyclophosphamide, imatinib, bevacizumab	Metastatic colorectal cancer	35	ORR: 20%	[62]
Tegafur, leucovorin, 5-Fluorouracil	Recurrent metastatic colorectal cancer	28	ORR: 45.5%	[63]

Abbreviations: CBR, clinical benefit rate; ORR, overall response rate.

on around 35 patients to determine the safety, MTD, and pharmacokinetics of cyclophosphamide, imatinib, and bevacizumab in patients with advanced colorectal cancer, which showed an overall response rate of 20%. Toxicities included neutropenia, hematuria, and nausea.

2.10 CONCLUSION

Traditional chemotherapy, while effective to an extent, has limitations due to its severe effects and the potential to develop resistance over time. This has led to the development of MCT, a novel approach that involves administering lower doses of chemotherapy continuously, without prolonged gaps. MCT offers several advantages including lowered side effects, improved quality of life, and a potentially more balanced therapeutic index. It also takes advantage of the immune-modulatory and anti-angiogenic properties of certain chemotherapeutic agents, ultimately aiming to enhance treatment outcomes.

The chapter highlights the successful application of MCT in breast, lung, prostate, and colorectal cancer. Researchers have explored various drugs, including those used in conventional chemotherapy, in the context of MCT. These studies have shown promising results in terms of clinical benefits and manageable toxicities. By providing insights into the principles and potential benefits of MCT, the chapter underscores the importance of moving beyond the one-size-fits-all approach of MTD chemotherapy. MCT's ability to maintain a continuous therapeutic effect, minimize drug resistance, and enhance the well-being of patients is the pursuit of more effective and patient-friendly cancer treatments.

In conclusion, the chapter emphasizes the significance of multitargeted therapies in cancer treatment and underscores the role of MCT as a viable multitargeted approach. Multitargeted therapies are designed to tackle various pathways and proteins that contribute to tumor progression. MCT aligns well with the principles of multitargeted therapy. It combines continuous drug delivery, low doses, and an immune system. By delivering low and continuous doses, MCT prevents tumor cells from recovering and developing resistance. Its immune-modulatory and anti-angiogenic effects contribute to its multitargeted nature.

In essence, the study highlights how MCT embodies the multitargeted therapy concept and offers a well-balanced strategy for treating cancer. As research continues to advance understanding of the intricate interactions between cancer cells, signaling pathways and tumor microenvironment become paramount. MCT's ability to target multiple aspects of tumor development and progression positions it as a promising avenue for improved cancer treatment outcomes while minimizing toxicities. It paved the way for innovative and patient-centered therapeutic approaches that hold great potential in the fight against cancer.

For further reading, some textbooks mentioned here also give a further understanding of MCT:

- *Progress in Anti-cancer Chemotherapy* by Gabriel N. Hortobagyi and David Khayat.
- *Triple-Negative Breast Cancer: Biomarkers, Emerging Therapeutic Strategies and Clinical Challenges* by Marion Montgomery.
- *Current Cancer Treatment: Novel Beyond Conventional Approaches* by Oner Ozdemir.
- *Breast Cancer: Current and Alternative Therapeutic Modalities* by Esra Gunduz and Mehmet Gunduz.

The following links for video lectures may also be helpful:

https://youtu.be/TAqO0Mq19JQ
https://youtu.be/6wEnYfvoBqk
https://youtu.be/A9fRRCUlMms
https://youtu.be/zIWvmCsKr34

For more insight about the topic, we would suggest detailed findings from the books of (Mir MA, 2022) https://doi.org/10.1016/C2021-0-02565-7, https://doi.org/10.1016/C2022-0-00074-X, (Mir MA, 2021) https://doi.org/10.52305/WXJL6770, and (Mir MA, 2015) https://doi.org/10.1016/C2014-0-02898-5 and from the cancer.net website on the following links:

https://www.cancer.net/cancer-types/breast-cancer/types-treatment
https://discovery.ucl.ac.uk/id/eprint/1472740/
https://www.jmedsciences.com/doi/JMEDS/pdf/10.5005/jp-journals-10045-00138

For diagrammatic illustrations and descriptive tables (Lazzeroni, 2012): http://www.eurekaselect.com/article/49928

See video links on the overall status of cancer, its various types, and current new possible treatment options available: https://www.sciencedirect.com/science/article/pii/S205970292032278X

REFERENCES

1. Sung, H., et al., Global cancer statistics 2020: GLOBOCAN estimates of incidence and mortality worldwide for 36 cancers in 185 countries. CA: a cancer journal for clinicians, 2021. **71**(3): p. 209–249.
2. Zugazagoitia, J., et al., Current challenges in cancer treatment. Clinical therapeutics, 2016. **38**(7): p. 1551–1566.
3. Maiti, R., *Metronomic chemotherapy.* Journal of pharmacology and pharmacotherapeutics, 2014. **5**(3): p. 186–192.
4. Brianna and S.H. Lee, Chemotherapy: how to reduce its adverse effects while maintaining the potency? Medical oncology, 2023. **40**(3): p. 88.
5. Kummar, S., et al., Drug development in oncology: classical cytotoxics and molecularly targeted agents. British journal of clinical pharmacology, 2006. **62**(1): p. 15–26.
6. Lien, K., et al., Low-dose metronomic chemotherapy: a systematic literature analysis. European journal of cancer, 2013. **49**(16): p. 3387–3395.
7. Galluzzi, L., et al., Immunological effects of conventional chemotherapy and targeted anticancer agents. Cancer cell, 2015. **28**(6): p. 690–714.
8. Takimoto, C.H. and E.K. Rowinsky, Dose-intense paclitaxel: deja vu all over again? Journal of clinical oncology, 2003. **21**(15): p. 2810–2814.
9. Lippman, M.E., *High-dose chemotherapy plus autologous bone marrow transplantation for metastatic breast cancer.* New England journal of medicine, 2000. **342**(15), p. 1119–1120.
10. DeVita Jr, V.T. and E. Chu, *A history of cancer chemotherapy.* Cancer research, 2008. **68**(21): p. 8643–8653.
11. Cazzaniga, M.E., et al., Metronomic oral vinorelbine in advanced breast cancer and non-small-cell lung cancer: current status and future development. Future oncology, 2016. **12**(3): p. 373–387.
12. Hanahan, D., G. Bergers, and E. Bergsland, *Less is more, regularly: metronomic dosing of cytotoxic drugs can target tumor angiogenesis in mice.* The journal of clinical investigation, 2000. **105**(8): p. 1045–1047.
13. Torimura, T., et al., *Metronomic chemotherapy: possible clinical application in advanced hepatocellular carcinoma.* Translational oncology, 2013. **6**(5): p. 511–519.
14. Bocci, G., et al., *Cyclophosphamide-methotrexate 'metronomic' chemotherapy for the palliative treatment of metastatic breast cancer. A comparative pharmacoeconomic evaluation.* Annals of oncology, 2005. **16**(8): p. 1243–1252.
15. Nygren, P., *What is cancer chemotherapy?* Acta oncologica, 2001. **40**(2-3): p. 166–174.
16. Thanki, K., et al., *Oral delivery of anticancer drugs: challenges and opportunities.* Journal of controlled release, 2013. **170**(1): p. 15–40.
17. Mahmud, F., et al., *Metronomic chemotherapy using orally active carboplatin/deoxycholate complex to maintain drug concentration within a tolerable range for effective cancer management.* Journal of controlled release, 2017. **249**: p. 42–52.
18. Moes, J., et al., *Development of an oral solid dispersion formulation for use in low-dose metronomic chemotherapy of paclitaxel.* European journal of pharmaceutics and biopharmaceutics, 2013. **83**(1): p. 87–94.

19. Kerbel, R.S. and B.A. Kamen, *The anti-angiogenic basis of metronomic chemotherapy.* Nature reviews. Cancer, 2004. **4**(6): p. 423–436.
20. Shaked, Y., et al., *Low-dose metronomic combined with intermittent bolus-dose cyclophosphamide is an effective long-term chemotherapy treatment strategy.* Cancer research, 2005. **65**(16): p. 7045–7051.
21. André, N., M. Carré, and E. Pasquier, *Metronomics: towards personalized chemotherapy?* Nature reviews. Clinical oncology, 2014. **11**(7): p. 413–431.
22. Kerbel, R.S. and Y. Shaked, *The potential clinical promise of 'multimodality' metronomic chemotherapy revealed by preclinical studies of metastatic disease.* Cancer letters, 2017. **400**: p. 293–304.
23. Gennari, A., et al., *ESMO Clinical Practice Guideline for the diagnosis, staging and treatment of patients with metastatic breast cancer.* Annals of oncology, 2021. **32**(12): p. 1475–1495.
24. Rajabi, M. and S.A. Mousa, *The role of angiogenesis in cancer treatment.* Biomedicines, 2017. **5**(2): p. 34.
25. Mackey, J.R., et al., *Controlling angiogenesis in breast cancer: a systematic review of anti-angiogenic trials.* Cancer treatment reviews, 2012. **38**(6): p. 673–688.
26. Cao, Y. and Q. Liu, *Therapeutic targets of multiple angiogenic factors for the treatment of cancer and metastasis.* Advances in cancer research, 2007. **97**: p. 203–224.
27. Chaudhary, B. and E. Elkord, *Regulatory T cells in the tumor microenvironment and cancer progression: role and therapeutic targeting.* Vaccines, 2016. **4**(3): p. 28.
28. Giordano, S. and A. Petrelli, *From single-to multi-target drugs in cancer therapy: when a specificity becomes an advantage.* Current medicinal chemistry, 2008. **15**(5): p. 422–432.
29. Broekman, F., E. Giovannetti, and G.J. Peters, *Tyrosine kinase inhibitors: Multi-targeted or single-targeted?* World journal of clinical oncology, 2011. **2**(2): p. 80–93.
30. El-Kenawi, A.E. and A.B. El-Remessy, *Angiogenesis inhibitors in cancer therapy: mechanistic perspective on classification and treatment rationales.* British journal of pharmacology, 2013. **170**(4): p. 712–729.
31. Andre, N., L. Padovani, and E. Pasquier, *Metronomic scheduling of anticancer treatment: the next generation of multitarget therapy?* Future oncology, 2011. **7**(3): p. 385–394.
32. Murciano-Goroff, Y.R., A.B. Warner, and J.D. Wolchok, *The future of cancer immunotherapy: microenvironment-targeting combinations.* Cell research, 2020. **30**(6): p. 507–519.
33. Block, K.I., et al. *Designing a broad-spectrum integrative approach for cancer prevention and treatment.* in Seminars in cancer biology. 2015. Elsevier.
34. Wuerth, R., et al., *Drug-repositioning opportunities for cancer therapy: novel molecular targets for known compounds.* Drug discovery today, 2016. **21**(1): p. 190–199.
35. Shen, P., et al., *Dual-target Janus kinase (JAK) inhibitors: Comprehensive review on the JAK-based strategies for treating solid or hematological malignancies and immune-related diseases.* European journal of medicinal chemistry, 2022. **239**: p. 114551.
36. Moserle, L., A. Amadori, and S. Indraccolo, *The angiogenic switch: implications in the regulation of tumor dormancy.* Current molecular medicine, 2009. **9**(8): p. 935–941.
37. Bergers, G. and L.E. Benjamin, *Tumorigenesis and the angiogenic switch.* Nature reviews cancer, 2003. **3**(6): p. 401–410.
38. Romiti, A., et al., *Metronomic chemotherapy for cancer treatment: a decade of clinical studies.* Cancer chemotherapy and pharmacology, 2013. **72**: p. 13–33.
39. Wildiers, H., et al., *Pertuzumab and trastuzumab with or without metronomic chemotherapy for older patients with HER2-positive metastatic breast cancer (EORTC 75111-10114): an open-label, randomised, phase 2 trial from the Elderly Task Force/Breast Cancer Group.* The Lancet oncology, 2018. **19**(3): p. 323–336.
40. Cazzaniga, M.E., et al., *Metronomic oral vinorelbine in advanced breast cancer and non-small-cell lung cancer: current status and future development.* Future oncology, 2016. **12**(3): p. 373–387.
41. Pasquier, E., N. Andre, and D. Braguer, *Targeting microtubules to inhibit angiogenesis and disrupt tumour vasculature: implications for cancer treatment.* Current cancer drug targets, 2007. **7**(6): p. 566–581.
42. Shaked, Y., et al., *Optimal biologic dose of metronomic chemotherapy regimens is associated with maximum antiangiogenic activity.* Blood, 2005. **106**(9): p. 3058–3061.
43. Fedele, P., et al., *Efficacy and safety of low-dose metronomic chemotherapy with capecitabine in heavily pretreated patients with metastatic breast cancer.* European journal of cancer, 2012. **48**(1): p. 24–29.
44. Munzone, E., et al., *Metronomic administration of pegylated liposomal-doxorubicin in extensively pretreated metastatic breast cancer patients: a mono-institutional case-series report.* The breast, 2010. **19**(1): p. 33–37.

45. Colleoni, M., et al., *Metronomic low-dose oral cyclophosphamide and methotrexate plus or minus thalidomide in metastatic breast cancer: antitumor activity and biological effects.* Annals of oncology, 2006. **17**(2): p. 232–238.
46. Gebbia, V., H. Boussen, and M.R. Valerio, *Oral metronomic cyclophosphamide with and without methotrexate as palliative treatment for patients with metastatic breast carcinoma.* Anticancer research, 2012. **32**(2): p. 529–536.
47. Dellapasqua, S., et al., *Metronomic cyclophosphamide and capecitabine combined with bevacizumab in advanced breast cancer.* Journal of clinical oncology, 2008. **26**(30): p. 4899–4905.
48. Addeo, R., et al., *Low-dose metronomic oral administration of vinorelbine in the first-line treatment of elderly patients with metastatic breast cancer.* Clinical breast cancer, 2010. **10**(4): p. 301–306.
49. Cazzaniga, M., et al., *Efficacy and safety of the all-oral schedule of metronomic vinorelbine and capecitabine in locally advanced or metastatic breast cancer patients: the phase I-II VICTOR-1 study.* International journal of breast cancer, 2014. **2014**: 769790.
50. Tannock, I.F., et al., *Chemotherapy with mitoxantrone plus prednisone or prednisone alone for symptomatic hormone-resistant prostate cancer: a Canadian randomized trial with palliative end points.* Journal of clinical oncology, 1996. **14**(6): p. 1756–1764.
51. André, N., et al., *Has the time come for metronomics in low-income and middle-income countries?* The Lancet oncology, 2013. **14**(6): p. e239–e248.
52. Jellvert, Å., et al., *Effective oral combination metronomic chemotherapy with low toxicity for the management of castration-resistant prostate cancer.* Experimental and therapeutic medicine, 2011. **2**(4): p. 579–584.
53. Khan, Z., et al., *Biology of Cox-2: an application in cancer therapeutics.* Current drug targets, 2011. **12**(7): p. 1082–1093.
54. Gebbia, V., et al., *Salvage therapy with oral metronomic cyclophosphamide and methotrexate for castration-refractory metastatic adenocarcinoma of the prostate resistant to docetaxel.* Urology, 2011. **78**(5): p. 1125–1130.
55. Millikan, R., et al., *Randomized, multicenter, phase II trial of two multicomponent regimens in androgen-independent prostate cancer.* Journal of clinical oncology, 2003. **21**(5): p. 878–883.
56. Siegel, R.L., K.D. Miller, and A. Jemal, *Cancer statistics, 2018.* CA: a cancer journal for clinicians, 2018. **68**(1): p. 7–30.
57. Hirsch, F.R., et al., *New and emerging targeted treatments in advanced non-small-cell lung cancer.* Lancet, 2016. **388**(10048): p. 1012–1024.
58. Del Conte, A., et al., *Pilot single institutional experience on tolerance and clinical efficacy of metronomic chemotherapy with oral vinorelbine in unfit patients with advanced or metastatic non-small cell lung cancer (NSCLC).* 2011.
59. Correale, P., et al., *Phase II trial of bevacizumab and dose/dense chemotherapy with cisplatin and metronomic daily oral etoposide in advanced non-small-cell-lung cancer patients.* Cancer biology & therapy, 2011. **12**(2): p. 112–118.
60. Hurwitz, H., et al., *Bevacizumab plus irinotecan, fluorouracil, and leucovorin for metastatic colorectal cancer.* The New England journal of medicine, 2004. **350**(23): p. 2335–2342.
61. Carreca, I., et al., *Metronomic (M), capecitabine (C), and oxaliplatin (O) plus bevacizumab (B) as treatment of advanced colorectal cancer (ACRC) in very elderly people (M-COB): Efficacy and safety (E&S) evaluation—A 2-year monitoring.* Journal of clinical oncology, 2011. **29**(15_suppl): p. e14086–e14086.
62. Kelley, R., et al., *A phase 1 trial of imatinib, bevacizumab, and metronomic cyclophosphamide in advanced colorectal cancer.* British journal of cancer, 2013. **109**(7): p. 1725–1734.
63. Lin, P.-C., et al., *Biweekly oxaliplatin plus 1-day infusional fluorouracil/leucovorin followed by metronomic chemotherapy with tegafur/uracil in pretreated metastatic colorectal cancer.* Cancer chemotherapy and pharmacology, 2007. **60**: p. 351–356.

3 Conventional Chemotherapy vs. Metronomic Chemotherapy

Manzoor Ahmad Mir, Nusrat Jan, and Hina Qayoom

3.1 INTRODUCTION

Globally, millions of individuals are affected and killed by the disease called cancer. Breast cancer (BC), in particular, is a serious medical condition, challenging to cure, and has a high incidence and fatality rate (1, 2). It is the second most prevalent form of carcinoma globally and the fifth most frequent reason for cancer-related death. It continues to be the leading reason for cancer-related fatality in females in underdeveloped areas. In developed countries, it is the second most common reason for carcinoma death in females after lung cancer (3). The prevalence of cancer is steadily rising in developed countries, especially in low- and middle-income countries (4). Nevertheless, the survival rate for carcinoma is improving in high-income nations. Prevention, early detection, and complete eradication are the three main concerns for addressing this disease. In particular, treating cancer is a complex issue, and radiation, surgery, and pharmacotherapy are the usual methods of treating cancer (1, 5). The cancer type, screening methods that are supported by evidence, early detection, and treatment and supportive care facilities are some of the variables that influence improvements (6, 7). The discovery of targeted treatment modalities for cancer has substantially advanced cancer therapy over the past few years (8, 9).

Despite the continued high occurrence of BC, there has been a decline in the rate of death over time, primarily as a result of advancements in diagnosis and treatment, which differ greatly across different geographical regions (10, 11). Although early-stage BC treatment has advanced, treating metastatic illness remains challenging (12, 13). Given that it is the most serious type of disease and has the highest level of genetic complexity, metastasis accounts for a large portion of BC patients' deaths (9, 14, 15). Therefore, the choice of treatment regimens for BC depends on a variety of factors, including the stage and subtype of the disease, the hormone receptor status, whether the carcinoma is HER2-positive or HER2-negative, the overall health of the patient (especially whether or not the patient has other diseases or a history of carcinoma), whether the female has undergone menopause, and whether or not the therapy is tolerated well enough to be changed to another course of action. Given this complexity, it is challenging to develop a widely acknowledged form of treatment, mostly because every case is different and the appropriate therapeutic plan should be customized for each individual. To determine the optimal course of treatment, highly qualified healthcare practitioners are crucial (3, 16, 17). Non-cancerous cells that are present in the tumor microenvironment should not be distinguished from cancer cells, as the search for more effective cancer treatments is leading to a shift in the way it is approached. In actuality, compound interactions between cancer cells and the non-cancerous cell microenvironment facilitate the emergence and growth of tumors. The extracellular matrix, tumor stroma, tumor vasculature, specialized mesenchymal cells, immune cells invading the tumor, cancer cells, and cancer stem cells are all potential targets that need to be attacked in the battle against cancer (18–20).

Chemotherapy is thought to be the most efficient and popular form of cancer treatment, whether it is given alone or in conjunction with radiotherapy. Traditional chemotherapy does have certain advantages, but there are also significant drawbacks, including chemoresistance, maximum dose requirements, poor absorption, side effects, therapeutic indices that are minimal, non-specific targeting, and the emergence of multidrug resistance (MDR) (21, 22). A novel therapeutic approach

DOI: 10.1201/9781032632674-3

called "metronomic chemotherapy" (MCT) has been created to overcome the limitations of traditional chemotherapy. The MCT treatment plan targets the activated endothelium cells in carcinomas by often administering low dosages of traditional chemotherapeutic drugs without longer rest intervals and has benefits that include minimal side effects and a low risk of developing acquired drug resistance (23, 24). Traditional cytotoxic chemotherapy has been changing in this new scenario. To increase the effectiveness and tolerability of chemotherapies, both new drug combinations and delivery methods have been in use (25, 26). Contrary to conventional chemotherapy, MCT has many therapeutic benefits, including reduced tumor vascularization, decreased drug resistance, improved cellular and molecular targets, host toxicity, and most importantly, improved antitumor immune responses and patient quality of life (QOL) (27).

In fact, "metronomic scheduling of anticancer treatment" is being investigated as a multitargeted therapy because it permits the continuous and simultaneous administration of several drugs with various anticancer properties. The potential of this strategy, which also offers a wide range of molecules that might be used in combination, still requires much experimentation and understanding. Currently, several anticancer processes are associated with the MCT regimen. Some of these activities have solid preclinical foundations, while others are theoretically sound but lack experimental data (28).

3.2 CONVENTIONAL/TRADITIONAL CHEMOTHERAPEUTICS

Although new metabolic and biochemical pathways have been discovered within the microenvironment of tumors, scientists and medical professionals are developing ways to block the network routes and signals of neovascularization, metastasis, activating apoptosis, and "awakening" the immune response (29). The dynamic and chaotic molecular structure of tumors makes this effort difficult since they can draw in stromal cells, make use of beneficial resources like organic metals and vitamins, and generate their blood supply through abnormal signaling. The cytotoxic drugs, chemotherapeutics used in conventional strategy, have helped oncology patients since the 1970s, but these strategies lack the desired selectivity (30).

Traditional chemotherapy employs chemical substances as a single drug or combination of drugs to treat illness, particularly carcinoma, administered at the maximum tolerable dose (MTD) to achieve maximum cytotoxicity. In the treatment of cancer, it can be combined with other methods like radiation therapy and surgery (31). The most common traditional carcinoma treatment modalities involve surgically removing the tumor, followed by radiotherapy, X-rays, and/or chemotherapy. Among these therapy options, surgery is incredibly effective during the early stages of disease progression. Just like chemotherapy, radiotherapy can harm healthy tissues, cells, and organs. Cytotoxic medicines are administered intravenously at the MTD to completely eradicate cancer cells while maintaining tolerable toxicity and manageable side effects (31–33). Almost all anticancer medications kill healthy cells, especially those that divide quickly and proliferate, even though chemotherapy lowers morbidity and mortality rates. Thus, developing strategies to enhance effectiveness and reduce negative effects will significantly improve the potential of cancer therapy (34). In many cases, tumor chemoresistance can be attributed to cancer stem cells (CSCs) (35). These self-renewing cells that initiate the tumor make up a minor portion of the heterogeneous tumor. The classic chemotherapeutic technique also has other limitations, including difficulties in choosing a dosage, poor precision, quick drug metabolism, and particularly severe adverse effects. Notably, metastatic relapse following chemotherapy has been hypothesized to be caused by therapeutic resistance, particularly in CSCs where apoptosis evasion allows tumor regression. Numerous processes involved in the regulation of the availability of drugs, epithelial-mesenchymal transitions (EMTs), and cancer signaling pathways enable tumor cells to withstand the harmful effects of medications (36). Additionally, through changes in the coding genes for apoptosis-inducing proteins, tumor cells have acquired the capacity to increase genes and alter gene expressions, leading to drug resistance. Additionally, modifications to the structure or function of proteins involved in drug delivery affect

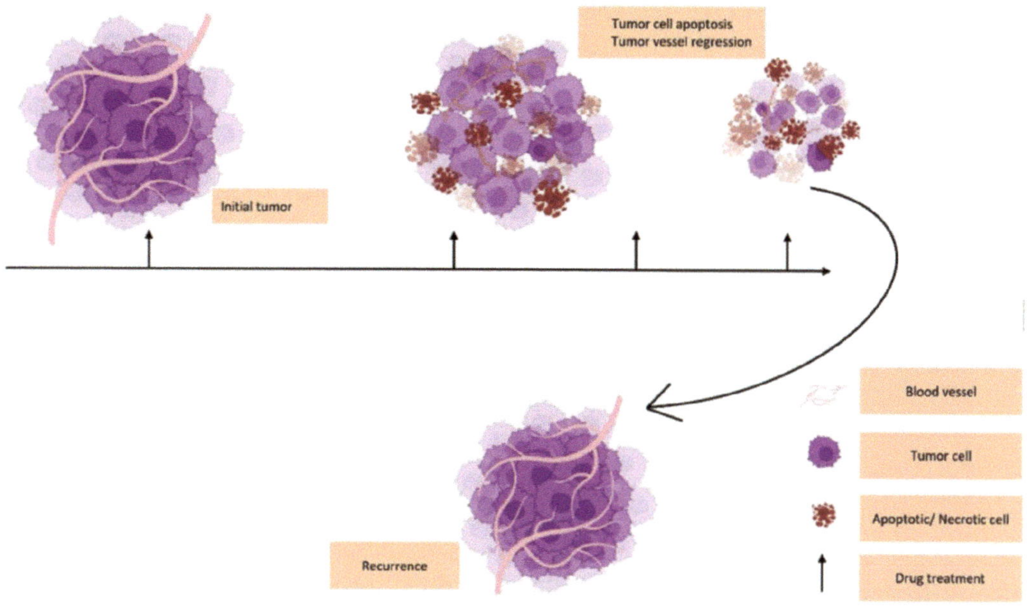

FIGURE 3.1 Traditional chemotherapy regimens are given irregularly with cytotoxic antineoplastic medications at MTD, followed by a long rest, for example, 4 weeks, resulting in tumor angiogenesis recurrence.

the quantity of medication that reaches carcinoma cells, exposing the cancer cells to lower drug concentrations and conferring resistance (37).

The administration of conventional chemotherapy often relies on the idea of the MTD, which includes the highest survivable dose and is geared to destroy as many tumor cells as feasible. But this approach, which had a very high cure rate in preclinical tests, needs periods without therapy to restart the creation of healthy host cells such as epithelial cells and the hematopoietic progenitor cells (38) (Figure 3.1). However, cancer patients in the majority, except for some hematological or germinal malignancies that may be treated with conventional chemotherapy, experience tumor growth after a period of disease regression or stabilization. The heterogeneity and genetic instability of cancer cells, the protective effects of the microenvironment, and the inhibition of antineoplastic immune responses all work to prevent the effectiveness of traditional regimens (39). Metastatic and high-risk tumors may experience recurrence after initial improvements. For many years, the justification and efficacy of traditional MTD-based chemotherapy schedules and the strategies to modify doses have been challenged, particularly in individuals with poor prognoses, and it was necessary to support the potential of an alternative treatment strategy with scientifically convincing research data. Furthermore, MTD chemotherapy does not effectively disrupt tumor-associated angiogenesis, a crucial stage in the growth and formation of tumors. The reason that traditional cytotoxic procedures have little effect on this compartment may be because endothelial cells reproduce more slowly than cancer cells (chronic angiogenesis). It has been intriguingly proposed that the recovery of endothelial cells during periods of no therapy encourages the regrowth of cancer cells (40, 41).

3.2.1 Cytotoxic Chemotherapy for BC

The disease-specific deaths have been reducing in numerous developed nations, despite an increase in the prevalence of BC (42). The early use of systemic medicine is undoubtedly one of the key causes of it. After the diagnosis of BC, the chance of a distant recurrence is highest during the first decade, but during post-diagnosis, it is interesting that it may be significant throughout the second decade. The major goals of systemic adjuvant therapy are to increase long-term overall survival, control any micro-metastatic illness, and lower the recurrence rate. Since the reduction in 15-year BC mortality caused by adjuvant chemotherapeutics, adjuvant radiotherapy, and hormonal therapy occurs after the first 5 years (43), there may be a 10-year or longer delay between any widespread changes in practice and the primary effects that these will eventually have on national BC death rates (44). Consequently, the rapid reduction in BC death rates of 25–30% observed in the United States and UK during the 1980s was mostly attributed to earlier diagnosis, increased use of adjuvant therapies, or both (45). Additionally, as a result of more cautious and thorough screening, modest advancements during the 1990s in local disease management and better systemic treatment of both early and advanced disease should help these decreases in national mortality rates to continue throughout the current decade (46). Despite advances in knowledge of how adjuvant medicines are used in the treatment of early-stage BC, treating metastatic illness continues to be very difficult. As more women with metastatic cancer present with exposure to anthracyclines and taxanes due to their usage in adjuvant therapy, the care of these patients has become more challenging. Although metastatic BC is incurable, it frequently retains chemosensitivity, allowing for symptom management and life extension. Long-term survival is still rare, and response time is still frustratingly short. Chemotherapeutics for BC have been evaluated in advanced metastatic disease, and after showing effective activity with tolerable toxicity effects, have moved into the adjuvant studies. Additional potential targets for cutting-edge treatments are provided by the continual advances in understanding the molecular biology of BC. The introduction of trastuzumab, a humanized monoclonal antibody against HER-2/neu, offers the first illustration of a targeted biological therapy for BC that was effectively evaluated in randomized clinical settings and is now universally recognized as standard therapy (47). Resistance to therapy still poses a problem in treating metastatic BC, despite these significant advancements (44). The increased understanding of the molecular biology behind BC offers potential targets for new treatments. The risk-benefit ratio of current treatments may be improved by therapies such as adjuvant or neoadjuvant chemotherapy for BC based on the tumor's molecular characteristics. The use of biomarkers will thereby direct us toward personalized therapies while also saving a subset of individuals who do not benefit from adjuvant chemotherapy. At the same time, a deeper comprehension of cancer biology will facilitate the creation of new regimens for patients who aren't likely to respond well to current treatments. As opposed to a single gene or gene product that might produce regimen-specific predictors, combinations of markers are anticipated to be more informative in predicting response. This theory can be tested using new molecular analytical methods, particularly transcriptional profiling (48).

The use of cytotoxic chemotherapy in both advanced and early stages of BC has evolved significantly over the past 10 years, and numerous studies have shown that novel therapies offer obvious survival advantages (12).

3.2.2 Neoadjuvant Therapy in Breast Cancer

A significant recent development with effectiveness for the management of early BC in the future is primary or neoadjuvant chemotherapy. The usual roles of surgery and chemotherapy are inverted in this method, and chemotherapy is administered before surgery (adjuvant treatment). The idea first came about as a result of experience with locally advanced, incurable BC, where standard practice now includes medical treatment, particularly chemotherapy given before local therapy (49). The management of individuals with extensive but treatable BC has been expanded to include this strategy, since the only standard treatment is mastectomy rather than conservative surgery in these cases. Regardless of local treatment, these individuals typically have a poor prognosis (50), and adjuvant chemotherapy is typically provided in any case. Initially, the main justification for chemotherapy before surgery was downstaging in the hopes of avoiding mastectomy. Furthermore, it was mentioned that preoperative medical therapy made it possible to use the tumor as an in vivo indicator of treatment efficacy. Preoperative medical care has become more widely accepted, and it is now being used more frequently in patients with small primary tumors in the hopes of achieving some survival benefit. Data on primary chemotherapy are currently accumulating quickly, but some important questions remain unanswered, including its long-term role in the treatment of early BC. With overall objective response rates ranging from 70% to more than 90%, preoperative chemotherapy is very effective in treating early BC. These rates are greater than those previously reported for individuals with metastatic disease. Patients underwent CHOP chemotherapy (CTX, doxorubicin, vincristine, and prednisolone), either as first-line therapy or after failing endocrine therapy, in one of the earliest studies to be released from Edinburgh. Their most recent findings revealed a response rate of 70% and a complete clinical remission rate of 26% (51).

3.2.3 Adjuvant Therapy in BC

Early BC is a type of carcinoma that is confined to only the breast or, in the case of node-positive patients, the local lymph nodes and the breast. All early BCs can be surgically removed. With more advanced surgical methods, it is no longer a problem to delay starting chemotherapy after surgery. If necessary, chemotherapy is often started within 6 weeks of surgery. But if left untreated, micrometastatic illness may persist locally or at distant places and, over the coming years, could become a potentially fatal clinical recurrence. The early strategy for treating BC was based on the idea that the illness propagated via locoregional spread in a systematic way and could be treated with vigorous surgery. As a result, radical mastectomy became the go-to surgical treatment for BC in the early 20th century (52). Randomized studies later revealed no advantage from radical mastectomy in comparison to less invasive surgical techniques and showed that distant recurrence remained a significant clinical issue regardless of the original surgical therapy (53, 54). Adjuvant systemic therapies came in more varieties with more applications when the method of local therapy changed from being more aggressive to being less aggressive. Adjuvant systemic chemotherapy, endocrine therapy, and anti-HER2–focused therapy all significantly decreased the likelihood of recurrence when combined with local therapy, according to numerous clinical trials carried out. These treatments also significantly increased overall survival (55).

Based on the improvements in outcomes connected to systemic therapies mentioned later, the available adjuvant therapy options are often customized to four phenotypic subtypes that are identified in a useful way by using data on estrogen receptor, progesterone receptor, and HER2 expression. The "intrinsic subtypes" revealed by gene expression profiling closely correspond to this practical phenotypic classification (56), even if the latter classification might offer more precise prognostic and predictive data (57, 58). The National Surgical Adjuvant Breast and Bowel Project (NSABP) B-01 trial, the first randomized study examining adjuvant chemotherapy for BC, was started in 1958 and reported in 1968 that an adjuvant alkylating agent (thiotepa) given after radical mastectomy significantly decreased recurrence rate in premenopausal women with four or more positive axillary lymph nodes. Another alkylating substance, L-phenylalinine mustard, was beneficial, according to a later randomized research study published in 1975. Other studies from the Istituto Nazionale Tumori in Milan, Italy, demonstrated that a combination chemotherapy regimen known as "CMF" that included an alkylating agent (CTX) and antimetabolites (MTX and 5-fluorouracil) significantly decreased the risk of recurrence (55). This marked the beginning of the modern era of adjuvant polychemotherapy regimens, which are now routinely utilized in clinical practice. These studies were among the first to demonstrate the value of adjuvant chemotherapy, initially in premenopausal women at the highest risk of recurrence and with axillary node-positive disease. Subsequent studies also demonstrated this benefit in postmenopausal women at lower risk and in individuals with axillary node-negative disease (55, 59). "Because adjuvant polychemotherapy improves survival, it should be recommended to the majority of women with localized breast cancer regardless of lymph node, menopausal, or hormone receptor status," said a National Institute of Health consensus panel in the United States in 2001.

Although the widespread use of more efficient systemic medicines helped in reducing BC death rates in the United States and around the world (60, 61), it also led to many individuals who might have been cured without chemotherapy being accidentally "overtreated" with chemotherapy. Recent research has demonstrated that numerous multiparameter gene expression assays can predict prognosis in individuals with ER-positive BC (62, 63) and can help determine which individual will benefit most from adjuvant chemotherapy (64, 65). Evidence-based recommendations for the use of adjuvant chemotherapy in particular contexts support the use of some of these assays. Hormone receptors are expressed in about 75% of breast carcinomas (66). Regardless of whether hormone therapy is given alone or in combination with chemotherapy, it lowers the chance of recurrence in diseases that express hormone receptors. Adjuvant tamoxifen increased the survival rate and decreased the chance of recurrence when administered for 2 years in 1982. In addition, 5 years of therapy was more effective than shorter durations, regardless of nodal metastasis; the proportional benefits were similar; and the benefits were only observed in individuals with ER-positive cancers. Moreover, the studies conducted later and meta-analyses of these studies and also supported a survival benefit. Later research revealed that tamoxifen was less effective in postmenopausal women than aromatase inhibitors (55).

Additionally longer adjuvant treatments, such as tamoxifen for up to 10 years (67) or subsequent tamoxifen followed by an aromatase inhibitor (68), are more beneficial than shorter treatments of up to 5 years. Finally, it has been demonstrated that ovarian suppression when combined with an aromatase inhibitor is superior (efficacious) to tamoxifen in premenopausal women who are at high risk for recurrence (69, 70).

HER2 genes are overexpressed in about 25% of all breast carcinomas (71). Trastuzumab, an anti-HER2 antibody, was added to adjuvant chemotherapy in 2005, either concurrently or sequentially, as shown by randomized studies conducted, and it was found that this significantly reduced the chance of recurrence in individuals with HER2 overexpressing node-positive or high-risk node-negative BC. A combination of trastuzumab with non-anthracycline regimens (such as carboplatin/docetaxel) was linked with decreased rates of cardiac toxicity, whereas adding trastuzumab to sequential anthracycline/CTX-taxane was associated with an approximately 3% risk of cardiac toxicity (72–75). Outstanding outcomes were shown by single-arm nonrandomized studies conducted in individuals with lower-risk node-positive carcinoma who were excluded from prior studies but were predicted to have greater recurrence rates in the absence of adjuvant trastuzumab (76, 77). Trastuzumab was found to be more beneficial for 1 year than it was for 6 months in subsequent investigations, but 2 years of treatment had no greater benefit than 1 year (78). Trastuzumab was given in combination with the HER2 tyrosine kinase inhibitor lapatinib, but the results were not improved (79).

3.2.4 Applications of Cytotoxic Chemotherapy in Clinical Settings

3.2.4.1 Hematological Malignancy

Several hematological malignancies respond favorably to conventional chemotherapy (cytotoxic) at first due to their high proliferation indices. With rigorous therapy, high-grade leukemia and lymphomas can be cured, whereas low-grade cancers can be managed for years with medication doses that allow for normal daily activities. However, the prognosis for recurrence is poor, and additional treatment may not provide long-lasting remission (80).

3.2.4.2 Advanced (Metastatic) Solid Cancers

The primary treatment choice for locally advanced solid tumors includes surgical procedures and radiotherapy, although many patients also have metastatic spread or recurring diseases that cannot be treated with these techniques. Individuals with advanced cancer frequently receive cytotoxic conventional treatment. It is rarely therapeutically effective. The best response to chemotherapy is usually a reduction in tumor size. This can greatly improve survival time and symptom control compared to the normal course of the disease. Drug resistance is the most frequent cause of treatment failure, and mutations (intrinsic or acquired) in carcinogenesis shield tumor cells from chemotherapy (81).

3.2.4.3 Early Cancer Chemotherapy as an Adjunct to Surgery

To increase cure rates after the initial surgery procedures, chemotherapeutics are frequently first evaluated in late disease and, if effective there, typically adopted for early disease treatment.

Neoadjuvant treatment is being utilized more and more as a treatment for localized carcinoma before the planned curative surgery. When compared to surgery alone, this method has been found to enhance overall survival (OS) in individuals having surgical treatment for localized esophageal carcinoma. It has also made it now possible to perform a more conservative surgical removal of large primary breast tumors.

Adjuvant cancer treatment is now regularly administered to individuals at greater risk of recurrence after surgical removal of, for example, primary BC or bowel carcinoma (82).

Radiotherapy and chemotherapy are sometimes more efficient when used together. 5-Fluorouracil and other radio-sensitization drugs can effectively shrink large esophageal and rectal carcinomas that may have been untreatable otherwise (83).

3.3 METRONOMIC CHEMOTHERAPEUTICS

In terms of delivery schedules, toxicity profiles, and ultimately anticancer effects, molecular-targeted medicines are distinct from conventional chemotherapeutic drugs (41). A novel therapeutic approach called MCT has been established to address the limitations of traditional chemotherapy. This therapeutic strategy addresses the persistent, evenly spaced administration of various chemotherapy medicines in low strengths without longer rest intervals. The uniqueness of this treatment approach also lies in a change in cell target, now concentrating on tumor-endothelial cells, in addition to its success against tumors with extremely low toxicity. The knowledge and experience gained in the preclinical MCT field, along with the enhanced experience gained in clinical settings, will enable a change in the formulation of therapeutic guidelines against carcinoma (84, 85). The MCT deals with the repetitive, relatively low-dose administration of anticancer drugs, frequently without allowing a longer interval of a drug-free condition. When it comes to controlling the progression of cancer, it was once believed that MCT targets angiogenesis; however, more recent research has revealed different processes, leading to the discovery of MCT as a component of a multitargeted treatment approach. A further scientific tenet of MCT is that traditional anticancer medications concentrate on enlarging vascular endothelial cells. However, because endothelial cells have a chance to recover while under treatment, the effect against angiogenesis cannot be extended. Furthermore, this may be overcome by therapy administered at low dosages repeatedly (23, 86). MCT has several targets since it impacts cancer cells both directly and indirectly and the environment around them. Inhibiting tumor angiogenesis, boosting the immune system's capacity to combat tumors, and even inducing tumor dormancy are all possible effects. MCT was previously proven to have antiangiogenic activity, but recently it has been proposed that it can enhance these effects by several interrelated pathways, making it different from conventional chemotherapy (87).

In the 2000s, Kerbel (88) and Hanahan (89) originally used the term "MCT" to describe the continuous administration of traditional chemotherapy medications at low dosages regularly without long drug-free breaks. The main mechanism of action does not directly inhibit tumor growth. In addition to stimulating the host's immune response against carcinoma and acting on stromal tissue, MCT mostly has indirect effects on cancer cells, particularly on their microenvironment, by inhibiting tumor angiogenesis (89). Preliminary research on MCT in several carcinoma types, particularly BC and lung carcinoma, however, revealed intriguing and encouraging findings: MCT remained restricted to palliative settings for a considerable amount of time since it was mistakenly believed to be devoid of anticancer potential. Since oral medications can be delivered for longer durations than conventional chemotherapy, they are chosen for use in a metronomic regimen (86).

Depending on the use context, the cancer type, and the preferences of the individual, different metronomic medication concentrations and regimens can have varied effects. The most researched medications for MCT in this context include cyclophosphamide (CTX), methotrexate (MTX), capecitabine (CAPE), and oral vinorelbine (VNR) for BC; CAPE for gastrointestinal malignancies; and oral VNR for non–small cell lung cancer. With the introduction of new medications, particularly immune checkpoint inhibitors, and a better understanding of the specific modes of action of MCT, it is advised that these regimens be employed more frequently in clinical settings (88).

Browder and colleagues presented such long-awaited preclinical evidence from Judah Folkman's laboratory 13 years ago, and it was verified in Robert Kerbel's laboratory. Both groups used transplantable tumors and xenograft models to demonstrate the antiangiogenic effectiveness of low-dose chemotherapy. The first trial found that a metronomic CTX regimen was more successful than standard chemotherapy and could overcome medication resistance. The second trial investigated the potential synergistic effects of anti-vascular endothelial growth factor (VEGF) receptor (VEGFR) therapy with continuous treatment of low-dose vinblastine therapy administered continuously (40, 90). The scientific rationale behind MCT is that while traditional anticancer drugs target vascular endothelial cell proliferation, their antiangiogenic effect is short-lived due to endothelial cell recovery during treatment breaks, which can be overcome by continuous treatment at low doses (89), and MCT is the regular delivery of chemotherapeutic medications at dosages lower than the MTD without a long period of drug-free rest. It therefore achieves a prolonged low blood level of the medication without major side effects (91) (Figure 3.2). The following are the primary characteristics of MCT (92):

- Using an optimized biological dose (OBD) instead of MTD for frequent (dose-dense) chemotherapy treatment.
- No use of hematopoietic growth factors.
- Preference for oral medications.
- Low prevalence of treatment-related side effects.
- Possibility of developing a delayed resistance.

Numerous investigations have revealed that the majority of clinical trials using MCT are well tolerated. Furthermore, as evaluated in multiple clinical trials, this therapy technique is regarded as an alternative treatment that has a cost benefit. The severe toxic consequences are either nonexistent or extremely infrequent. The side effects generally associated with this therapy include anemia, mild to severe neutropenia, leukopenia, lymphopenia, and vomiting, as well as modest fatigue. Moreover, MCT offers exceptional therapeutic progress, including an improvement in QOL with low harm. To supplement MCT use in clinical settings, however, a well-conducted phase III trial demonstrating the significant importance of metronomic dosage is required. Furthermore, well-evaluated randomized phase II trials would provide convincing data to support the use of low-dose chemotherapy. The individuals to be investigated should be proportionate and associated with patients from the real world (93, 94). So far, locating clinical studies that are both randomized and statistically significant has been challenging.

BC is among the well-researched cancers for MCT. The therapy is mainly aimed at individuals suffering from metastatic disease to improve their QOL and alleviate their symptoms. MCT is the ideal alternative to conventional treatment in terms of adverse effects and efficacy. CTX, MTX, VRB, and CAPE were the most frequently studied cytotoxic medicines in metronomic trials. CAPE and VRB have been the most thoroughly investigated anticancer drugs as monotherapy (86).

Taguchi and colleagues studied the effectiveness of CAPE at a low dose as first-line treatment in 33 individuals with metastatic BC in 2010. This medication was given regularly according to the following schedule: 825 mg/m^2 two times daily for 21 days out of 28. The median OS and progression-free survival (PFS) (95) were 24.8 and 6.9 months, respectively. This study found that CAPE was more tolerable and effective on this schedule (96).

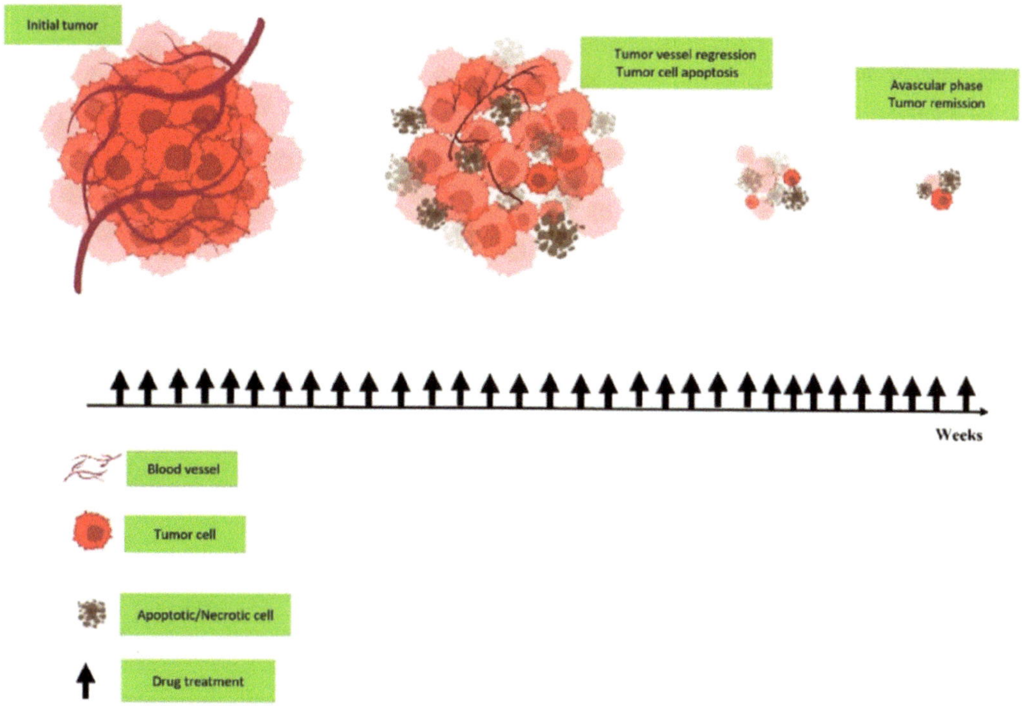

FIGURE 3.2 MCT is given regularly, for example, every 4 days or daily, utilizing fractionated MTD of chemotherapy medications for a time period without long drug-free breaks, leading to tumor angiogenesis inhibition and inhibition of tumor growth and resulting in disease remission.

Fedele et al. investigated the efficacy of CAPE given regularly to pretreated 60 metastatic BC individuals by providing 1500 mg once daily. According to this study, the median time to progression, OS, and clinical benefit rate were 7%, 17%, and 62%, respectively. Furthermore, VRB was measured at 70 mg/m^2 three times a week in 34 senior individuals, followed by a 1-week respite. This MCT program was tolerated well (6% grade 3 neutropenia) with an OS and median PFS of 15.9 and 7.7 months, respectively (95). Furthermore, De Iuliis and colleagues evaluated VRB in 32 individuals at a dose of 30 mg orally every other day. The safety profile was optimal in this trial (no grade 3 or 4 accidents), and the clinical benefit rate (CBR) was 50%. Similarly, 63 individuals received a modest dose of CTX and MTX in a combinational oral administration, yielding an overall CBR and overall response rate (ORR) of 32% and 10%, respectively (97) (Table 3.1).

3.4 DIFFERENCE FROM CONVENTIONAL CHEMOTHERAPY

In conventional chemotherapy, cytotoxic medicines are often administered in cycles close to or at the MTD and replace a significant drug-free interval to allow the individual to recuperate from severe drug responses. The effectiveness of this method slows the illness process in many pediatric and adult patients, although it has significant drawbacks. Furthermore, while most cancers improve in the early stages, recurrence is a common issue in metastatic and dangerous malignancies (23, 98). MCT, on the other hand, refers to the repetitive and continuous use of traditional chemotherapy drugs at low doses, which is an advanced alternative to conventional chemotherapy (99). Furthermore, its suitable toxicity profile, cheaper price, and ease of application are critical advantages over traditional treatments (100) (Table 3.2).

TABLE 3.1
MCT for Breast Cancer

Type of Cancer	MCT	Findings	Reference
BC	*CAPE*: 825 mg/m² two times daily for 21 out of 28 days.	The OS and MPFS were 24.8 and 6.9 months, respectively. This study showed the excellent tolerability and effectiveness of CAPE in this schedule.	(96)
BC	*VNR*: In 34 elderly patients, 70 mg/m² was administered thrice per week, followed by a 1-week rest.	The schedule was tolerated well (grade 3 neutropenia 6%), and OS and MPFS were 15.9 and 7.7 months, respectively.	(95)
BC	63 patients received a low-dose combinational oral administration of CPM and MTX.	The overall CBR and overall objective response rate were 32% and 19% respectively.	(97)

Abbreviations: BC, Breast cancer; CAPE, capecitabine; CBR, clinical benefit rate; MPFS, median progression-free survival; OS, overall survival; VNR, vinorelbine.

TABLE 3.2
Distinguishing Features between Conventional Chemotherapy and Metronomic Chemotherapy

S. No.	Parameters	Traditional/Conventional Chemotherapy	Metronomic Chemotherapy (MCT)	References
1	Dose	Established to deliver at MTD	Low doses below the MTD	(91)
2	Dosing frequency	Given at intervals (weekly, fortnightly, three-weekly)	regularly, every alternate day, weekly	(23)
3	Concentration of plasma	Altered	controlled	(23, 101)
4	Cells targeted	Proliferating cancer cells	Endothelial cells in the tumor's developing vasculature that have been angiogenically stimulated	(102)
5	Purpose of administration	Directly treating cancer by blocking or removing rapidly developing tumor cells	Controlling cancer by focusing on angiogenesis	(23)
6	Toxicity	Toxicity is a big problem as a result of medications being used at maximum tolerated doses	Because of the achievement of a continual low level of medicines in the bloodstream, no significant harmful side effects were seen	(91)
7	Effectiveness	Much more effective in the primary tumor than in metastasis	Advanced cancer	(23)
8	Quality of life	Poor	Improved	(27)

Abbreviation: MTD, maximally tolerable dose.

3.5 CONCLUSION

MCT in cancer treatment began silently, restricted to later lines of therapy and entirely in palliative care. Since the early twenties, the use of MCT has grown in popularity, with BC serving as the forerunner, followed by other types of carcinomas, particularly lung cancer and pediatric cancers. Oncologists currently recommend MCT for a variety of cancers, typically outside of clinical trials and following the failure of at least two prior lines of therapy. Contrary to MTD therapies, the dosage of which was established in well-defined phase I trials, MCT dosages and frequency have typically been developed empirically to determine the OBD, the dose that reduces the tumor volume greatly with no or minimal side effects.

MCT's principles include its simple administration, its low frequency of adverse effects, and its numerous modal methods of action. Angiogenesis suppression is the most well-known effect of MCT, but interesting new findings show that it has activity on immune system activation. Different preclinical findings have shown the potential combination of MCT with antiangiogenic medications (e.g., bevacizumab, sunitinib, erlotinib) or anti-PD-L1 therapies. Additionally, some clinical data are also available, but with inconsistent outcomes and seemingly no added benefit for MCT. Further investigation is needed for these combinations in both preclinical and clinical settings for a deep knowledge of the complicated interplay stimulated or suppressed by the combination of these drugs to properly investigate the role of MCT. In terms of BC, several areas have already been established. According to real-life studies, MCT is a beneficial treatment, particularly in advanced HR-positive/HER2-negative BC patients without visceral crises and with minimal tumor load, as well as in elderly patients. In clinical settings, some areas of interest remain unresolved, with little information available, particularly in maintenance treatment in high-risk metastatic triple-negative breast cancer (TNBC) patients and prolonged adjuvant treatment in individuals with early TNBC who received radical resection and were at high risk of relapse.

Preclinical and clinical research has established MCT as a new treatment option for specific forms of cancer. Even after a decade of clinical research, optimizing a metronomic anticancer therapy remains a difficult issue. As a result, future cancer research should seek to identify the optimal agents to utilize based on cancer type, calculate the doses of each agent to be used alone or in combination, and specify medication administration time. In oncology clinics, new methods are being developed to combine MCT with traditional chemotherapy, radiation, and/or targeted therapy, and with these methods, an unlimited number of possible combinations can be explored. In-depth pharmacogenetic and pharmacoproteomic studies on tumor endothelium cells, however, are now required to assess their susceptibility to MCT and to establish the most effective drug combination to utilize in the clinic.

MCT has mostly only been used in palliative settings until now. It is now appropriate to apply its application to numerous other fields with the increasing significance of preclinical and clinical evidence. It should be taken into consideration for early-stage BC, including other types of cancer. Treatment of patients in nations with few resources could be greatly aided using oral treatment regimens that can be provided at home and which also have very favorable safety profiles.

For further reading, some textbooks mentioned here also give further understanding of MCT:

- *Progress in Anti-cancer Chemotherapy* by Gabriel N. Hortobagyi and Khayat, David.
- *Triple-Negative Breast Cancer: Biomarkers, Emerging Therapeutic Strategies and Clinical Challenges* by Marion Montgomery.
- *Current Cancer Treatment: Novel Beyond Conventional Approaches* by Oner Ozdemir.
- *Breast Cancer: Current and Alternative Therapeutic Modalities* by Esra Gunduz and Mehmet Gunduz.

The following links for video lectures may also be helpful:

https://youtu.be/TAqO0Mq19JQ

https://youtu.be/6wEnYfvoBqk
https://youtu.be/A9fRRCUlMms
https://youtu.be/zIWvmCsKr34

For more insight about the topic, we would suggest detailed findings from the books of (Mir MA, 2022) https://doi.org/10.1016/C2021-0-02565-7, https://doi.org/10.1016/C2022-0-00074-X, (Mir MA, 2021) https://doi.org/10.52305/WXJL6770, and (Mir MA, 2015) https://doi.org/10.1016/C2014-0-02898-5 and from the cancer.net website on the following mentioned links:

https://www.cancer.net/cancer-types/breast-cancer/types-treatment
https://discovery.ucl.ac.uk/id/eprint/1472740/
https://www.jmedsciences.com/doi/JMEDS/pdf/10.5005/jp-journals-10045-00138

For diagrammatic illustrations and descriptive tables (Lazzeroni, 2012): http://www.eurekaselect.com/article/49928

See video links on the overall status of cancer, its various types, and current and new possible treatment options available: https://www.sciencedirect.com/science/article/pii/S205970292032278X

REFERENCES

1. Correia A, Silva D, Correia A, Vilanova M, Gärtner F, Vale N. Study of new therapeutic strategies to combat breast cancer using drug combinations. Biomolecules. 2018;8(4):175.
2. Mir MA, Qayoom H, Mehraj U, Nisar S, Bhat B, Wani NA. Targeting different pathways using novel combination therapy in triple negative breast cancer. Current cancer drug targets. 2020;20(8):586–602.
3. Correia AS, Gärtner F, Vale N. Drug combination and repurposing for cancer therapy: The example of breast cancer. Heliyon. 2021;7(1):e05948.
4. Allemani C, Matsuda T, Di Carlo V, Harewood R, Matz M, Nikšić M, et al. Global surveillance of trends in cancer survival 2000–14 (CONCORD-3): Analysis of individual records for 37 513 025 patients diagnosed with one of 18 cancers from 322 population-based registries in 71 countries. The Lancet. 2018;391(10125):1023–75.
5. Qayoom H, Bhat BA. U Mehraj U, Mir MA. Rising trends of cancers in Kashmir valley: Distribution pattern, incidence and causes. Journal of oncology research and therapy. 2020;5(150):2.
6. Gurney J, Stanley J, McLeod M, Koea J, Jackson C, Sarfati D. Disparities in Cancer-specific survival between Māori and non-Māori New Zealanders, 2007-2016. JCO global oncology. 2020;6:766–74.
7. Sofi S, Mehraj U, Qayoom H, Aisha S, Asdaq SMB, Almilaibary A, et al. Cyclin-dependent kinases in breast cancer: Expression pattern and therapeutic implications. Medical oncology. 2022;39(6):106.
8. Krause DS, Van Etten RA. Tyrosine kinases as targets for cancer therapy. New England journal of medicine. 2005;353(2):172–87.
9. Mehraj U, Dar AH, Wani NA, Mir MA. Tumor microenvironment promotes breast cancer chemoresistance. Cancer chemotherapy and pharmacology. 2021;87:147–58.
10. Anjum F, Razvi N, Masood MA. Breast cancer therapy: A mini review. MOJ drug design development & therapy. 2017;1(2):00006.
11. Mehraj U, Mushtaq U, Mir MA, Saleem A, Macha MA, Lone MN, et al., editors. Chemokines in triple-negative breast cancer heterogeneity: New challenges for clinical implications 2022: Elsevier.
12. Hassan MSU, Ansari J, Spooner D, Hussain SA. Chemotherapy for breast cancer. Oncology reports. 2010;24(5):1121–31.
13. Sofi S, Qayoom H, Jan N, Khaliq N, Almilaibary A, Mir MA. A comprehensive analysis of Notch signalling genes in breast cancer: Expression pattern and prognostic significance. Advances in cancer biology-metastasis. 2023;7:100104.
14. Bertucci F, Ng CKY, Patsouris A, Droin N, Piscuoglio S, Carbuccia N, et al. Genomic characterization of metastatic breast cancers. Nature. 2019;569(7757):560–4.
15. Qayoom H, Wani NA, Alshehri B, Mir MA. An insight into the cancer stem cell survival pathways involved in chemoresistance in triple-negative breast cancer. Future oncology. 2021;17(31):4185–206.
16. Mir MA, Hamdani SS, Sheikh BA, Mehraj U. Recent advances in metabolites from medicinal plants in cancer prevention and treatment. Current immunology reviews. 2019;15(2):185–201.

17. Mehraj U, Qayoom H, Mir MA. Prognostic significance and targeting tumor-associated macrophages in cancer: New insights and future perspectives. Breast cancer. 2021;28:539–55.
18. Pasquier E, Kavallaris M, André N. Metronomic chemotherapy: New rationale for new directions. Nature reviews clinical oncology. 2010;7(8):455–65.
19. Mir MA, Mehraj U. Double-crosser of the immune system: macrophages in tumor progression and metastasis. Current immunology reviews. 2019;15(2):172–84.
20. Mehraj U, Ganai RA, Macha MA, Hamid A, Zargar MA, Bhat AA, et al. The tumor microenvironment as driver of stemness and therapeutic resistance in breast cancer: New challenges and therapeutic opportunities. Cellular oncology. 2021;44(6):1–21.
21. Panthi VK, Dua K, Singh SK, Gupta G, Hansbro PM, Paudel KR. Nanoformulations-Based Metronomic Chemotherapy: Mechanism, Challenges, Recent Advances, and Future Perspectives. Pharmaceutics. 2023;15(4):1192.
22. Mehraj U, Wani NA, Hamid A, Alkhanani M, Almilaibary A, Mir MA. Adapalene inhibits the growth of triple-negative breast cancer cells by S-phase arrest and potentiates the antitumor efficacy of GDC-0941. Frontiers in pharmacology. 2022;13:958443.
23. Maiti R. Metronomic chemotherapy. Journal of pharmacology and pharmacotherapeutics. 2014;5(3):186–92.
24. Mehraj U, Mir IA, Alkhanani M, Wani NA, Mir MA. Adapalene and doxorubicin synergistically promote apoptosis of TNBC cells by hyperactivation of the ERK1/2 pathway through ROS induction. Frontiers in oncology. 2022;12:938052.
25. Romiti A, Cox MC, Sarcina I, Di Rocco R, D'Antonio C, Barucca V, et al. Metronomic chemotherapy for cancer treatment: A decade of clinical studies. Cancer chemotherapy and pharmacology. 2013;72:13–33.
26. Qayoom H, Mehraj U, Sofi S, Aisha S, Almilaibary A, Alkhanani M, et al. Expression patterns and therapeutic implications of CDK4 across multiple carcinomas: a molecular docking and MD simulation study. Medical oncology. 2022;39(10):158.
27. Kim JY, Kim Y-M. Tumor endothelial cells as a potential target of metronomic chemotherapy. Archives of pharmacal research. 2019;42:1–13.
28. Andre N, Padovani L, Pasquier E. Metronomic scheduling of anticancer treatment: the next generation of multitarget therapy? Future oncology. 2011;7(3):385–94.
29. O'Meara T, Safonov A, Casadevall D, Qing T, Silber A, Killelea B, et al. Immune microenvironment of triple-negative breast cancer in African-American and Caucasian women. Breast cancer research and treatment. 2019;175:247–59.
30. Ventola CL. Progress in nanomedicine: Approved and investigational nanodrugs. Pharmacy and therapeutics. 2017;42(12):742.
31. Frei Iii E, Canellos GP. Dose: A critical factor in cancer chemotherapy. The American journal of medicine. 1980;69(4):585–94.
32. Hanna N, Einhorn LH. Testicular cancer: A reflection on 50 years of discovery. Journal of clinical oncology. 2014;32(28):3085–92.
33. May T, Goldstein DP, Berkowitz RS. Current chemotherapeutic management of patients with gestational trophoblastic neoplasia. Chemotherapy research and practice. 2011;2011:806256. doi: 10.1155/2011/806256.
34. Pangeni R, Panthi VK, Yoon I-S, Park JW. Preparation, characterization, and in vivo evaluation of an oral multiple nanoemulsive system for co-delivery of pemetrexed and quercetin. Pharmaceutics. 2018;10(3):158.
35. Wang X, Zhang H, Chen X. Drug resistance and combating drug resistance in cancer. Cancer drug resistance. 2019;2(2):141.
36. Phi LTH, Sari IN, Yang Y-G, Lee S-H, Jun N, Kim KS, et al. Cancer stem cells (CSCs) in drug resistance and their therapeutic implications in cancer treatment. Stem cells international. 2018;2018:5416923. doi: 10.1155/2018/5416923.
37. Peetla C, Vijayaraghavalu S, Labhasetwar V. Biophysics of cell membrane lipids in cancer drug resistance: Implications for drug transport and drug delivery with nanoparticles. Advanced drug delivery reviews. 2013;65(13-14):1686–98.
38. He S. Experimental evaluation of potential anticancer agents. XIII. On the criteria and kinetics associated with "curability" of experimental leukemia. Cancer chemotherapy reports. 1964;61: 1307–17.
39. Gasparini G. Metronomic scheduling: The future of chemotherapy? The Lancet oncology. 2001;2(12):733–40.

40. Browder T, Butterfield CE, Kräling BM, Shi B, Marshall B, O'Reilly MS, et al. Antiangiogenic scheduling of chemotherapy improves efficacy against experimental drug-resistant cancer. Cancer research. 2000;60(7):1878–86.
41. Kerbel RS, Kamen BA. The anti-angiogenic basis of metronomic chemotherapy. Nature reviews Cancer. 2004;4(6):423–36.
42. Guarneri V, Conte PF. The curability of breast cancer and the treatment of advanced disease. European journal of nuclear medicine and molecular imaging. 2004;31:S149–S61.
43. Early Breast Cancer Trialists' Collaborative G. Favourable and unfavourable effects on long-term survival of radiotherapy for early breast cancer: An overview of the randomised trials. The Lancet. 2000;355(9217):1757–70.
44. Hussain SA, Palmer DH, Stevens A, Spooner D, Poole CJ, Rea DW. Role of chemotherapy in breast cancer. Expert review of anticancer therapy. 2005;5(6):1095–110.
45. Peto R, Boreham J, Clarke M, Davies C, Beral V. UK and USA breast cancer deaths down 25% in year 2000 at ages 20-69 years. Lancet. 2000;355(9217):1822.
46. Early Breast Cancer Trialists' Collaborative G. Effects of chemotherapy and hormonal therapy for early breast cancer on recurrence and 15-year survival: An overview of the randomised trials. The Lancet. 2005;365(9472):1687–717.
47. Henson ES, Hu X, Gibson SB. Herceptin sensitizes ErbB2–overexpressing cells to apoptosis by reducing antiapoptotic Mcl-1 expression. Clinical cancer research. 2006;12(3):845–53.
48. Pusztai L, Symmans FW, Hortobagyi GN. Development of pharmacogenomic markers to select preoperative chemotherapy for breast cancer. Breast cancer. 2005;12(2):73–85.
49. Bartelink H, Rubens RD, Van der Schueren E, Sylvester R. Hormonal therapy prolongs survival in irradiated locally advanced breast cancer: A European Organization for Research and Treatment of Cancer Randomized Phase III Trial. Journal of clinical oncology. 1997;15(1):207–15.
50. Haagensen CD, Bodian C. A personal experience with Halsted's radical mastectomy. Annals of surgery. 1984;199(2):143.
51. Sapunar F, Smith IE. Neoadjuvant chemotherapy for breast cancer. Annals of medicine. 2000;32(1):43–50.
52. Cotlar AM, Dubose JJ, Rose DM. History of surgery for breast cancer: Radical to the sublime. Current surgery. 2003;60(3):329–37.
53. Fisher B, Anderson S, Bryant J, Margolese RG, Deutsch M, Fisher ER, et al. Twenty-year follow-up of a randomized trial comparing total mastectomy, lumpectomy, and lumpectomy plus irradiation for the treatment of invasive breast cancer. New England journal of medicine. 2002;347(16):1233–41.
54. Fisher B, Jeong J-H, Anderson S, Bryant J, Fisher ER, Wolmark N. Twenty-five-year follow-up of a randomized trial comparing radical mastectomy, total mastectomy, and total mastectomy followed by irradiation. New England journal of medicine. 2002;347(8):567–75.
55. Anampa J, Makower D, Sparano JA. Progress in adjuvant chemotherapy for breast cancer: an overview. BMC medicine. 2015;13(1):1–13.
56. Sørlie T, Tibshirani R, Parker J, Hastie T, Marron JS, Nobel A, et al. Repeated observation of breast tumor subtypes in independent gene expression data sets. Proceedings of the national academy of sciences. 2003;100(14):8418–23.
57. Bastien RRL, Rodríguez-Lescure Á, Ebbert MTW, Prat A, Munárriz B, Rowe L, et al. PAM50 breast cancer subtyping by RT-qPCR and concordance with standard clinical molecular markers. BMC medical genomics. 2012;5(1):1–12.
58. Parker JS, Mullins M, Cheang MCU, Leung S, Voduc D, Vickery T, et al. Supervised risk predictor of breast cancer based on intrinsic subtypes. Journal of clinical oncology. 2009;27(8):1160.
59. Albain KS, Barlow WE, Ravdin PM, Farrar WB, Burton GV, Ketchel SJ, et al. Adjuvant chemotherapy and timing of tamoxifen in postmenopausal patients with endocrine-responsive, node-positive breast cancer: A phase 3, open-label, randomised controlled trial. The Lancet. 2009;374(9707):2055–63.
60. Jemal A, Center MM, DeSantis C, Ward EM. Global patterns of cancer incidence and mortality rates and trends. Cancer epidemiology, biomarkers & prevention. 2010;19(8):1893–907.
61. Munoz D, Near AM, Van Ravesteyn NT, Lee SJ, Schechter CB, Alagoz O, et al. Effects of screening and systemic adjuvant therapy on ER-specific US breast cancer mortality. Journal of the National Cancer Institute. 2014;106(11):dju289.
62. Sparano JA, Fazzari M, Kenny PA. Clinical application of gene expression profiling in breast cancer. Surgical oncology clinics. 2010;19(3):581–606.
63. Sotiriou C, Pusztai L. Gene-expression signatures in breast cancer. New England journal of medicine. 2009;360(8):790–800.

64. Paik S, Tang G, Shak S, Kim C, Baker J, Kim W, et al. Gene expression and benefit of chemotherapy in women with node-negative, estrogen receptor–positive breast cancer. Journal of clinical oncology. 2006;24(23):3726–34.
65. Albain KS, Barlow WE, Shak S, Hortobagyi GN, Livingston RB, Yeh IT, et al. Breast Cancer Intergroup of North America Prognostic and predictive value of the 21-gene recurrence score assay in postmenopausal women with node-positive, oestrogen-receptor-positive breast cancer on chemotherapy: A retrospective analysis of a randomised trial. The Lancet oncology. 2010;11(1):55–65.
66. Hammond MEH, Hayes DF, Dowsett M, Allred DC, Hagerty KL, Badve S, et al. American Society of Clinical Oncology/College of American Pathologists guideline recommendations for immunohistochemical testing of estrogen and progesterone receptors in breast cancer (unabridged version). Archives of pathology & laboratory medicine. 2010;134(7):e48–e72.
67. Goss PE, Ingle JN, Martino S, Robert NJ, Muss HB, Piccart MJ, et al. A randomized trial of letrozole in postmenopausal women after five years of tamoxifen therapy for early-stage breast cancer. New England journal of medicine. 2003;349(19):1793–802.
68. Davies C, Pan H, Godwin J, Gray R, Arriagada R, Raina V, et al. Long-term effects of continuing adjuvant tamoxifen to 10 years versus stopping at 5 years after diagnosis of oestrogen receptor-positive breast cancer: ATLAS, a randomised trial. The Lancet. 2013;381(9869):805–16.
69. Pagani O, Regan MM, Francis PA. Exemestane with ovarian suppression in premenopausal breast cancer. The New England journal of medicine. 2014;371(14):1358–9.
70. Pagani O, Regan MM, Walley BA, Fleming GF, Colleoni M, Láng I, et al. Adjuvant exemestane with ovarian suppression in premenopausal breast cancer. New England journal of medicine. 2014;371(2):107–18.
71. Wolff AC, Hammond MEH, Hicks DG, Dowsett M, McShane LM, Allison KH, et al. Recommendations for human epidermal growth factor receptor 2 testing in breast cancer: American Society of Clinical Oncology/College of American Pathologists clinical practice guideline update. Archives of pathology and laboratory medicine. 2014;138(2):241–56.
72. Romond EH, Perez EA, Bryant J, Suman VJ, Geyer Jr CE, Davidson NE, et al. Trastuzumab plus adjuvant chemotherapy for operable HER2-positive breast cancer. New England journal of medicine. 2005;353(16):1673–84.
73. Joensuu H, Kellokumpu-Lehtinen P-L, Bono P, Alanko T, Kataja V, Asola R, et al. Adjuvant docetaxel or vinorelbine with or without trastuzumab for breast cancer. New England journal of medicine. 2006;354(8):809–20.
74. Piccart-Gebhart MJ, Procter M, Leyland-Jones B, Goldhirsch A, Untch M, Smith I, et al. Trastuzumab after adjuvant chemotherapy in HER2-positive breast cancer. New England journal of medicine. 2005;353(16):1659–72.
75. Slamon D, Eiermann W, Robert N, Pienkowski T, Martin M, Press M, et al. Adjuvant trastuzumab in HER2-positive breast cancer. New England journal of medicine. 2011;365(14):1273–83.
76. Tolaney SM, Barry WT, Dang CT, Yardley DA, Moy B, Marcom PK, et al. Adjuvant paclitaxel and trastuzumab for node-negative, HER2-positive breast cancer. New England journal of medicine. 2015;372(2):134–41.
77. Jones SE, Collea R, Paul D, Sedlacek S, Favret AM, Gore I, et al. Adjuvant docetaxel and cyclophosphamide plus trastuzumab in patients with HER2-amplified early stage breast cancer: A single-group, open-label, phase 2 study. The Lancet oncology. 2013;14(11):1121–8.
78. Goldhirsch A, Gelber RD, Piccart-Gebhart MJ, de Azambuja E, Procter M, Suter TM, et al. Herceptin Adjuvant (HERA) Trial Study Team. 2 years versus 1 year of adjuvant trastuzumab for HER2-positive breast cancer (HERA): An open-label, randomised controlled trial. Lancet. 2013;382(9897):1021–8.
79. Hutchinson L. ALTTO: Wake-up call for setting up clinical trials. Nature reviews clinical oncology. 2013;10(3):121.
80. Corrie PG. Cytotoxic chemotherapy: Clinical aspects. Medicine. 2008;36(1):24–8.
81. Corrie PG. Cytotoxic chemotherapy: Clinical aspects. Medicine. 2004;32(3):25–9.
82. Working MRCOC. Surgical resection with or without preoperative chemotherapy in oesophageal cancer: A randomised controlled trial. The Lancet. 2002;359(9319):1727–33.
83. Corrie PG. Cytotoxic chemotherapy: Clinical aspects. Medicine. 2011;39(12):717–22.
84. Scharovsky OG, Mainetti LE, Rozados VR. Metronomic chemotherapy: Changing the paradigm that more is better. Current oncology. 2009;16(2):7–15.
85. Liu J, He M, Wang Z, Li Q, Xu B. Current research status of metronomic chemotherapy in combination treatment of breast cancer. Oncology research and treatment. 2022;45(11):681–92.
86. Cazzaniga ME, Cordani N, Capici S, Cogliati V, Riva F, Cerrito MG. Metronomic chemotherapy. Cancers. 2021;13(9):2236.

87. Lupo G, Caporarello N, Olivieri M, Cristaldi M, Motta C, Bramanti V, et al. Anti-angiogenic therapy in cancer: Downsides and new pivots for precision medicine. Frontiers in pharmacology. 2017;7:519.
88. Kerbel R, Folkman J. Clinical translation of angiogenesis inhibitors. Nature reviews cancer. 2002;2(10):727–39.
89. Hanahan D, Bergers G, Bergsland E. Less is more, regularly: Metronomic dosing of cytotoxic drugs can target tumor angiogenesis in mice. The journal of clinical investigation. 2000;105(8):1045–7.
90. Klement G, Baruchel S, Rak J, Man S, Clark K, Hicklin DJ, et al. Continuous low-dose therapy with vinblastine and VEGF receptor-2 antibody induces sustained tumor regression without overt toxicity. The journal of clinical investigation. 2000;105(8):R15–R24.
91. Bahl A, Bakhshi S. Metronomic chemotherapy in progressive pediatric malignancies: Old drugs in new package. The Indian journal of pediatrics. 2012;79:1617–22.
92. Mross K, Steinbild S. Metronomic anti-cancer therapy–an ongoing treatment option for advanced cancer patients. Journal of cancer therapeutics and research. 2012;1(1):32.
93. Wichmann V, Eigeliene N, Saarenheimo J, Jekunen A. Recent clinical evidence on metronomic dosing in controlled clinical trials: A systematic literature review. Acta oncologica. 2020;59(7):775–85.
94. Malik PS, Raina V, André N. Metronomics as maintenance treatment in oncology: Time for chemo-switch. Frontiers in oncology. 2014;4:76.
95. Fedele P, Marino A, Orlando L, Schiavone P, Nacci A, Sponziello F, et al. Efficacy and safety of low-dose metronomic chemotherapy with capecitabine in heavily pretreated patients with metastatic breast cancer. European journal of cancer. 2012;48(1):24–9.
96. Taguchi T, Nakayama T, Masuda N, Yoshidome K, Akagi K, Nishida Y, et al. Study of low-dose capecitabine monotherapy for metastatic breast cancer. Chemotherapy. 2010;56(2):166–70.
97. De Iuliis F, Salerno G, Taglieri L, Lanza R, Scarpa S. On and off metronomic oral vinorelbine in elderly women with advanced breast cancer. Tumori journal. 2015;101(1):30–5.
98. Weitman SD, Glatstein E, Kamen BA. Back to the basics: The importance of concentration× time in oncology. Journal of clinical oncology. 1993;11(5):820–1.
99. Lien K, Georgsdottir S, Sivanathan L, Chan K, Emmenegger U. Low-dose metronomic chemotherapy: A systematic literature analysis. European journal of cancer. 2013;49(16):3387–95.
100. Simsek C, Esin E, Yalcin S. Metronomic chemotherapy: A systematic review of the literature and clinical experience. Journal of oncology. 2019;2019.
101. Pasquier E, Andre N, Braguer D. Targeting microtubules to inhibit angiogenesis and disrupt tumour vasculature: Implications for cancer treatment. Current cancer drug targets. 2007;7(6):566–81.
102. Bocci G, Nicolaou KC, Kerbel RS. Protracted low-dose effects on human endothelial cell proliferation and survival in vitro reveal a selective antiangiogenic window for various chemotherapeutic drugs. Cancer research. 2002;62(23):6938–43.

4 The Antiangiogenic Basis of Metronomic Chemotherapy

Manzoor Ahmad Mir, Asma Jan, Aisha Shabir, and Burhan Ul Haq

4.1 INTRODUCTION

Cytotoxic chemotherapeutics have been the gold standard of systemic cancer treatment for almost 50 years. The majority of such medications are inhibitors of microtubules or substances that damage DNA intended to slow down or destroy quickly dividing cells. The maximum tolerated dose (MTD) is the largest dose that may be given to a patient without developing toxicity that poses a life-threatening risk. It is frequently given in single doses or brief courses of therapy. Between each cycle of MTD therapy, there must be lengthy pauses (usually lasting 2–3 weeks). Despite the abundance of these chemotherapeutic agents and the enormous number of clinical trials conducted to evaluate them, little advancement has been made in regard to treatment or considerably extending the longevity of cancer survivors, specifically individuals who have metastatic or late-stage illness [1, 2]. Furthermore, considering the harsh adverse effects that are commonly linked to MTD-based chemotherapy, the advancements that have been made in the treatment of some types of cancer frequently come at a hefty cost. These include short-term myelosuppression, mucositis, nausea, intestinal mucosa damage, and hair loss along with long-term effects on the nervous system, kidneys, heart, and reproductive system. Antinausea medications and growth factors that are given to cancer survivors to lessen the intensity of or hasten healing from the side effects of chemotherapy are a large part of recent pharmacological breakthroughs in oncology treatment. These "supportive-care" medications come with their own set of adverse effects and can dramatically increase the cost of chemotherapy for cancer.

A reassessment is currently in progress regarding the optimal methods for delivering chemotherapy. Rather than relying solely on brief periods of intense toxic chemotherapy followed by extended intervals for recuperation from adverse side effects, there is now a changing perspective favoring more condensed or accelerated treatment schedules using significantly lower individual doses than the MTD. This approach aims to be more efficient not just in decreasing specific side effects but potentially enhancing the antitumor effects as well [3–6]. Furthermore, certain approaches for dosing and scheduling are highly compatible with the combination of chemotherapy and emerging targeted therapies, which are comparatively less toxic in their anticancer effects and are currently being developed. The latest advancement in this idea is known as "metronomic" chemotherapy, and it involves giving chemotherapeutic agents often at dosages much below the MTD, with no extended intervals between treatments.

4.2 METRONOMIC THERAPY

The idea that chemotherapy is best administered frequently rather than seldom and that "less is more" has been supported by a variety of considerations. Initially, the alternative strategy, which involves "dose-dense" chemotherapy followed by a transplant of the patient's bone marrow stem cells, intended to restore the damaged stem cells, has failed to yield the anticipated survival advantages, particularly when used with individuals who have metastatic breast cancer [7, 8]. Further, this strategy is extremely toxic and very costly. Additionally, "dose-dense" chemotherapy, which involves the administration of one or more chemotherapeutics at more regular intervals (i.e., every

DOI: 10.1201/9781032632674-4

other week), has demonstrated definite advantages in randomized phase III clinical trials [9–11]. Typically, this approach is intended to ensure the administration of at least an equivalent amount, and often a higher cumulative dosage of the drug throughout the treatment duration [9].

Hence, if the effectiveness of treatment increases by administering chemotherapy every other week instead of every 3 weeks, it raises the question of why not consider administering it on a weekly or even daily basis. There is a growing trend toward administering taxane drugs to specific cancer patients, like those with breast cancer, every week [12–15]. This approach allows for increased dose density, facilitated by the use of externally administered supportive care measures such as transfusion medicine, antibiotics, and growth factors. In certain aspects, this dose density aligns with the concept of metronomic therapy, which will be further elaborated upon in the subsequent discussion.

The term "metronomic chemotherapy" refers to a type of high-dose treatment, although the total dosage in metronomic therapy is often considerably lower than in chemotherapy based on the MTD [15, 16]. By reducing toxicity levels, metronomic therapy diminishes or eliminates the requirement for growth-factor support to expedite myelosuppression healing. Additionally, although using low total dosages of medication, this method may potentially have better antitumor effects than traditional MTD regimens in terms of extending survival periods, particularly in some preclinical models [17–19]. Mathematical modelling studies have also provided additional support for the efficacy of metronomic therapy [20, 21]. In contrast to high-dose chemotherapy, which primarily focuses on proliferating tumor cells, metronomic chemotherapy is used often or continuously and primarily targets the endothelial cells present in the developing vasculature of the tumor [22]. Essentially, chemotherapeutic agents are utilized as inhibitors of angiogenesis, thereby reshaping the conventional chemotherapeutic goal [23]. This shift in perspective is also the rationale behind the term coined by Browder et al. [22], known as "anti-angiogenic chemotherapy," to characterize this particular approach of treatment. Metronomic chemotherapy has the added benefit of being compatible with antiangiogenic medications, various forms of aimed treatments like antitumor vaccinations or drugs that focus on certain signal transduction mechanisms, and other sorts of targeted therapies. Ironically, the original intent behind the development of targeted medicines was to displace chemotherapy and lessen the severe morbidities related to conventional MTD or high-dose chemotherapy. However, despite their lower toxicity, it was discovered that most of these rationally designed medications only have limited effectiveness when taken alone to treat people in the late stages of an illness. As a result, they are typically employed in tandem with more conventional treatments like conventional chemotherapy or radiotherapy procedures. Bevacizumab (Avastin) is an example of a humanized monoclonal antibody used for the management of metastatic colorectal cancer. It works by targeting vascular endothelial cell growth factor (VEGF), as shown in (Figure 4.1). Bevacizumab is typically administered in conjunction with 5-fluorouracil (5-FU), irinotecan, and leucovorin [24, 25]. The second instance is trastuzumab (Herceptin), which is a humanized monoclonal antibody used for the treatment of metastatic breast cancer. It specifically targets the ERBB2 oncoprotein. Trastuzumab is commonly used in combination with paclitaxel or an alkylating substance in metastatic breast cancer management [26].

The expected advantages of specific treatment were thought to have fewer side effects and enhancement in the overall well-being of patients. Nonetheless, when these medications are used in conjunction with MTDs of chemotherapy, these advantages are not achieved. Considering that chemotherapy is likely to remain the primary approach for systemic cancer treatment in the foreseeable future, it will be progressively vital to develop more efficient methods of administration, as well as potential therapeutic combinations with the latest class of molecule-specific medications.

4.3 ANTIANGIOGENIC MECHANISMS

A significant amount of evidence, primarily derived from laboratory studies (in vitro), suggests that the "activated" endothelial cells present in newly developing capillaries of blood vessels exhibit a high and specific sensitivity to extremely low dosages of different chemotherapy medications

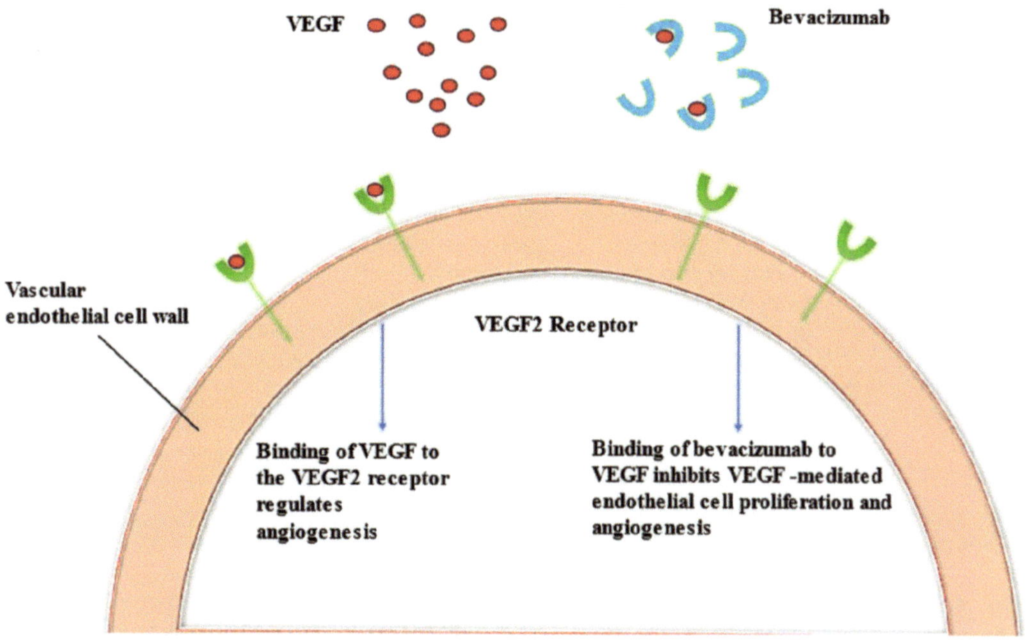

FIGURE 4.1 Bevacizumab's therapeutic effect on the body. Bevacizumab decreases vascular permeability by blocking VEGF-mediated endothelial proliferation and angiogenesis.

[27–32]. Numerous research investigations have been conducted to assess the impact of picomolar dosages of these medications on several kinds of human cells, such as tumor cells, epithelial cells, lymphocytes, and fibroblasts from different tissues, as well as endothelium cells, either of the micro- or macro-vascular variety. These studies have explored the antiproliferative, migration-inhibitory, and sometimes cytotoxic effects of these chemotherapeutic drugs. A summary of selected studies is provided in (Table 4.1). Several intriguing studies have focused on different microtubule inhibitors, including vinblastine, paclitaxel, and docetaxel. These investigations have revealed that extremely low concentrations of these drugs can specifically hinder the migration of endothelial cells or proliferation while leaving other examined kinds of cells unaffected. In one instance, Wang et al. discovered that inhibiting the growth and mammary epithelial cells, movement of human astrocytes, fibroblasts, prostate epithelial cells, keratinocytes, or smooth muscle cells required concentrations of paclitaxel that were 10–100,000 times higher compared to epithelial cells [31]. However, taxanes require special formulations in order to be injected to prevent their interaction with serum proteins. In medically relevant quantities, these carriers or binding proteins can drastically reduce the antiangiogenic action of taxanes, necessitating larger dosages of injectable taxanes to elicit antiangiogenic effects in vivo [33].

The majority of these research studies did not observe any cytotoxic effects. However, it was revealed that when endothelial cells are consistently subjected to a lower dosage of medications like paclitaxel for 6 days, mimicking metronomic therapy, only the endothelial cells, and not dermal fibroblasts or tumor cells, experience programmed cell death (apoptosis) within approximately 5 days [29]. This concept is visually depicted in (Figure 4.2), illustrating the relationship between low-dose metronomic chemotherapy and the specific mechanisms involved in the response of vascular endothelial cells.

According to two recent studies, thrombospondin 1 (TSP1) has been identified as a possible intermediary of the metronomic cyclophosphamide's consequences. According to a single research study, endothelial cells of the blood vessels (other cell types were not examined) were exposed for 5 days to low doses of several chemotherapeutic medicines, and this led to substantially elevated

TABLE 4.1
Sensitivity of Metronomic Therapy in Human Vascular Endothelial Cells

Medine Used	Assay	Effects	References
Vinblastine and paclitaxel	Inhibiting the proliferation of cancer cells or endothelial cells in tumor spheroid or monolayer cultures (for cancer cells).	To inhibit cancer cells, an IC_{50} of 3,434–10,084 nM was observed in spheroid culture and 2–27 nM was seen in monolayer culture. To inhibit human umbilical endothelial cells, an IC_{50} of 0.4–0.5 nM was observed.	[34]
Vinblastine	Proliferation, movement, and the production of metalloproteinases are all inhibited.	Concentrations as low as 0.1–1 pM/l were able to inhibit human endothelial cells but did not affect cancer cells, fibroblasts, or leukocytes.	[30]
Paclitaxel	Inhibiting the movement, development of cords and tubes, and proliferation in endothelial cells.	The invasiveness and chemotoxic effects of endothelial cells were inhibited by drug doses even at 10 pM (IC_{50}* = 0.5–4 nM) after many hours of incubation.	[28]
Methotrexate	Inhibiting the proliferation of cells.	Inhibition of human endothelial cells at 5×10^{-9}M concentration of the drug.	[27]
Docetaxel and paclitaxel	Movement, capillary sprouting, and inhibiting proliferation of cells.	The efficacy of docetaxel is 10 times greater than that of paclitaxel; cancer cells are 10–100 times less sensitive as compared with endothelial cells.	[32]
Paclitaxel, cyclophosphamide, and epothilone B	Cancer cells, human endothelial cells, and fibroblasts were tested for inducing apoptosis and inhibiting the proliferation of cells.	Human endothelial cell proliferation was inhibited with IC_{50} values between 50 and 100 pM after being exposed to medications for 6 days regularly. When compared to endothelial cells, IC_{50} values for fibroblasts or cancer cells were typically at least 10 times higher; only endothelial cells showed signs of apoptotic induction.	[29]

levels of proteins and TSP1 messenger RNA (mRNA). A well-known endogenous inhibitor of angiogenesis is TSP1, a part of the extracellular matrix that is released as well as detected in the circulation [35, 36]. It appears to function primarily through attaching to endothelial cells' CD36 receptors [37]. Endothelial cells are affected by this contact, which inhibits their ability to proliferate and causes them to undergo apoptosis [38, 39]. CD36-negative cells, like cells from hair follicles

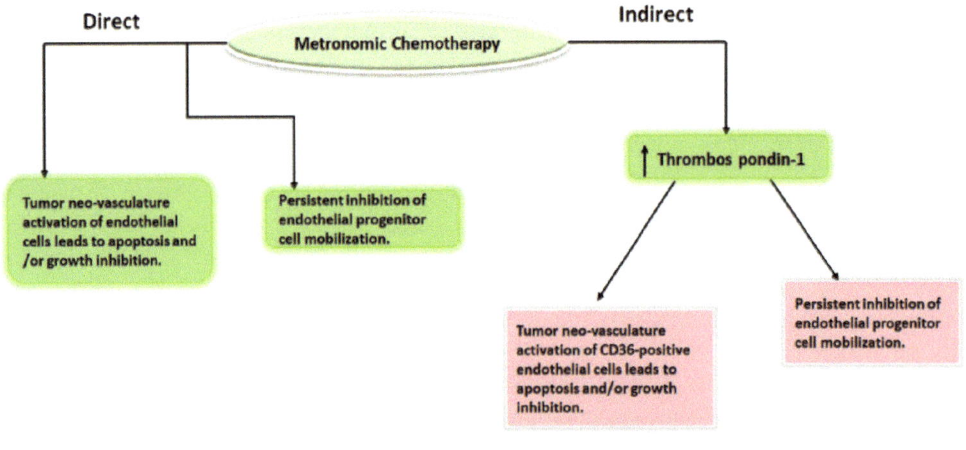

FIGURE 4.2 Metronomic chemotherapy: A possible mechanism for its antiangiogenic effects.

and hematopoietic stem cells produced from bone marrow, would not be predicted to be affected by this interaction. Additionally, VEGF can be bound to by TSP1 and sequestered, which inhibits its proangiogenic function [40].

4.4 CLINICAL TRIALS OF METRONOMIC CHEMOTHERAPY

Metronomic chemotherapy regimens, in particular when paired with an antiangiogenic medication, have sparked interest and prompted the initiation of phase II clinical studies to investigate their potential advantages. (Table 4.2) is a summary of some of these tests. Most of them entail continuous dosing of oral cyclophosphamide for up to 2 years as part of chemotherapy treatment. Some patients additionally get low-dose methotrexate orally twice weekly. One example of a targeted medicine is celecoxib, a daily oral inhibitor of cyclooxygenase-2 (COX2), and another is bevacizumab, an intravenous monoclonal antibody against VEGF that is injected every 2 weeks. It was decided to include celecoxib in the experiment because of its widespread availability, low risk of side effects, and potential antiangiogenic properties; all of these factors contributed to the drug's selection for inclusion in the trial [41, 42].

A clinical study of cyclophosphamide plus methotrexate was conducted in Italy, and its success prompted the current investigation of this combination in another 49 studies. Low-dose oral cyclophosphamide and oral methotrexate were administered on a daily and weekly basis, respectively, to 64 women with progressing, advanced, or refractory breast cancer. In addition to having had prior first-, second-, or third-line therapy, the majority of study participants had progressing metastatic illness at the outset. The response rate was 32%, with 2 complete responders, 10 partial responders, and 12 patients experiencing stable illness for 6 months or longer [48]. Although several patients had received prior chemotherapy, no serious side effects were reported. From a toxicity standpoint, this is preferable to the usual third-line chemotherapy protocols employed in this scenario. About US$10 per month [48] was projected to be spent on this form of outpatient treatment.

TABLE 4.2
Metronomic Chemotherapeutic Studies in Humans

Medication Therapy	Disease Treated	References
Combination of repeated chronic dosing of low-dose oral cyclophosphamide for 3 weeks, then low-dose oral etoposide for 3 weeks with low-dose celecoxib and thalidomide, all taken orally once a day	Pediatric and adult patients with refractory solid tumors	[43]
Twice-daily 400 mg celecoxib and 50 mg orally low-dose cyclophosphamide	Metastatic renal cell carcinoma	[44]
Treatment with a low dose of 2.5 mg oral methotrexate daily and twice-daily 400 mg celecoxib	Recurrent and metastatic chemoresistant cancerous squamous cells of the neck and head	[45]
Low doses of 1 mg dexamethasone and 50 mg oral cyclophosphamide are given daily	Refractory and advanced prostate cancer; phase II clinical study involving 32 individuals	[46]
25 mg of rofecoxib (Vioxx) and a lower dosage of oral treosulfan 500 mg every day	Refractory and advanced melanoma; a pilot trial involving 12 individuals	[47]
Twice-weekly lower dosage of oral methotrexate and 50 mg of oral cyclophosphamide every day	Refractory and advanced breast cancer; phase II clinical study involving 64 individuals	[48]

Androgen-independent metastatic prostate cancer was also treated in a recent trial by Glode et al., which involved 32 patients. Oral cyclophosphamide and dexamethasone were given daily to these individuals [46]. The antiangiogenic effects of dexamethasone have been described. Seventy percent of the participants had a 50% or greater reduction in blood prostate-specific antigen levels in this short research study. Such promising early findings warrant confirmation in bigger, prospective, controlled clinical trials [49].

The antiangiogenic effects of these medicines, which may contribute to their purported antitumor effectiveness, are likewise extremely difficult to establish with any degree of certainty. Serum VEGF levels were shown to be lower in individuals who improved while receiving treatment, as reported by Colleoni et al. The anti-VEGF monoclonal antibody bevacizumab has been shown to decrease the number of bone marrow–derived circulating endothelial progenitors in the blood of patients with rectal carcinoma [50], and Bertolini et al. reported a similar effect in patients with lymphoma and breast cancer who received daily low-dose cyclophosphamide therapy [51]. Drugs like celecoxib and dexamethasone, among others, have side effects that contribute to their therapeutic success, but this is only one aspect. Activated endothelial cells and cancer cells both express the COX2 that is inhibited by celecoxib. It has also been shown that low-dose chemotherapy can boost the immune system [52, 53], making it a promising adjunct to cancer vaccines and other forms of immunotherapy. Some research suggests that the effectiveness of immunotherapeutic vaccinations can be improved with cyclophosphamide in a metronomic chemotherapy regimen [54, 55].

Even though there is no concrete proof, metronomic treatment might have direct effects on cancer cells, such as inducing cell differentiation. The "repopulation" of cancer cells, which can occur during the downtime between MTD chemotherapy sessions, may be avoided using continuous chemotherapy regimens. To avoid the development of "kinetic drug resistance," it may be necessary to reduce or eliminate the drug-free break periods between MTD pulsatile chemotherapy cycles, as repopulation kinetics tend to grow more aggressive with each cycle [56].

Several effective pediatric chemotherapy techniques are similar to metronomic therapy in that they include continuous administration of low doses of cytotoxic medicines over lengthy periods, as so-called "maintenance" therapies, and have little toxicity (Table 4.3), as stated earlier. Pediatric oncologists have found that combining weekly dosing with vinca alkaloid medications like vincristine or vinblastine with daily dosing of cyclophosphamide is the most effective treatment for disorders like neuroblastoma [57]. Treatment plans for Wilms tumor and rhabdomyosarcoma often include weekly delivery of vinca alkaloids. Pediatric CHOP (cyclophosphamide, Adriamycin, vincristine, and prednisone) treatment for patients with non-Hodgkin lymphoma is conceptually similar to metronomic therapy. It has been shown that low-dose vinblastine and methotrexate therapy can successfully manage aggressive fibromatosis [58]. Treatments for babies with localized, unresectable neuroblastoma [57] are being tried using low-dose chemotherapy regimens, such as low-dose cyclophosphamide treatment and low-dose vincristine therapy.

TABLE 4.3
Pediatric Cancer Treatment with Metronomic Chemotherapy

Medicine Used	Syndrome	References
Cyclophosphamide, vincristine, actinomycin, vinblastine, and actinomycin	Rhabdomyosarcoma	[59]
Vincristine	Wilms tumor	[60]
6-Mercaptopurine and methotrexate	Acute lymphatic leukemia	[61]
CHOP (cyclophosphamide, doxorubicin, vincristine, and prednisone)	Lymphoma	[62]

4.5 EFFECTS AGAINST ANGIOGENESIS

Chemotherapeutics don't explicitly target tumor cells; instead, they disrupt the division of cells by, for example, blocking microtubules or topoisomerases, thymidylate synthase, and other enzymes essential for the metabolism and replication of DNA. As a result, such medications also harm hair follicle cells and the routinely dividing cells of quickly regenerating organs including bone marrow and gut mucosa. Because of this, host toxicity frequently has a small relative difference to antitumor efficacy, leading to a narrow therapeutic index. However, there may be some light at the end of the otherwise gloomy tunnel because dividing endothelial cells, like other normally dividing cells, should be vulnerable to chemotherapeutics [63]. These cells are located in the expanding blood arteries observed in tumors [64]. There would likely be an antiangiogenic effect if the endothelial cells in the process of mitosis were eliminated or their division was inhibited. Furthermore, because endothelial cells of the host blood vessels are believed to be genetically stable and free of the various genetic flaws that tumor cells exhibit and which result in drug resistance, chemotherapy's potential effects may last longer even after treatment is continued. For instance, myelosuppression can occur after each cycle of chemotherapy based on MTD, but the severity does not significantly change over time [63]. Myelosuppression would gradually decrease and vanish if normal progenitor cells in the bone marrow developed chemoresistance, as highly changeable, genetically unstable cancer cells do. Accordingly, it was initially postulated in 1991 that tumor cells that have developed resistance to a certain chemotherapeutic treatment may respond to that therapy in a roundabout way, via a side effect – damage or loss to its linked vasculature [65]. The majority of chemotherapeutic drug classes have been shown to have antiangiogenic or antivascular effects in a variety of in vitro and in vivo studies in literature dating back to the middle of the 1980s [23].

Although many tumors initially respond to chemotherapy treatments, they quickly develop drug resistance or are fundamentally resistant to drugs. Therefore, it would appear that chemotherapy has only minimal or insignificant antiangiogenic effects. Why is this the case? It's possible that there aren't enough dividing endothelial cells in the blood arteries connected to tumors for chemotherapy to have a meaningful therapeutic effect. Elevated local concentrations of endothelial cell survival factors like basic fibroblast growth factor (BFGF), VEGF, and angiopoietin 1 may also protect the endothelial cells against chemotherapy-induced cell death [66, 67]. Judah Folkman's ground-breaking research [22] suggests a third reason: The conventional administration of chemotherapy obscures and diminishes its antiangiogenic effects. In this instance, the antiangiogenic effects of the medications are diminished by the lengthy gaps between medication administrations required to let the patient recuperate from the toxic side effects of the treatment, notably myelosuppression.

The antiangiogenic and antitumor effects of the alkylating drug cyclophosphamide were examined by Browder and colleagues in immune-competent syngeneic mice that had received subcutaneous injections of several tumor types [22]. They discovered that this medication, when given at the MTD, led to endothelial cell death in the developing tumor microvessels [22]. The endothelial cells in the tumor were the first to experience apoptosis, according to a thorough temporal analysis [22]. It appears that the damaged tumor vasculature was fully restored over the extensive rest/recovery intervals (2–3 weeks) during subsequent cycles of MTD-based treatment, preventing this antiangiogenic impact from translating into a major therapeutic advantage.

Accordingly, it was suggested that when cyclophosphamide was administered very often, most likely every week with no long intervals in between, consequently chemotherapy's antiangiogenic impact would build up irreversibly, and the injured endothelium would have much less chance to heal (Figure 4.3). Naturally, this calls for reducing the dosage of the medication given with each injection. Several mouse tumor cell lines produced subcutaneously in syngeneic mice were used by Browder et al. to demonstrate the significant antitumor and antiangiogenic benefits of this low-dose treatment that is administered regularly and more often, provided at one-third of the MTD [22]. This method made it possible for even extremely large, established subcutaneous tumors to respond

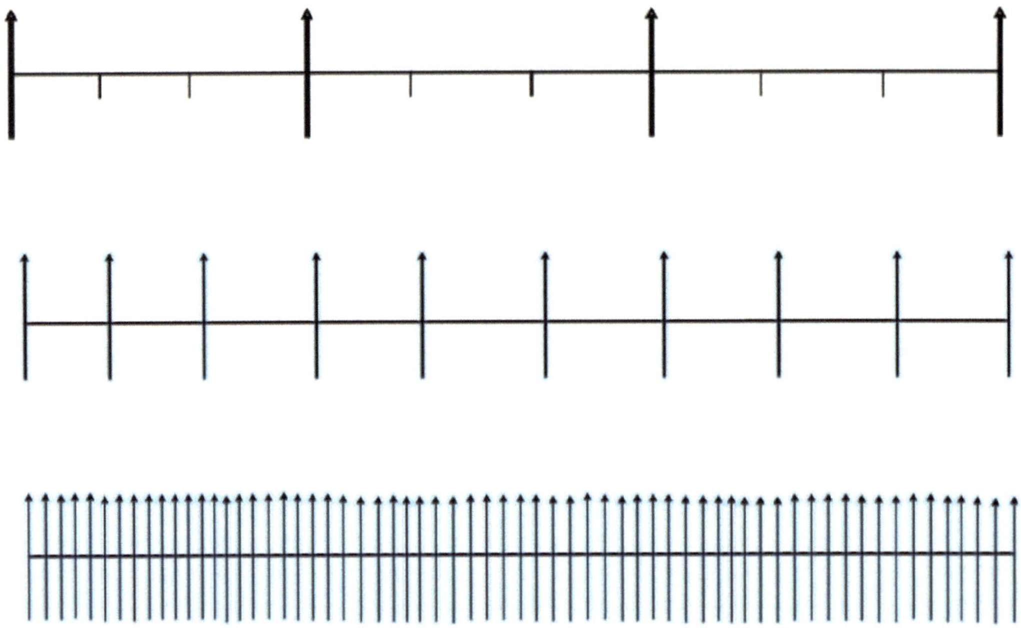

FIGURE 4.3 Various treatment schedules of metronomic chemotherapy.

to the same medication and virtually fully retreat after being previously chosen in vivo for cyclophosphamide resistance that developed while utilizing the standard MTD treatment. In summary, it's possible to easily reverse a condition of developed resistance to drugs by seemingly redirecting treatment efforts far from the population of drug-resistant cancer cells and toward the drug-sensitive endothelium lining tumors [22].

These findings have also been changed with daily oral medication delivery in the form of consumable water, which appears to be safer than the regular weekly dose [19, 68]. These modifications have been supported by additional research groups [69, 70]. Lymphopenia was the sole harmful adverse effect observed in a recent in-depth study of cyclophosphamide, given regularly at a low dose, for an extended period of time, which demonstrated no substantial damage to cells or tissues ordinarily impacted by MTD compared to regimens of the same medication [71].

4.6 POSSIBLE ADVANTAGES OF EMPLOYING ANTIANGIOGENIC/METRONOMIC CHEMOTHERAPY

Antiangiogenic chemotherapy may offer a variety of benefits. With this strategy, numerous cheap, commercially available medications with well-defined mechanisms of action are now ready for use right away. Even in patients who have been highly pretreated and are in the latter stages of their disease, the therapy is anticipated to be well tolerated at the necessary doses. Instead of focusing on a specific cytokine route like VEGF, antiangiogenic chemotherapy targets the growth of endothelial and pericyte cells. This ought to boost this method's potential efficacy, regardless of the specific cytokine expression patterns of a given tumor. Because they have a normal genetic makeup, tumor-induced endothelium cells aren't capable of the fast mutations that make cancer cells resistant to chemotherapy. This suggests that individuals can receive treatment with metronomic agents, despite the prior utilization of these identical medications for the initial tumor. Ultimately, the favorable tolerability of antiangiogenic therapy renders it exceptionally compatible with combined strategies involving other antiangiogenic drugs, biological inhibitors that target signaling pathways, conventionally administered chemotherapy, and/or radiation. Furthermore, they may be well-suited as adjunctive therapy after the induction of remission, aiming to impede the reactivation of the angiogenic switch by residual tumor cells.

4.7 HOW THE CHEMOTHERAPY APPROACH KNOWN AS "ANTIANGIOGENIC" EVOLVED INTO A TREATMENT STRATEGY CALLED "METRONOMIC" CHEMOTHERAPY

Hanahan and colleagues introduced the term "metronomic" dosage based on the idea of antiangiogenic chemotherapy being administered frequently, rhythmically, and at low doses [3]. Regardless of the anticancer mechanism involved, chemotherapy is referred to as "metronomic therapy" when it is administered more frequently. Antiangiogenic chemotherapy is now more easily understood thanks to this new phrase, but it has also generated a lot of confusion. Numerous chemotherapy drugs can be given in tiny doses every day or once a week, resulting in an overall dose increase that is comparable to the maximally tolerated levels that are given for a few days at the start of a 3- to 4-week cycle. In children with recurring or progressing brain tumors, for instance, temozolomide, which has been administered in a low-dose daily formulation, has shown tolerance and activity. When temozolomide is delivered on a metronomic schedule, the total dose can reach at least 1.5 times the level seen with regular administration therapy (the usual schedule is 200 mg/m^2 over 5 days, or 1000 mg/m^2, compared to the metronomic schedule, which is 90 mg/m^2 over 21 days, or 1,890 mg/m^2) [72]. Since many "metronomic" treatments may have an improved anticancer impact, it is unable to determine the antiangiogenic effect separately from that effect. Antiangiogenic chemotherapy's original definition suggests a particular mode of operation. It is simple to distinguish between anticancer and antiendothelial effects in preclinical mice. However, in humans, this differentiation is considerably more challenging. Simultaneously, in order to advance in this field, it is crucial to acquire a comprehensive understanding of the mechanisms and causes behind the effectiveness or ineffectiveness of a specific tumor and model system when choosing the most suitable combination of antiangiogenic treatments for each type of tumor and the individual patient. Therefore, even if it is also known as "metronomic" chemotherapy, it is essential to remember that not every low-dose, often-administered chemotherapy targets angiogenesis.

4.8 ANTIANGIOGENIC AND METRONOMIC CHEMOTHERAPY DRUGS

Different chemotherapeutic drugs' capacity to effectively block endothelium and pericyte proliferative, migratory, and tube-forming processes cannot be inferred based solely on their anticancer activity. In their initial publication, Browder et al. examined a variety of substances for their action, including doxorubicin, 5-FU, and 6-mercaptopurine, and they identified several schedules and "metronomic" dosages that offered the best antiangiogenic therapy in preclinical models [22]. Nonetheless, in extensively researched medications like cyclophosphamide, identifying the precise metabolites responsible for achieving the most effective antiendothelial cell capabilities can prove challenging [73]. Klement et al. demonstrated that administering vinblastine at a dosage ranging from one-tenth to one-twentieth of the MTD resulted in a robust induction of antiangiogenic impact in two cancer cell lines implanted subcutaneously [16]. These results have been supported by other studies [74], which also show that antiangiogenesis occurs at even picomolar doses [30], indicating that for this drug, action may be present over many log values below the MTD. There have also been reports on the antiangiogenic potential of numerous additional medications, including 5-FU [75], gemcitabine [76], trofosfamide [77], Adriamycin, idarubicin, 5-FU, paclitaxel, etoposide [78], and temozolomide in an orthotopic glioma model [79].

When this strategy was paired with other antiangiogenic pathways in the initial in vivo preclinical models of antiangiogenic chemotherapy, tumor response was developed. Consequently, the impact of metronomic cyclophosphamide was elevated when combined with TNP-470 [22]. A similar outcome was observed when vinblastine was administered alongside DC101 (a murine anti-VEGFR2 antibody) [16], along with various low-dose chemotherapy agents like cisplatin and paclitaxel [34]. Preclinical investigations have revealed that the combination of metronomic chemotherapy and various antiangiogenic therapies leads to increased effectiveness either through additive or synergistic effects. These therapies encompass mTOR inhibition using RAD001 (everolimus) [80], VEGF inhibition using SU5416 (semaxinib) [81], axitinib [82], sunitinib, lenalidomide [83], platelet-derived growth factor receptor (PDGFR) inhibition with imatinib or sunitinib [84], AEE788 (a dual kinase inhibitor targeting VEGF/endothelial growth factor receptor [EGFR]) [85], and ABT-510 (a peptide mimicking thrombospondin) [86]. In the case of brain tumors, the combination of metronomic cyclophosphamide and DC101, a murine antibody targeting VEGFR2, demonstrated significant antitumor activity in preclinical models [87]. Likewise, combining metronomic chemotherapy with traditional chemotherapeutic agents like Adriamycin [88], 5-FU [89], full-dose cyclophosphamide [90], and the hypoxia agent tirapazamine [91], as well as immune modulators [54] has shown additive or synergistic effects.

4.9 CONCLUSION

A variety of pro- and antiangiogenic signals are involved in the complex biological mechanism known as angiogenesis. The signal pathways that regulate neovascularization are always evolving, and new regulators of angiogenesis are still being found. Although the first antiangiogenic medication was approved less than 5 years ago, biologists and physicians still find it difficult to keep up with the ever-growing complexity of the rules governing this essential process. We first overestimated the potential therapeutic effect of these new domains because we lacked a thorough knowledge of how complex they were, as is the case with many new fields in biology and medicine. Many prematurely gave up the field whenever an immediate cure was unavailable. Thankfully, the pharmaceutical sector and a large group of basic biologists and physicians have recognized the long-term potential of this discipline, both for cancer and a variety of other diseases. Both adult and pediatric patients, including those with end-stage cancer, appear to respond to antiangiogenic chemotherapy, which is now accessible, affordable, and efficacious. The majority of recently created antiangiogenic medications are still in clinical testing, and even when they do become commercially available, the majority of people around the globe will not be able to afford them. Improved medication combinations with suitable marker platforms can be quickly introduced into well-organized clinical trials and should become quickly attainable by concentrating on the underlying biology of targeting endothelial cell proliferation.

For further reading, some textbooks mentioned here also give a further understanding of metronomic chemotherapy:

- *Progress in Anti-cancer Chemotherapy* by Gabriel N. Hortobagyi and David Khayat.
- *Triple-Negative Breast Cancer: Biomarkers, Emerging Therapeutic Strategies and Clinical Challenges* by Marion Montgomery.
- *Current Cancer Treatment: Novel Beyond Conventional Approaches* by Oner Ozdemir.
- *Breast Cancer: Current and Alternative Therapeutic Modalities* by Esra Gunduz and Mehmet Gunduz.

The following links for video lectures may also be helpful:

https://youtu.be/TAqO0Mq19JQ
https://youtu.be/6wEnYfvoBqk
https://youtu.be/A9fRRCUlMms
https://youtu.be/zIWvmCsKr34

For more insight about the topic, we would suggest detailed findings from the books of (Mir MA, 2022) https://doi.org/10.1016/C2021-0-02565-7, https://doi.org/10.1016/C2022-0-00074-X, (Mir MA, 2021) https://doi.org/10.52305/WXJL6770, and (Mir MA, 2015) https://doi.org/10.1016/C2014-0-02898-5 and from the cancer.net website on the following mentioned links:

https://www.cancer.net/cancer-types/breast-cancer/types-treatment
https://discovery.ucl.ac.uk/id/eprint/1472740/
https://www.jmedsciences.com/doi/JMEDS/pdf/10.5005/jp-journals-10045-00138

For diagrammatic illustrations and descriptive tables (Lazzeroni, 2012): http://www.eurekaselect.com/article/49928
See video links on the overall status of cancer, its various types, and current and new possible treatment options available: https://www.sciencedirect.com/science/article/pii/S205970292032278X

REFERENCES

1. Schiller, J.H., et al., *Comparison of four chemotherapy regimens for advanced non–small-cell lung cancer.* New England journal of medicine, 2002. **346**(2): p. 92–98.
2. Leaf, C., *Why we're losing the war on cancer (and how to win it).* Fortune, 2004. **149**(5): p. 42–55.
3. Hanahan, D., G. Bergers, and E. Bergsland, *Less is more, regularly: metronomic dosing of cytotoxic drugs can target tumor angiogenesis in mice.* The journal of clinical investigation, 2000. **105**(8): p. 1045–1047.
4. Gasparini, G., *Metronomic scheduling: the future of chemotherapy?* The Lancet oncology, 2001. **2**(12): p. 733–740.
5. Kamen, B.A., et al., *High-time chemotherapy or high time for low dose.* Journal of clinical oncology, 2000. **18**(16): p. 2935–2937.
6. Kerbel, R., et al., *Continuous low-dose anti-angiogenic/metronomic chemotherapy: from the research laboratory into the oncologyclinic.* Annals of oncology, 2002. **13**(1): p. 12–15.
7. Nieto, Y., *The verdict is not in yet. Analysis of the randomized trials of high-dose chemotherapy for breast cancer.* Haematologica, 2003. **88**(2): p. 201–211.
8. Rochéa, H., et al., *High-dose chemotherapy for breast cancer: the French PEGASE experience.* Cancer control, 2003. **10**(1): p. 42–47.
9. Piccart-Gebhart, M.J., *Mathematics and oncology: a match for life?* Journal of clinical oncology, 2003. **21**(8): p. 1425–1428.
10. Tuma, R.S., *Dosing study seen as victory for clinical trials, mathematical models.* Journal of the National Cancer Institute, 2003. **95**(4): p. 254–255.
11. Citron, M.L., et al., *Randomized trial of dose-dense versus conventionally scheduled and sequential versus concurrent combination chemotherapy as postoperative adjuvant treatment of node-positive primary breast cancer: first report of Intergroup Trial C9741/Cancer and Leukemia Group B Trial 9741.* Journal of clinical oncology, 2003. **21**(8): p. 1431–1439.
12. Lokich, J., *Phase I clinical trial of weekly combined paclitaxel plus docetaxel in patients with solid tumors.* Cancer, 2000. **89**(11): p. 2309–2314.
13. Burstein, H.J., et al., *Docetaxel administered on a weekly basis for metastatic breast cancer.* Journal of clinical oncology, 2000. **18**(6): p. 1212–1219.
14. Aihara, T., Y. Kim, and Y. Takatsuka, *Phase II study of weekly docetaxel in patients with metastatic breast cancer.* Annals of oncology, 2002. **13**(2): p. 286–292.
15. Tulpule, A., et al., *Multicenter trial of low-dose paclitaxel in patients with advanced AIDS-related Kaposi sarcoma.* Cancer, 2002. **95**(1): p. 147–154.
16. Klement, G., et al., *Continuous low-dose therapy with vinblastine and VEGF receptor-2 antibody induces sustained tumor regression without overt toxicity.* The journal of clinical investigation, 2000. **105**(8): p. R15–R24.
17. Bertolini, F., et al., *Maximum tolerable dose and low-dose metronomic chemotherapy have opposite effects on the mobilization and viability of circulating endothelial progenitor cells.* Cancer research, 2003. **63**(15): p. 4342–4346.

18. Bello, L., et al., *Low-dose chemotherapy combined with an antiangiogenic drug reduces human glioma growth in vivo.* Cancer research, 2001. **61**(20): p. 7501–7506.
19. Man, S., et al., *Antitumor effects in mice of low-dose (metronomic) cyclophosphamide administered continuously through the drinking water.* Cancer research, 2002. **62**(10): p. 2731–2735.
20. Hahnfeldt, P., J. Folkman, and L. Hlatky, *Minimizing long-term tumor burden: the logic for metronomic chemotherapeutic dosing and its antiangiogenic basis.* Journal of theoretical biology, 2003. **220**(4): p. 545–554.
21. Stoll, B.R., et al., *A mathematical model of the contribution of endothelial progenitor cells to angiogenesis in tumors: implications for antiangiogenic therapy.* Blood, 2003. **102**(7): p. 2555–2561.
22. Browder, T., et al., *Antiangiogenic scheduling of chemotherapy improves efficacy against experimental drug-resistant cancer.* Cancer research, 2000. **60**(7): p. 1878–1886.
23. Miller, K.D., C.J. Sweeney, and G.W. Sledge Jr, *Redefining the target: chemotherapeutics as antiangiogenics.* Journal of clinical oncology, 2001. **19**(4): p. 1195–1206.
24. Hurwitz, H. *Bevacizumab (Avastin, a Monoclonal Antibody to Vascular Endothelial Growth Factor) Prolongs Survival in First-Line Colorectal Cancer (CRC): Results of a Phase III trial of Bevacizumab in Combination with IFL (irinotecan, 5-fluorouracil, leucovorin) as first-line therapy in subjects with metastatic CRC.* in *Proc Am Assoc Clin Oncol.* 2003.
25. Hurwitz, H., et al., *Addition of bevacizumab (rhuMab VEGF) to bolus IFL in the first line treatment of patients with metastatic colorectal cancer: results of a randomized phase III trial.* New England journal of medicine, 2004. **350**(23): p. 2335–2342.
26. Slamon, D.J., et al., *Use of chemotherapy plus a monoclonal antibody against HER2 for metastatic breast cancer that overexpresses HER2.* New England journal of medicine, 2001. **344**(11): p. 783–792.
27. Hirata, S., et al., *Inhibition of in vitro vascular endothelial cell proliferation and in vivo neovascularization by low-dose methotrexate.* Arthritis & rheumatism, 1989. **32**(9): p. 1065–1073.
28. Belotti, D., et al., *The microtubule-affecting drug paclitaxel has antiangiogenic activity.* Clinical cancer research, 1996. **2**(11): p. 1843–1849.
29. Bocci, G., K. Nicolaou, and R.S. Kerbel, *Protracted low-dose effects on human endothelial cell proliferation and survival in vitro reveal a selective antiangiogenic window for various chemotherapeutic drugs.* Cancer research, 2002. **62**(23): p. 6938–6943.
30. Vacca, A., et al., *Antiangiogenesis is produced by nontoxic doses of vinblastine.* Blood, 1999. **94**(12): p. 4143–4155.
31. Wang, J., et al., *Paclitaxel at ultra low concentrations inhibits angiogenesis without affecting cellular microtubule assembly.* Anti-cancer drugs, 2003. **14**(1): p. 13–19.
32. Grant, D.S., et al., *Comparison of antiangiogenic activities using paclitaxel (taxol) and docetaxel (taxotere).* International journal of cancer, 2003. **104**(1): p. 121–129.
33. Ng, S.S., W.D. Figg, and A. Sparreboom, *Taxane-mediated antiangiogenesis in vitro: influence of formulation vehicles and binding proteins.* Cancer research, 2004. **64**(3): p. 821–824.
34. Klement, G., et al., *Differences in therapeutic indexes of combination metronomic chemotherapy and an anti-VEGFR-2 antibody in multidrug-resistant human breast cancer xenografts.* Clinical cancer research, 2002. **8**(1): p. 221–232.
35. de Fraipont, F., et al., *Thrombospondins and tumor angiogenesis.* Trends in molecular medicine, 2001. **7**(9): p. 401–407.
36. Lawler, J., *Thrombospondin-1 as an endogenous inhibitor of angiogenesis and tumor growth.* Journal of cellular and molecular medicine, 2002. **6**(1): p. 1–12.
37. Dawson, D.W., et al., *CD36 mediates the in vitro inhibitory effects of thrombospondin-1 on endothelial cells.* The journal of cell biology, 1997. **138**(3): p. 707–717.
38. Guo, N.-h., et al., *Thrombospondin 1 and type I repeat peptides of thrombospondin 1 specifically induce apoptosis of endothelial cells.* Cancer research, 1997. **57**(9): p. 1735–1742.
39. Jiménez, B., et al., *Signals leading to apoptosis-dependent inhibition of neovascularization by thrombospondin-1.* Nature medicine, 2000. **6**(1): p. 41–48.
40. Gupta, K., et al., *Binding and displacement of vascular endothelial growth factor (VEGF) by thrombospondin: effect on human microvascular endothelial cell proliferation and angiogenesis.* Angiogenesis, 1999. **3**(2): p. 147–58.
41. Masferrer, J.L., et al., *Antiangiogenic and antitumor activities of cyclooxygenase-2 inhibitors.* Cancer research, 2000. **60**(5): p. 1306–11.
42. Gately, S. and R. Kerbel, *Therapeutic potential of selective cyclooxygenase-2 inhibitors in the management of tumor angiogenesis.* Progress in experimental tumor research, 2003. **37**: p. 179–192.

43. Kieran, M.W., Anti-angiogenic chemotherapy in central nervous system tumors. *Cancer Treat* Research. 2004. 117: p. 337–349. doi: 10.1007/978-1-4419-8871-3_19
44. Bjarnason, G., et al., *Phase II trial of continuous low dose cyclophosphamide and celecoxib in patients with progressing advanced renal cell carcinoma (RCC).* in *Proc. Am. Soc. Clin. Oncol.* 2003.
45. Glück, S., et al., *Metronomic therapy in recurrent and metastatic chemoresistant SCCHN: data from a pilot study.* Biomedicine & pharmacotherapy, 2005. **59**: p. S319.
46. Glode, L.M., et al., *Metronomic therapy with cyclophosphamide and dexamethasone for prostate carcinoma.* Cancer, 2003. **98**(8): p. 1643–1648.
47. Spieth, K., R. Kaufmann, and J. Gille, *Metronomic oral low-dose treosulfan chemotherapy combined with cyclooxygenase-2 inhibitor in pretreated advanced melanoma: a pilot study.* Cancer chemotherapy and pharmacology, 2003. **52**: p. 377–382.
48. Colleoni, M., et al., *Low-dose oral methotrexate and cyclophosphamide in metastatic breast cancer: antitumor activity and correlation with vascular endothelial growth factor levels.* Annals of oncology, 2002. **13**(1): p. 73–80.
49. DiPaola, R.S., H.J. Durivage, and B.A. Kamen, High time for low-dose prospective clinical trials. Cancer, 2003. **98**(8): p. 1559–1561.
50. Willett, C.G., et al., *Direct evidence that the VEGF-specific antibody bevacizumab has antivascular effects in human rectal cancer.* Nature medicine, 2004. **10**(2): p. 145–147.
51. Mancuso, P. *Circulating endothelial cells in preclinical cancer models and in cancer patients: origin, kinetics and viability after conventional-dose or metronomic chemotherapy.* in *Proc. Am. Assoc. Cancer Res. A.* 2003.
52. Ben-Efraim, S., *Immunomodulating anticancer alkylating drugs: targets and mechanisms of activity.* Current drug targets, 2001. **2**(2): p. 197–212.
53. Matar, P., et al., *Antimetastatic effect of a single-low dose of cyclophosphamide on a rat lymphoma.* Journal of experimental & clinical cancer research, 1995. **14**(1): p. 59–63.
54. Hermans, I.F., et al., *Synergistic effect of metronomic dosing of cyclophosphamide combined with specific antitumor immunotherapy in a murine melanoma model.* Cancer research, 2003. **63**(23): p. 8408–8413.
55. Dunussi-Joannopoulos, K., *The combination of chemotherapy and systemic immunotherapy and the concept of cure in murine leukemia and lymphoma.* Leukemia & lymphoma, 2002. **43**(11): p. 2075–2082.
56. Wu, L. and I.F. Tannock, *Repopulation in murine breast tumors during and after sequential treatments with cyclophosphamide and 5-fluorouracil.* Cancer research, 2003. **63**(9): p. 2134–2138.
57. Rubie, H., et al., *Localised and unresectable neuroblastoma in infants: excellent outcome with low-dose primary chemotherapy.* British journal of cancer, 2003. **89**(9): p. 1605–1609.
58. Azzarelli, A., et al., *Low-dose chemotherapy with methotrexate and vinblastine for patients with advanced aggressive fibromatosis.* Cancer, 2001. **92**(5): p. 1259–1264.
59. Crist, W.M., et al., *Intergroup rhabdomyosarcoma study-IV: results for patients with nonmetastatic disease.* Journal of clinical oncology, 2001. **19**(12): p. 3091–3102.
60. Grundy, P., Infectious Complications in pediatric cancer patients. In Principles and practice of pediatric oncology (eds Pizzo, PA & Poplack, DG) 865–893. 2002, Lippincott Williams and Wilkins, Philadelphia.
61. Kamen, B. Why more 6-mercaptopurine? Seminars in hematology, 1991. **28**(3): p. 12–14..
62. Link, M.P., et al., *Treatment of children and young adults with early-stage non-Hodgkin's lymphoma.* New England journal of medicine, 1997. **337**(18): p. 1259–1266.
63. Crawford, J., et al., *Reduction by granulocyte colony-stimulating factor of fever and neutropenia induced by chemotherapy in patients with small-cell lung cancer.* New England journal of medicine, 1991. **325**(3): p. 164–170.
64. Eberhard, A., et al., *Heterogeneity of angiogenesis and blood vessel maturation in human tumors: implications for antiangiogenic tumor therapies.* Cancer research, 2000. **60**(5): p. 1388–1393.
65. Kerbel, R.S., *Inhibition of tumor angiogenesis as a strategy to circumvent acquired resistance to anti-cancer therapeutic agents.* Bioessays, 1991. **13**(1): p. 31–36.
66. Sweeney, C.J., et al., *The antiangiogenic property of docetaxel is synergistic with a recombinant humanized monoclonal antibody against vascular endothelial growth factor or 2-methoxyestradiol but antagonized by endothelial growth factors.* Cancer research, 2001. **61**(8): p. 3369–3372.

67. Tran, J., et al., *A role for survivin in chemoresistance of endothelial cells mediated by VEGF.* Proceedings of the National Academy of Sciences, 2002. **99**(7): p. 4349–4354.
68. Bocci, G., et al., *Thrombospondin 1, a mediator of the antiangiogenic effects of low-dose metronomic chemotherapy.* Proceedings of the National Academy of Sciences, 2003. **100**(22): p. 12917–12922.
69. Takahashi, N., et al., *Antiangiogenic therapy of established tumors in human skin/severe combined immunodeficiency mouse chimeras by anti-endoglin (CD105) monoclonal antibodies, and synergy between anti-endoglin antibody and cyclophosphamide.* Cancer research, 2001. **61**(21): p. 7846–7854.
70. Hamano, Y., et al., *Thrombospondin-1 associated with tumor microenvironment contributes to low-dose cyclophosphamide-mediated endothelial cell apoptosis and tumor growth suppression.* Cancer research, 2004. **64**(5): p. 1570–1574.
71. Emmenegger, U., et al., *A comparative analysis of low-dose metronomic cyclophosphamide reveals absent or low-grade toxicity on tissues highly sensitive to the toxic effects of maximum tolerated dose regimens.* Cancer research, 2004. **64**(11): p. 3994–4000.
72. Baruchel, S., et al., *Safety and pharmacokinetics of temozolomide using a dose-escalation, metronomic schedule in recurrent paediatric brain tumours.* European journal of cancer, 2006. **42**(14): p. 2335–2342.
73. Günther, M., E. Wagner, and M. Ogris, *Acrolein: unwanted side product or contribution to antiangiogenic properties of metronomic cyclophosphamide therapy?* Journal of cellular and molecular medicine, 2008. **12**(6b): p. 2704–2716.
74. Albertsson, P., B. Lennernäs, and K. Norrby, *Dose effects of continuous vinblastine chemotherapy on mammalian angiogenesis mediated by VEGF-A.* Acta Oncologica, 2008. **47**(2): p. 293–300.
75. Ooyama, A., et al., *Anti-angiogenic effect of 5-Fluorouracil-based drugs against human colon cancer xenografts.* Cancer letters, 2008. **267**(1): p. 26–36.
76. Laquente, B., et al., *Antiangiogenic effect of gemcitabine following metronomic administration in a pancreas cancer model.* Molecular cancer therapeutics, 2008. **7**(3): p. 638–647.
77. Klink, T., et al., *Metronomic trofosfamide inhibits progression of human lung cancer xenografts by exerting anti-angiogenic effects.* Journal of cancer research and clinical oncology, 2006. **132**: p. 643–652.
78. Drevs, J., et al., *Antiangiogenic potency of various chemotherapeutic drugs for metronomic chemotherapy.* Anticancer research, 2004. **24**(3A): p. 1759–1764.
79. Kim, J.T., et al., *Metronomic treatment of temozolomide inhibits tumor cell growth through reduction of angiogenesis and augmentation of apoptosis in orthotopic models of gliomas.* Oncology reports, 2006. **16**(1): p. 33–39.
80. Cejka, D., et al., *Everolimus (RAD001) and anti-angiogenic cyclophosphamide show long-term control of gastric cancer growth in vivo.* Cancer biology & therapy, 2008. **7**(9): p. 1377–1385.
81. Bocci, G., et al., *Antiangiogenic and anticolorectal cancer effects of metronomic irinotecan chemotherapy alone and in combination with semaxinib.* British journal of cancer, 2008. **98**(10): p. 1619–1629.
82. Ma, J. and D.J. Waxman, *Modulation of the antitumor activity of metronomic cyclophosphamide by the angiogenesis inhibitor axitinib.* Molecular cancer therapeutics, 2008. **7**(1): p. 79–89.
83. Blansfield, J.A., et al., *Combining agents that target the tumor microenvironment improves the efficacy of anticancer therapy.* Clinical cancer research, 2008. **14**(1): p. 270–280.
84. Pietras, K. and D. Hanahan, *A multitargeted, metronomic, and maximum-tolerated dose "chemo-switch" regimen is antiangiogenic, producing objective responses and survival benefit in a mouse model of cancer.* Journal of clinical oncology, 2005. **23**(5): p. 939–952.
85. Kamat, A.A., et al., *Metronomic chemotherapy enhances the efficacy of antivascular therapy in ovarian cancer.* Cancer research, 2007. **67**(1): p. 281–288.
86. Yap, R., et al., *Metronomic low-dose chemotherapy boosts CD95-dependent antiangiogenic effect of the thrombospondin peptide ABT-510: a complementation antiangiogenic strategy.* Clinical cancer research, 2005. **11**(18): p. 6678–6685.
87. Folkins, C., et al., *Anticancer therapies combining antiangiogenic and tumor cell cytotoxic effects reduce the tumor stem-like cell fraction in glioma xenograft tumors.* Cancer research, 2007. **67**(8): p. 3560–3564.
88. Shiraga, E., et al., *A metronomic schedule of cyclophosphamide combined with PEGylated liposomal doxorubicin has a highly antitumor effect in an experimental pulmonary metastatic mouse model.* International journal of pharmaceutics, 2008. **353**(1-2): p. 65–73.

89. Munoz, R., et al., *Highly efficacious nontoxic preclinical treatment for advanced metastatic breast cancer using combination oral UFT-cyclophosphamide metronomic chemotherapy.* Cancer research, 2006. **66**(7): p. 3386–3391.
90. Shaked, Y., et al., *Low-dose metronomic combined with intermittent bolus-dose cyclophosphamide is an effective long-term chemotherapy treatment strategy.* Cancer research, 2005. **65**(16): p. 7045–7051.
91. Emmenegger, U., et al., *Low-dose metronomic daily cyclophosphamide and weekly tirapazamine: a well-tolerated combination regimen with enhanced efficacy that exploits tumor hypoxia.* Cancer research, 2006. **66**(3): p. 1664–1674.

5 Imaging Techniques, Biomarkers, and Methods for Metronomic Chemotherapy

Manzoor Ahmad Mir, Tabish Javed, Aisha Shabir, and Nusrat Jan

5.1 INTRODUCTION

Through tremendous advancements in cancer therapeutic concepts in recent years, people with incurable stages of the disease have witnessed increased survival rates [1, 2]. It is generally known that the microenvironment that supports and protects cancer cells from potential harm and death must also be addressed for therapy to be successful. The fight against cancer was launched in the 1970s with the use of cytotoxins, which are drugs that are often given in combination regimens with other drugs that have distinct toxicities and are delivered at or near the maximally tolerated dosage (MTD) [3]. The MTD strategy tries to stop cancer cells' overactive mitotic processes, which make them more vulnerable to these medications than healthy, quiescent cells [4]. However, this approach is rarely effective in malignancies with complicated genetic composition, and it is widely known that such tumors rely on the microenvironment to thrive.

Metronomic chemotherapy treatment (MCT) describes the uninterrupted application of low-dose chemotherapy that may maintain high levels of active medication in the blood for a long period while producing outstanding tolerability [4]. It may be viewed as a multitarget treatment since it affects tumor cells directly in addition to the microenvironment of the cell by suppressing tumor angiogenesis or enhancing immune response. The most common definition, which is schedule-based, relates to chronic chemotherapy treatment at low, minimally harmful dosages on continuous close intervals [5]. The least physiologically effective dosage of a chemotherapeutic agent, when administered according to a regular dosing schedule and without a protracted drug-free interval, results in anti-tumor activity, which is how Klement and Kamen defined MCT [6].

Molecularly targeted medicines have different delivery schedules, toxicity profiles, and anticancer efficacy from conventional chemotherapy drugs. The term "metronomic chemotherapy" was initially used in the 2000s by Kerbel and Hanahan in two separate studies to describe the recurrent application of conventional chemotherapeutic medications at lower dosages with no extended drug-free pauses [7, 8]. The main mechanism of action is not a direct antitumor effect. The majority of the effects of MCT on tumor cells, especially on their microenvironment, are indirect and therefore inhibit tumor angiogenesis. It also affects stromal tissue and stimulates the host's immune system to combat malignancy [9]. For a long time, MCT was only used in palliative conditions since it was mistakenly believed to have no antitumor effects [10]. Oral medications are favored for use in a metronomic regimen because they may be delivered for longer periods than conventional chemotherapy. Depending on the usage context, the kind of cancer, and the preferences of the patient, different metronomic medication doses and regimens can have varied effects. These consist of oral vinorelbine (VNR), capecitabine (CAPE), methotrexate (MTX), and cyclophosphamide (CTX) for breast cancer; oral VNR for non–small cell lung cancer (NSCLC); and CAPE for gastrointestinal malignancies. The increased use of innovative pharmaceuticals and a deeper understanding of the specific mechanisms of action of MCT indicate that these treatment protocols should be more often employed in clinical settings.

 DOI: 10.1201/9781032632674-5

5.2 METRONOMIC CHEMOTHERAPY IN BREAST CANCER

One of the tumors for MCT that has likely received the most research attention is breast cancer (Figure 5.1). Treatment goals for people with metastatic illness often include enhancing quality of life and easing disease symptoms. In terms of effectiveness, MCT offers a superior option

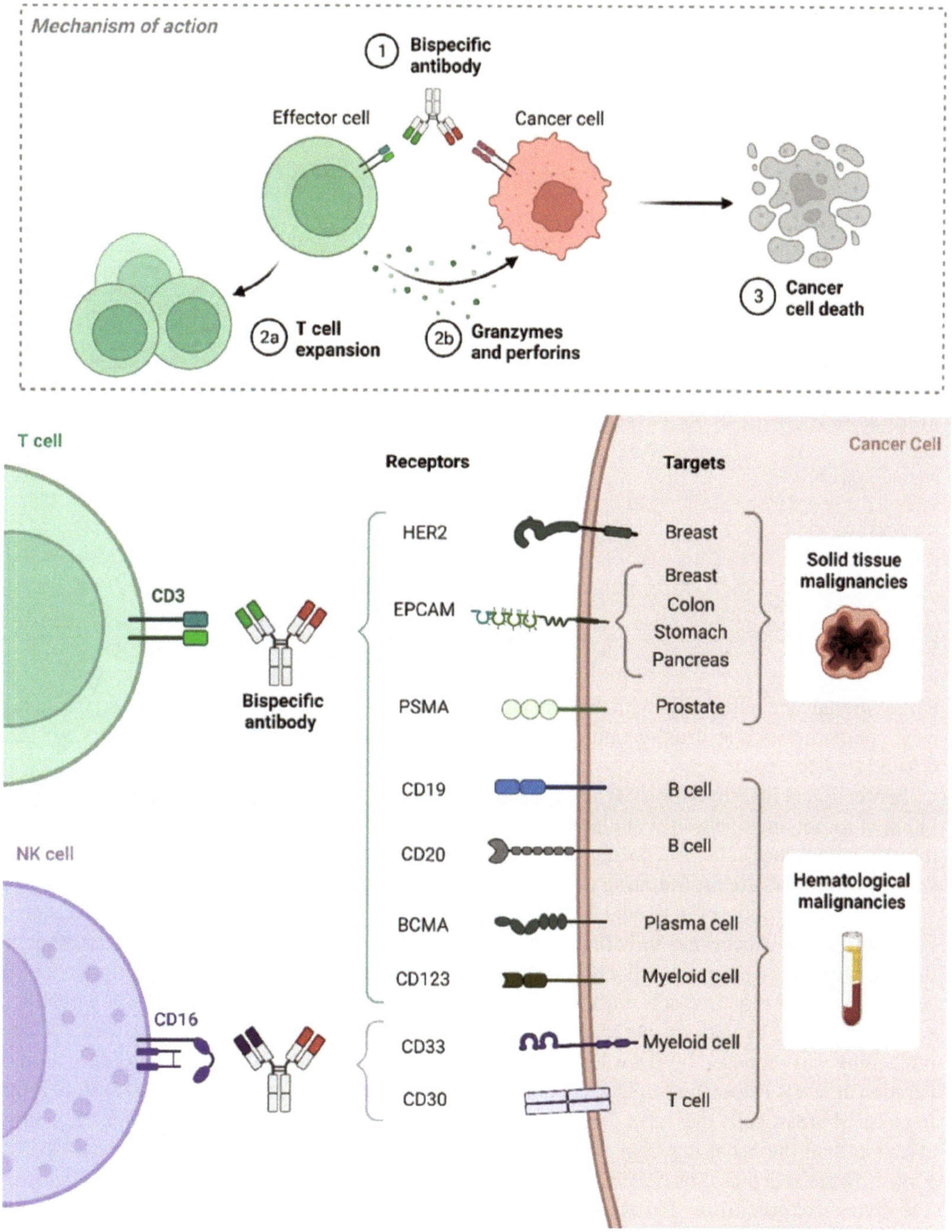

FIGURE 5.1 Mechanism involved in cancer cell death; receptors involved in targeting the specific cancer types with solid tissue malignancies and hematological malignancies.

TABLE 5.1
Result Summary of MCT Trials in Breast Cancers

Breast Cancer	CBR %	PFS (m)	OS (m)	Grade of Toxicity	Reference
Phase II Pretreated pts 22	63.6	7.5	13.6	Grade 3–4 (Hypertension 4%)	[19]
Phase II Metastatic pts 46	68	NA	NA	Grade 3 (Neutropenia 4% hypertension 17%)	[20]
Phase II; I line pts 34	38	7.7	15.9	Grade 3 (Neutropenia 9% anemia 9%)	[21]
Phase II; I line pts 33	42	6.9	24.8	Grade 3 (Neutropenia 6% HFS 15%)	[22]
Phase II Pretreated patients pts 58	62	NA	17	Grade 3 (HFS 5%)	[23]
Phase II; I and II line pts 45	44.4	12.3	NA	Grade 3 (Neutropenia 16%)	[24]
Case-cohort report Metastatic pts 33	56	NA	NA	Grade 3 (Transaminases toxicity 3%)	[25]
Phase II Pretreated patients pts 68	53.0	NA	16.9	Grade 3–4	[26]
Phase II Metastatic pts 41	NA	14.9	28.6	Grade 3 (HFS 7.3%)	[27]
Phase II Metastatic pts 80	48.8	NA	NA	Grade 3–4 (Neutropenia 4% HFS 10%)	[28]
Phase II Metastatic pts 32	50	NA	NA	Grade 3–4 (Neutropenia 4% HFS 10%)	[29]

Abbreviations: CBR, clinical benefit rate; HFS, hand-foot skin reaction; OS, overall survival; PFS, progression-free survival.

to traditional chemotherapy with fewer side effects (Tables 5.1 and 5.2). The most frequently used chemotherapeutic drugs examined in metronomic studies include CTX, MTX, VNR, and CAPE [11, 12].

In preclinical investigations, MCT has demonstrated various mechanisms of action, such as inhibition of tumor angiogenesis, cellular senescence, and immune system modulation, which interacts directly with tumor cells. According to several preclinical investigations, the main targets of MCT therapy for tumors are proliferating microvascular endothelial cells (ECs). Low-dose chemotherapy stops the proliferation of activated ECs and causes them to die, which directly affects the tumor vasculature [10]. According to Miller et al. (2001), it was hypothesized that some chemotherapy drugs could directly impact actively proliferating cancer cells, as well as indirectly influence the formation of new blood vessels. Furthermore, these effects were shown to diminish when treatment was discontinued by various methods [13]. An approach to boost the antiangiogenic effects of chemotherapy was proposed by Browder and colleagues based on this idea of continuous drug administration at levels below the MTD with short breaks in between cycles [14]. They demonstrated that in cultured breast cells that have developed drug resistance, a constant low dosage of CTX is more effective than the usual regimen. Using lower doses of vinblastine and anti–vascular endothelial growth factor receptor-2 (VEGFR-2) antibodies has been shown to induce cancer regression [15]. The division of infantile endothelial cells and the modification of pro- and antiangiogenic molecules were later demonstrated to be prevented by MCT, as well as the proliferation and movement of circulating endothelial cells (CECs) and endothelial progenitor cells (CEPs) [16]. Metronomic therapies have the potential to disrupt the equilibrium between proangiogenic and antiangiogenic factors.

TABLE 5.2
A Sample of Metronomic Schedule Investigations with Single-Agent Cytotoxics

Sample Size	Dose	ORR %	CBR%	Grade of Toxicity (3–4)	Reference
73	VRL at 30 mg, 40 mg, 50 mg thrice weekly	5	20	Neutropenia 11%–18%	[35]
34	VRL 70 mg/m^2 thrice weekly for 3 out of 4 weeks	38	68	Neutropenia 9%, febrile neutropenia 6%	[36]
43	VRL 50 mg thrice weekly	18.6	58	None	[23]
45	Cape 1000 mg/m^2 daily 4 out of 5 wk	21	51	Neutropenia 9%	[37]
60	Cape 1500 mg daily	24	62	Hand/foot 5%	[38]
107	Cape 650 mg/m^2 twice daily continuously	20	50	Thrombocytopenia 3%, anemia 3%, neutropenia 0.9%, hand/foot 16%	[39]
26	CTX 100 mg twice daily	27	69	Anemia 7.7%, thrombocytopenia 7.7%, lymphopenia 80.7%	[40]
22	CTX 50 mg daily	14	55	Leukopenia 5%	[41]
54	CTX 50 mg daily	20.4	40.7	Anemia 2%, asthenia 2%	[42]

Abbreviations: CBR, clinical benefit rate; ORR, objective response rate.

This disruption can lead to the generation and release of antiangiogenic factors in laboratory settings, specifically on ECs that have been subjected to various anticancer agents following the MCT regimen. Additionally, in vivo experiments involving mice treated with low daily doses of CTX have demonstrated similar effects [17]. The study showed that the production of thrombospondin-1 (TSP-1) is enhanced in ECs exposed to modest and sustained dosages of medicines and directly triggers growth arrest and apoptosis. The study by Bocci et al. (2003) demonstrated the suppression of angiogenesis in mice given modest daily dosages of CTX that had human tumors growing on them. They specifically noticed a rise in circulating TSP-1 and a decrease in the recruitment of proangiogenic factors. Later, Bocci and other researchers proved that MCT suppresses angiogenesis and tumor growth by upregulating TSP-1 expression and production in mice with various tumor types. In conjunction with TSP-1, MCT has also been shown to increase several endogenous angiogenesis inhibitors, including soluble VEGF receptors (sVEGFRs), pigment epithelium-derived factor (PEDF), and endostatin [18].

In contrast, MCT delivery during or after MCT administration reduces proangiogenic factors such as VEGF and hypoxia inducible factor (HIF-1). The efficacy of low dosages of CAPE as first-line treatment in 33 patients with metastatic breast cancer was investigated by Taguchi and colleagues. The following protocol was implemented to provide CAPE continuously. Take 825 mg/m^2 twice a day for 21 out of the 28 days. The medians for overall survival (OS) and progression-free survival (PFS) were 6.9 and 24.8 months, respectively. The results of this investigation showed that CAPE was effective and well-tolerated in this regimen [22]. In a second phase II study, 323 individuals with metastatic breast cancer were randomized to receive regular CMF (i.e. cyclophosphamide, methotrexate, and fluorouracil), CAPE 650 mg twice a day, or CAPE 1000 mg twice daily for 14 of every 21 days. While both CAPE regimens were equally potent as CMF when given routinely, they had fewer side effects [30]. The effectiveness of continuing CAPE at a dosage of 1500 mg once a day was examined by Fedele et al. in 60 patients with metastatic breast cancer who had already received treatment. According to their research, the median time to progression (TTP) was 7 months, the median OS was 17 months, and the clinical benefit rate (CBR) was 62% [23]. Proangiogenic factors including VEGF, VEGFR2, basic fibroblast growth factor (bFGF), and stromal cell-derived factor

1 (SDF1) are inhibited by MCT, whereas angiogenesis inhibitors such as TSP-1, platelet factor 4, and endostatin are produced by both stromal and cancer cells. Additionally, endothelial progenitor cell (EPC) mobilization is inhibited by MCT, which also causes death in circulating ECs [31]. Additionally, a combination of antiangiogenic and cytotoxic medication therapy slowed the growth of tumors and lengthened life in animal models with tumors [32]. A first study of 63 patients using the combination of two oral medications, CTX and MTX (CM), showed a 32% overall CBR and a 19% overall objective response rate (ORR) [33].

Two phase I–II experiments known as VICTOR-1 and VICTOR-2 studied the metronomic application of VNR and CAPE, revealing percentages of CBR of 48%–58% [11, 28]. The largest retrospective study presently available, VICTOR-6, summarizes the application of MCT in 584 individuals with metastatic breast cancer. First-line VNR-based regimens had noteworthy results, with ORRs fluctuating between 36.7% and 44% and disease control rates from 82.4% to 88% [34].

5.3 IMAGING TECHNIQUES

One of the main deficiencies in assessing the efficacy of MCT stems from its distinctive modes of action, which often do not result in tumor shrinkage, akin to other treatments using the MTD. This behavior has also been seen in more contemporary therapeutic approaches, such as immune checkpoint inhibitors. Prior research has shown that diffusion MRI (dMRI) and dynamic contrast-enhanced (DCE) MRI have been proposed as possible modalities for evaluating the therapeutic effectiveness of MCT in the 4T1 mammary tumor model of locally advanced breast cancer [3]. The transition to imaging at 3T favors DCE MRI, a developing technique for diagnosing breast illness. It offers significant capacities for cancer diagnosis, detection, and monitoring [43]. The non-invasive and three-dimensional nature of this technique has the advantage of facilitating the visualization of disease severity and angiogenic characteristics, as well as the capacity to anticipate treatment outcomes before therapy initiation or in the early stages of treatment. Furthermore, the evaluation of patients at high risk and the identification of high-grade ductal carcinoma in situ are emerging as efficacious applications of DCE MRI [44]. Nevertheless, there exist notable drawbacks in this context, such as the similarity in augmentation patterns observed in both benign and malignant diseases. Additionally, addressing microscopic disease, especially in the neoadjuvant setting, poses a considerable challenge. Furthermore, the prognostic value of the enhancement trend for clinical outcomes is inconsistent. The technological needs of each examination should be carefully considered, as should the necessity for postprocessing processes to be automated in order to effectively handle the expanding amount of data collected (Figure 5.2). Research is still being done, with an emphasis on employing greater field strengths with better spatial and temporal resolution data, understanding how cells increase contrast, and creating macromolecular and tailored contrast agents. The most recent publications from investigations of preoperative patients cited sensitivity values of 93% [45] as evidence of the high sensitivity of DCE MRI for the diagnosis of breast cancer. More erratic statistics have been released for specificity, with results ranging from 37% to 89%. One of the key concerns of MR breast experts is the development of methods to increase specificity without compromising whole-breast coverage [46].

The use of functional imaging has been thoroughly examined in many research studies (like DCE MRI). DCE MRI has been utilized as an effective tool for evaluating tumor vasculature in early-stage clinical studies of vascular targeting drugs [47, 48]. Using the DCE MRI approach, a paramagnetic short molecular weight contrast agent is given intravenously, and its passage through the tumor's blood arteries and into the extravascular and extracellular area is tracked over several minutes with several pictures [49]. T1- and T2-weighted sequences can be used to measure vascular parameters [50]. Therefore, it is important to presume that variations in levels during antiangiogenic therapy reflect comparable changes in other unmeasured lesions. Because only one or two relevant

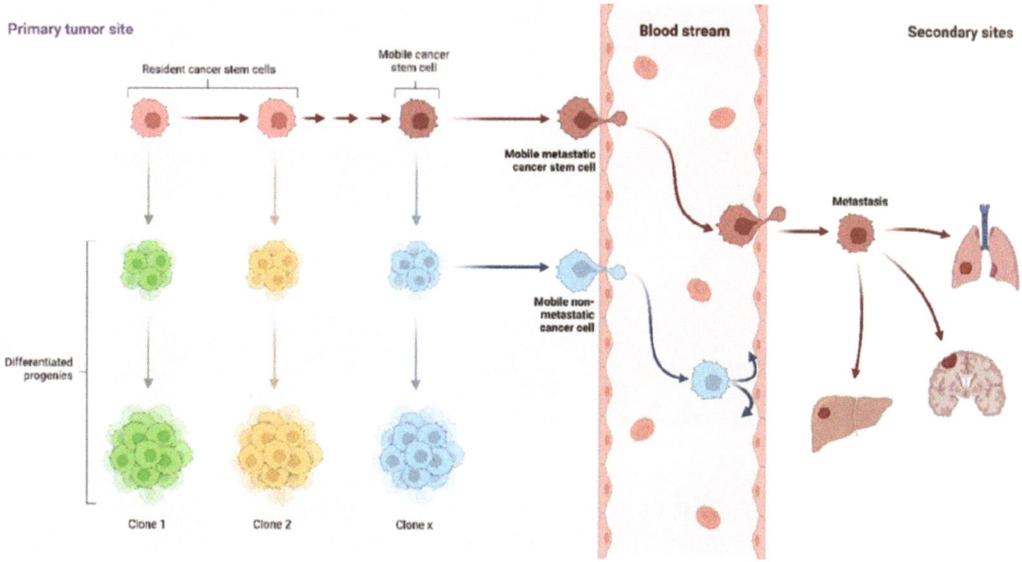

FIGURE 5.2 Differentiated progenies in primary tumor site bloodstream and secondary sites.

lesions from a single cancer location may be chosen for testing, functional imaging is restricted. Functional imaging therefore does not account for any potential behavioral variability of the distributed lesions. This characteristic might be the cause of functional imaging's inconsistent ability to predict benefits in clinical trials [51].

5.4 BIOMARKERS AND METRONOMIC DOSING

Studies have addressed the significance of biomarkers to direct medication development [52, 53] (Figure 5.3). To calculate the optimal metronomic dosage (OMD) of cytotoxins, biomarkers that show the accomplishment of pharmacodynamic effects are crucial. While MTD determination may be made with regular laboratory testing and clinical evaluations, OMD determination is significantly more difficult. Two key pharmacodynamic characteristics of interest in MCT antiangiogenic and immunological effects are possibly deserving of further investigation as biomarkers among the various known mechanisms of MCT's anticancer activities. These have probably been the subject of most research in medicines that target immune and angiogenesis regulation, but lack of validation prevents their frequent usage in clinical settings. Although lacking sensitivity, specificity, and repeatability for predicting reactions or survival outcomes in routine clinical settings, these methods can still be beneficial in aiding researchers in obtaining valuable insights into the metronomic activity of relevant cytotoxic agents, thereby facilitating appropriate metronomic dosing strategies. In the evaluation of antiangiogenic drugs, a comprehensive array of potential indicators was examined to demonstrate the efficacy of inhibiting angiogenesis targets. The components of these diagnostic tools include blood biomarkers that circulate within the bloodstream, such as cytokines like VEGF, TSP-1/2, and circulating ECs. Additionally, immunohistochemistry techniques are used to examine the presence of VEGF and TSP-2. Functional imaging methods, such as DCE MRI or DCE CT, are also utilized. Furthermore, single-nucleotide polymorphisms, such as VEGF, interleukin (IL)6, IL8, and others, are considered in the analysis [54]. The standard stages and variations have not been reliably proven to connect with response or survival results, even though changes may occur concurrently with the delivery of the medication.

It has been noted that MCT has a variety of effects on the tumor's immune biology. It does not, however, consistently alter tumor immune biology within or across tumor types, similar to immune

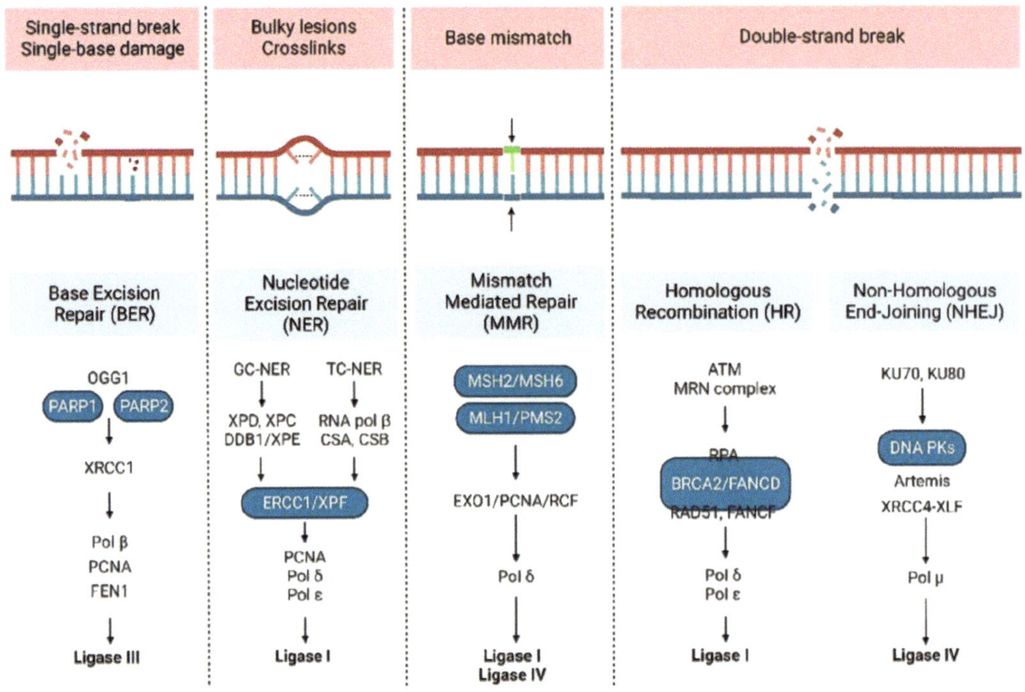

FIGURE 5.3 DNA repair mechanisms utilizing PARP1, PARP2, ERCC1/XPF/MSH2/MSH6, BRCA2/FANCD, and DNA PKs.

checkpoint inhibitors. A comprehensive characterization is necessary to get a complete understanding of how metronomic dosages of cytotoxics impact the immune-stimulatory and immunosuppressive elements that determine the inflammatory state of the tumor microenvironment. Significant progress has been made in the field of high throughput technologies, enabling a thorough examination of antibody responses, T-cell size, cytotoxic activity, and T-cell receptor repertoire. The progress made in MCT research has exceeded the previous accomplishments in this field. A more thorough understanding may be attained by establishing a connection between the impact of MCT and both the immune cells and the tumor's gene signature, mutant antigen profile, and epigenetic alterations [55]. Along with raising the standard for oncology practices dramatically, this may also lead to a better knowledge of how to mix MCT with immune checkpoint inhibitors. To comprehensively investigate the effects of MCT treatment, it is necessary to conduct meticulously designed research that involves the systematic collection of blood and tumor biospecimens throughout the treatment process. The existing literature on predictive biomarkers exhibits considerable inconsistency and complexity, likely because of variations in the selection of biomarkers, laboratory protocols, and patient categorization methodologies used for assessment [56]. Although variations in levels or measurements of some of the biomarkers that are currently accessible have been somewhat reproducibly demonstrated to suggest target action, they are still insufficiently reliable for therapeutic usefulness [57]. Although it would be desirable, the existing technological limitations prevent the establishment of a mechanistic definition for MCT. Nevertheless, considering the significant progress and advancements in the comprehension of biology and laboratory techniques, it may be plausible in the foreseeable future to establish a mechanistic-based strategy for determining the OMD of lethal substances, which can be universally applicable. The "proof of concept" (POC) represents the initial stage in the drug development process, during which a substantial body of evidence indicates a reasonable likelihood of the presence of essential elements for success and the absence of critical elements leading to failure [58]. The veracity of achieving POC status for the idea of MCT, as shown by the most recent outcomes from clinical studies, is open to scrutiny. In light of the lack of established biomarkers that have been validated to direct the development of MCT in terms of anti-angiogenic and immunological regulation, it may be appropriate to investigate potential surrogate biomarkers. The ability to assess cancer burden in real time by identifying mutation load in cell-free DNA from patients receiving treatment has been made feasible due to the progress in molecular diagnostics, including droplet digital polymerase chain reaction (PCR) and next-generation sequencing [59]. Regardless of the target, the ultimate objective of every cancer therapy is to reduce the tumor burden. Therefore, if this type of tumor biomarker that assesses tumor burden can be shown to be reliable and repeatable, it may be able to overcome the challenges associated with assessing the targets of angiogenesis and immune regulation. At this point, it is important to emphasize the significant role that pharmacokinetics plays in the development of MCT. To advance the field of metronomics from its current empirical state to a more precise and informed applied science, the incorporation of pharmacodynamics and pharmacokinetics biomarkers is imperative. In order to effectively target the appropriate patients with this therapeutic method, it is imperative that future studies use sample sizes of enough magnitude [60]. Studies done to date have not yet addressed this. Small studies that incorporate biomarkers lack the sample size capacity to identify the best biomarkers to advance, while large randomized trials don't include biomarkers to support the validation of possible biomarker candidates.

5.5 METHODS OF METRONOMIC CHEMOTHERAPY

MCT affects tumor cells and their surrounding tissue both directly and indirectly. In addition to inducing tumor dormancy, it can decrease tumor angiogenesis and boost the immune system's ability to fight cancer. MCT primarily inhibits tumor angiogenesis to achieve its anticancer effects (Figure 5.4). It has been demonstrated that using a metronomic medication delivery regimen considerably increases the antiangiogenic effects of several chemotherapeutic medicines, such as cyclophosphamide (CPA) and taxanes, in vitro [61]. MCT's antiangiogenic activity and mechanisms of action have been shown in vivo [62]. These include the selective suppression of EC migration, TSP-1, an endogenous inhibitor of angiogenesis, is expressed more often. EPCs produced from bone marrow have steadily decreased in quantity and viability [18].

Anticancer medication often results in endothelial damage. A drug's antiangiogenic dosage should not be construed as antiangiogenic if it exceeds the amount necessary to eradicate malignant cells. Several clearly stated requirements must be followed to classify chemotherapy with anticancer drugs as metronomic [32]. The key factors for implementing an antiangiogenic strategy in MCT involve the presence of significant variations in cytotoxicity between cancer cells

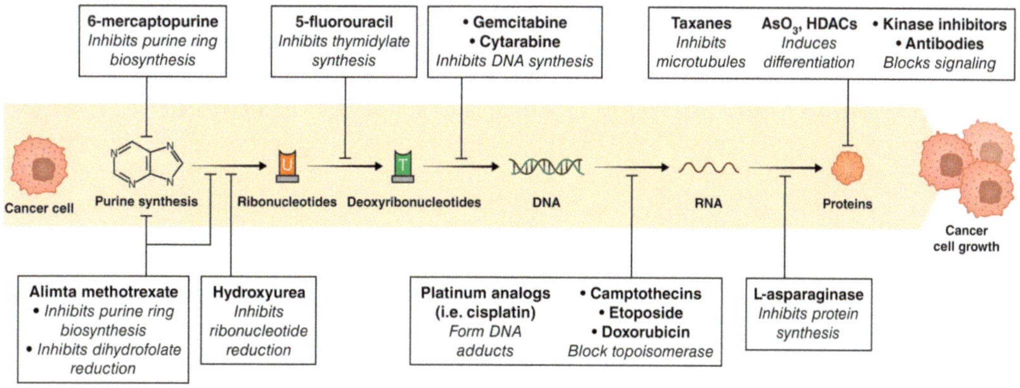

FIGURE 5.4 Methods of chemotherapy and mechanisms of action.

and ECs, the utilization of DCE MRI or contrast-enhanced ultrasonic examinations to identify changes in tumor permeability and blood flow that indicate altered functioning of ECs, and the observation of mechanistic effects that result in modifications [63]. The start of immunity and the genesis and management of cancer are significantly influenced by the innate and adaptive immune systems. The immune system is affected by neutropenia and lymphopenia. However, research has shown that certain cytotoxic drugs have immunostimulatory properties [64]. The impact on regulatory T cells (Tregs) is significant for metronomic therapies in this setting. Regulatory T cells (Tregs) are a subset of immune cells characterized by the expression of CD4, CD25, and Foxp3. These cells possess the ability to dampen or inhibit the immune system's response to a specific antigen via both cytokine signaling and direct cell-to-cell interaction. Treg cells can impede the antitumor immune response by suppressing many types of effector cells, including tumor-specific effector cells such as natural killer (NK) cells and NK T cells, as well as tumor-unspecific effector cells like CD8+ cytotoxic T lymphocytes and CD4+ T helper cells) [65]. Numerous investigations, both preclinical and clinical, have shown what low-dosage CTX does to Treg cells. It boosts lymphocyte proliferation and memory T cells while decreasing the amount of Treg cells and suppressing their activity [66]. This treatment also promotes dendritic cell development, which stimulates the immune system. Tanaka et al. found that MCT regimens of several chemotherapy drugs may promote dendritic cell maturation at low hazardous doses [67], establishing tumor dormancy cell-cycle arrest or a dynamic equilibrium condition in which the activation of apoptosis balances cell growth results in tumor dormancy. Tumor dormancy can be seen in the early stages of cancer development as well as during the remission phase that follows the end of anticancer therapies when tumors might begin growing as a result of any lingering illness [68]. The three basic mechanisms by which MCT causes tumor dormancy are the suppression of angiogenesis, death of cancerous cells, and immune surveillance [69]. The beginning of senescence is accompanied by antiproliferative responses that might begin when chemotherapeutic drugs cause damage of a lower grade. There is no activation of the caspase activity cascades that cause cellular death. Doxorubicin at a dosage of 250 nM induces prostate cells to undergo apoptosis, but at a level of 25 nM, senescence is induced. Many senescence-inducing medicines cause deoxyribonucleic acid damage that results in single- and double-strand breaks [70]. Repetitive, low-dose cytostatic medication regimens are capable of inducing senescence in tumors [71].

5.6 DISCUSSION

The utilization of MCT in cancer treatment has been limited and mostly employed as a last option and predominantly in palliative care environments. The use of MCT has shown a consistent upward trend since the early 1920s, first spearheaded by breast cancer but subsequently extending to include other malignancies such as lung cancer and pediatric cancers.

The three primary principles of mobile health technology include its straightforward administration approach, limited occurrence of negative outcomes, and diverse range of mechanisms of action. The collective influence of these characteristics has had a notable effect on the extensive use of MCT across several cancer types, including head and neck, ovarian, acute myeloid leukemia (AML), and glioblastoma multiforme (GBM). The use of magnetic resonance–guided high-intensity focused ultrasound (MCT) has been endorsed in the Advanced Breast Cancer–European Society for Medical Oncology (ABC-ESMO) metastatic breast cancer recommendations. Additionally, it has been included in national guidelines that have more recently adopted the systematic use of this technique.

MCT may be a good alternative for older patients, who frequently receive insufficient care merely because of their advanced age, regardless of hormone receptor (HR) status. The specialists highly advise that individuals with triple-negative illnesses or tumors that express the HR are the best candidates for MCT. The most well-known function of MCT is its ability to suppress angiogenesis, but fascinating new information about its possible role in immune system activation is about to be released. Diverse preclinical evidence points to the possible combination of MCT with antiangiogenic medications with anti–PD-L1 medicines. Additionally, certain clinical data are also available, even if they provide mixed outcomes [72] and appear to offer no additional benefit for MCT. We firmly feel that these combinations should be the subject of additional research in both preclinical and clinical settings. To properly examine the function of MCT, it is crucial to have a deeper knowledge of the intricate interplay that these combinations stimulate or suppress. Several regions have previously been identified concerning breast cancer: Real-world trials have shown the effectiveness of MCT, notably for elderly patients and HR+/HER2-positive metastatic breast cancer patients without visceral crises and with low tumor loads [73]. Considering the limitations imposed on more assertive treatment protocols and the incorporation of immune checkpoint inhibitors, we assert that the outcomes have been effectively consolidated among appropriately progressed patients, for whom modified conventional chemotherapy (MCT) may sometimes represent the only viable therapeutic alternative [74]. The maintenance context is probably the best place to apply MCT for colorectal cancer; however, other situations, such as elderly or physically unfit individuals, still warrant more research [75]. The significant findings from several trials on pediatric tumors also mark a significant advancement, particularly for kids living in low- and middle-income countries (LMICs) for whom every new medicine might mean the difference between a cure and palliative care. Even while several trials, mostly using oral MCT-based regimens, showed encouraging preliminary results, there hasn't been much information available for a while about the proper amount of each medication to be utilized during metronomic administration. Unfortunately, credible exploratory biomarker endpoints in well-designed and powered research are still missing. We are all aware that these studies are often quite expensive and require extensive center collaboration. In our perspective, future research in MCT may be jeopardized by the lack of support from significant pharmaceutical firms, financing for fundamental research, and innovative trial design. This serves as a vital reminder that certain parts of the globe still lack access to affordable, effective medications and that it is our moral obligation to support their development.

5.7 CONCLUSION

The alternative technique of metronomic therapy is sufficiently interesting to be pursued, but it struggles from a lack of resources for research and development. The attraction of this strategy for healthcare practitioners stems from its many focused qualities, compatibility with certain drugs and immunotherapies, cost-effectiveness, and convenient oral administration. Gaining a comprehensive understanding of the intricate processes behind the effects of metronomic cytotoxic agents might facilitate the identification of optimal pharmaceutical agents to be used concerning certain cancer types, patient cohorts, and therapeutic contexts.

For further reading, some textbooks mentioned here also give a further understanding of MCT:

- *Progress in Anti-cancer Chemotherapy* by Gabriel N. Hortobagyi and David Khayat.
- *Triple-Negative Breast Cancer: Biomarkers, Emerging Therapeutic Strategies and Clinical Challenges* by Marion Montgomery.
- *Current Cancer Treatment: Novel Beyond Conventional Approaches* by Oner Ozdemir.
- *Breast Cancer: Current and Alternative Therapeutic Modalities* by Esra Gunduz and Mehmet Gunduz.

The following links for video lectures may also be helpful:

https://youtu.be/TAqO0Mq19JQ
https://youtu.be/6wEnYfvoBqk
https://youtu.be/A9fRRCUlMms
https://youtu.be/zIWvmCsKr34

For more insight about the topic, we would suggest detailed findings from the books of (Mir MA, 2022) https://doi.org/10.1016/C2021-0-02565-7, https://doi.org/10.1016/C2022-0-00074-X, (Mir MA, 2021) https://doi.org/10.52305/WXJL6770, and (Mir MA, 2015) https://doi.org/10.1016/C2014-0-02898-5 and from the cancer.net website on the following mentioned links:

https://www.cancer.net/cancer-types/breast-cancer/types-treatment
https://discovery.ucl.ac.uk/id/eprint/1472740/
https://www.jmedsciences.com/doi/JMEDS/pdf/10.5005/jp-journals-10045-00138

For diagrammatic illustrations and descriptive tables (Lazzeroni, 2012): http://www.eurekaselect.com/article/49928

See video links on the overall status of cancer, its various types, and current and new possible treatment options available: https://www.sciencedirect.com/science/article/pii/S205970292032278X

REFERENCES

1. Leowattana, W., P. Leowattana, and T. Leowattana, *Systemic treatment for metastatic colorectal cancer.* World journal of gastroenterology, 2023. **29**(10): p. 1569.
2. Ganesan, P. and L.M. Kulik, *Hepatocellular carcinoma: new developments.* Clinics in liver disease, 2023. **27**(1): p. 85–102.
3. Cazzaniga, M.E., et al., *Metronomic chemotherapy.* Cancers, 2021. **13**(9): p. 2236.
4. Dehingia, K., H.K. Sarmah, and M.B. Jeelani, *A brief review on cancer research and its treatment through mathematical modelling.* Annals of cancer research and therapy, 2021. **29**(1): p. 34–40.
5. Rajasekaran, T., et al., *Metronomic chemotherapy: A relook at its basis and rationale.* Cancer letters, 2017. **388**: p. 328–333.
6. Klement, G.L. and B.A. Kamen, *Nontoxic, fiscally responsible, future of oncology: could it be beginning in the Third World?* Journal of pediatric hematology/oncology, 2011. **33**(1): p. 1.
7. Kerbel, R.S. and B.A. Kamen, *The anti-angiogenic basis of metronomic chemotherapy.* Nature reviews cancer, 2004. **4**(6): p. 423–436.
8. Hanahan, D., G. Bergers, and E. Bergsland, *Less is more, regularly: metronomic dosing of cytotoxic drugs can target tumor angiogenesis in mice.* The journal of clinical investigation, 2000. **105**(8): p. 1045–1047.
9. Janssen, N., Predicting clinical outcome in melanoma and breast cancer using blood and tumour-based immune biomarkers: a means to monitor anti-cancer immune responses. 2020, Universität Tübingen.
10. Krajnak, S., et al., *Metronomic chemotherapy for metastatic breast cancer.* Oncology research and treatment, 2022. **45**(1-2): p. 12–17.
11. Cazzaniga, M., et al., *Metronomic chemotherapy for advanced breast cancer patients in the real world practice: Final results of the VICTOR-6 study.* The breast, 2019. **48**: p. 7–16.
12. Romiti, A., et al., *Current achievements and future perspectives of metronomic chemotherapy.* Investigational new drugs, 2017. **35**: p. 359–374.
13. Miller, K.D., C.J. Sweeney, and G.W. Sledge Jr, *Redefining the target: chemotherapeutics as antiangiogenics.* Journal of clinical oncology, 2001. **19**(4): p. 1195–1206.
14. Browder, T., et al., *Antiangiogenic scheduling of chemotherapy improves efficacy against experimental drug-resistant cancer.* Cancer research, 2000. **60**(7): p. 1878–1886.
15. Klement, G., et al., *Continuous low-dose therapy with vinblastine and VEGF receptor-2 antibody induces sustained tumor regression without overt toxicity.* The journal of clinical investigation, 2000. **105**(8): p. R15–R24.
16. Simsek, C., E. Esin, and S. Yalcin, *Metronomic chemotherapy: a systematic review of the literature and clinical experience.* Journal of oncology, 2019. **2019**: p. 5483791.
17. Bocci, G., et al., *Thrombospondin 1, a mediator of the antiangiogenic effects of low-dose metronomic chemotherapy.* Proceedings of the National Academy of Sciences, 2003. **100**(22): p. 12917–12922.
18. Natale, G. and G. Bocci, *Does metronomic chemotherapy induce tumor angiogenic dormancy? A review of available preclinical and clinical data.* Cancer letters, 2018. **432**: p. 28–37.
19. Garcia-Saenz, J., et al., *Bevacizumab in combination with metronomic chemotherapy in patients with anthracycline-and taxane-refractory breast cancer.* Journal of chemotherapy, 2008. **20**(5): p. 632–639.

20. Dellapasqua, S., et al., *Metronomic cyclophosphamide and capecitabine combined with bevacizumab in advanced breast cancer.* Journal of clinical oncology, 2008. **26**(30): p. 4899–4905.
21. Addeo, R., et al., *Low-dose metronomic oral administration of vinorelbine in the first-line treatment of elderly patients with metastatic breast cancer.* Clinical breast cancer, 2010. **10**(4): p. 301–306.
22. Taguchi, T., et al., *Study of low-dose capecitabine monotherapy for metastatic breast cancer.* Chemotherapy, 2010. **56**(2): p. 166–170.
23. Fedele, P., et al., *Efficacy and safety of low-dose metronomic chemotherapy with capecitabine in heavily pretreated patients with metastatic breast cancer.* European journal of cancer, 2012. **48**(1): p. 24–29.
24. Yoshimoto, M., et al., *Metronomic oral combination chemotherapy with capecitabine and cyclophosphamide: a phase II study in patients with HER2-negative metastatic breast cancer.* Cancer chemotherapy and pharmacology, 2012. **70**: p. 331–338.
25. Aurilio, G., et al., *Oral metronomic cyclophosphamide and methotrexate plus fulvestrant in advanced breast cancer patients: A mono-institutional case-cohort report.* The breast journal, 2012. **18**(5): p. 470–474.
26. Wang, Z., et al., *An all-oral combination of metronomic cyclophosphamide plus capecitabine in patients with anthracycline-and taxane-pretreated metastatic breast cancer: a phase II study.* Cancer chemotherapy and pharmacology, 2012. **69**: p. 515–522.
27. Schwartzberg, L.S., et al., *Phase II trial of fulvestrant with metronomic capecitabine for postmenopausal women with hormone receptor-positive, HER2-negative metastatic breast cancer.* Clinical breast cancer, 2014. **14**(1): p. 13–19.
28. Cazzaniga, M., et al., *Metronomic chemotherapy with oral vinorelbine (mVNR) and capecitabine (mCAPE) in advanced HER2-negative breast cancer patients: is it a way to optimize disease control? Final results of the VICTOR-2 study.* Breast cancer research and treatment, 2016. **160**: p. 501–509.
29. De Iuliis, F., et al., *On and off metronomic oral vinorelbine in elderly women with advanced breast cancer.* Tumori journal, 2015. **101**(1): p. 30–35.
30. Stockler, M.R., et al., *Capecitabine versus classical cyclophosphamide, methotrexate, and fluorouracil as first-line chemotherapy for advanced breast cancer.* Journal of clinical oncology, 2011. **29**(34): p. 4498–4504.
31. Liu, J., et al., *Current research status of metronomic chemotherapy in combination treatment of breast cancer.* Oncology research and treatment, 2022. **45**(11): p. 681–692.
32. Hadjicharalambous, M., et al., *Combined anti-angiogenic and cytotoxic treatment of a solid tumour: in silico investigation of a xenograft animal model's digital twin.* Journal of theoretical biology, 2022. **553**: p. 111246.
33. Orlando, L., et al., *Prolonged clinical benefit with metronomic chemotherapy in patients with metastatic breast cancer.* Anti-cancer drugs, 2006. **17**(8): p. 961–967.
34. Montagna, E., et al., *Metronomic chemotherapy for first-line treatment of metastatic triple-negative breast cancer: a phase II trial.* Breast care, 2018. **13**(3): p. 177–181.
35. Camerini, A., et al., *Metronomic oral vinorelbine as first-line treatment in elderly patients with advanced non-small cell lung cancer: results of a phase II trial (MOVE trial).* BmC cancer, 2015. **15**(1): p. 1–6.
36. He, S., et al., *Capecitabine "metronomic" chemotherapy for palliative treatment of elderly patients with advanced gastric cancer after fluoropyrimidine-based chemotherapy.* Medical oncology, 2012. **29**: p. 100–106.
37. Brandi, G., et al., *Metronomic capecitabine in advanced hepatocellular carcinoma patients: a phase II study.* The oncologist, 2013. **18**(12): p. 1256–1257.
38. Glode, L.M., et al., *Metronomic therapy with cyclophosphamide and dexamethasone for prostate carcinoma.* Cancer, 2003. **98**(8): p. 1643–1648.
39. Penel, N., et al., *Megestrol acetate versus metronomic cyclophosphamide in patients having exhausted all effective therapies under standard care.* British journal of cancer, 2010. **102**(8): p. 1207–1212.
40. Kato, H., et al., *A randomized trial of adjuvant chemotherapy with uracil–tegafur for adenocarcinoma of the lung.* New England journal of medicine, 2004. **350**(17): p. 1713–1721.
41. Watanabe, T., et al., *Oral uracil and tegafur compared with classic cyclophosphamide, methotrexate, fluorouracil as postoperative chemotherapy in patients with node-negative, high-risk breast cancer: National Surgical Adjuvant Study for Breast Cancer 01 Trial.* Journal of clinical oncology, 2009. **27**(9): p. 1368–1374.
42. Pantziarka, P., et al., *Repurposing drugs in oncology (ReDO)—Cimetidine as an anti-cancer agent.* ecancermedicalscience, 2014. **8**.

43. McLaughlin, R. and N. Hylton, *MRI in breast cancer therapy monitoring.* NMR in biomedicine, 2011. **24**(6): p. 712–720.
44. Gilbert, F.J. and K. Pinker-Domenig, Diagnosis and staging of breast cancer: when and how to use mammography, tomosynthesis, ultrasound, contrast-enhanced mammography, and magnetic resonance imaging. In: Hodler J., Kubik-Huch R.A., von Schulthess G.K., editors, Diseases of the Chest, Breast, Heart and Vessels 2019-2022: Diagnostic and Interventional Imaging, Cham (CH): Springer, 2019: p. 155–166.
45. Wells, P.S., et al., *Accuracy of ultrasound for the diagnosis of deep venous thrombosis in asymptomatic patients after orthopedic surgery: a meta-analysis.* Annals of internal medicine, 1995. **122**(1): p. 47–53.
46. Schünemann, H.J., et al., *Breast cancer screening and diagnosis: a synopsis of the European Breast Guidelines.* Annals of internal medicine, 2020. **172**(1): p. 46–56.
47. Leach, M., et al., *Imaging vascular function for early stage clinical trials using dynamic contrast-enhanced magnetic resonance imaging.* European radiology, 2012. **22**: p. 1451–1464.
48. O'connor, J.P., et al., *Dynamic contrast-enhanced MRI in clinical trials of antivascular therapies.* Nature reviews. Clinical oncology, 2012. **9**(3): p. 167–177.
49. Padhani, A.R., *Dynamic contrast-enhanced MRI in clinical oncology: current status and future directions.* Journal of magnetic resonance imaging, 2002. **16**(4): p. 407–422.
50. Jacobs, I., *Magnetic resonance imaging biomarkers of cancer: with applications in high intensity focused ultrasound treatment evaluation.* [PhD Thesis 1 (Research TU/e / Graduation TU/e), Biomedical Engineering]. Technische Universiteit Eindhoven. 2015.
51. Eisenhauer, E.A., et al., *New response evaluation criteria in solid tumours: revised RECIST guideline (version 1.1).* European journal of cancer, 2009. **45**(2): p. 228–247.
52. Workman, P., et al., *Minimally invasive pharmacokinetic and pharmacodynamic technologies in hypothesis-testing clinical trials of innovative therapies.* Journal of the National Cancer Institute, 2006. **98**(9): p. 580–598.
53. O'connor, J.P., et al., *Imaging biomarker roadmap for cancer studies.* Nature reviews. Clinical oncology, 2017. **14**(3): p. 169–186.
54. Andre, N., et al., *Maintenance chemotherapy in children with ALL exerts metronomic-like thrombospondin-1 associated anti-endothelial effect.* Oncotarget, 2015. **6**(26): p. 23008.
55. Yuan, J., et al., *Novel technologies and emerging biomarkers for personalized cancer immunotherapy.* Journal for immunotherapy of cancer, 2016. **4**: p. 1–25.
56. Malottki, K., et al., *Problems of variable biomarker evaluation in stratified medicine research—A case study of ERCC1 in non-small-cell lung cancer.* Lung cancer, 2016. **92**: p. 1–7.
57. Tan, E.H., et al., *Metronomic vinorelbine (oral) in combination with sorafenib in advanced non-small cell lung cancer.* Lung cancer, 2015. **88**(3): p. 289–296.
58. Zhao, X., et al., *Biomarkers in pharmaceutical research.* Clinical chemistry, 2015. **61**(11): p. 1343–1353.
59. Levy, B., et al., *Clinical utility of liquid diagnostic platforms in non-small cell lung cancer.* The oncologist, 2016. **21**(9): p. 1121–1130.
60. Bocci, G. and R.S. Kerbel, *Pharmacokinetics of metronomic chemotherapy: a neglected but crucial aspect.* Nature reviews. Clinical oncology, 2016. **13**(11): p. 659–673.
61. Cesca, M., et al., *Tumor delivery of chemotherapy combined with inhibitors of angiogenesis and vascular targeting agents.* Frontiers in oncology, 2013. **3**: p. 259.
62. Cejka, D., et al., *Everolimus (RAD001) and anti-angiogenic cyclophosphamide show long-term control of gastric cancer growth in vivo.* Cancer biology & therapy, 2008. **7**(9): p. 1377–1385.
63. Maiti, R., *Metronomic chemotherapy.* Journal of pharmacology and pharmacotherapeutics, 2014. **5**(3): p. 186–192.
64. Innao, V., et al., *New frontiers about the role of human microbiota in immunotherapy: the immune checkpoint inhibitors and CAR T-cell therapy era.* International journal of molecular sciences, 2020. **21**(23): p. 8902.
65. Krishnamoorthy, M., L. Gerhardt, and S. Maleki Vareki, *Immunosuppressive effects of myeloid-derived suppressor cells in cancer and immunotherapy.* Cells, 2021. **10**(5): p. 1170.
66. Golubovskaya, V. and L. Wu, *Different subsets of T cells, memory, effector functions, and CAR-T immunotherapy.* Cancers, 2016. **8**(3): p. 36.
67. Tanaka, H., et al., *Classification of chemotherapeutic agents based on their differential in vitro effects on dendritic cells.* Cancer research, 2009. **69**(17): p. 6978–6986.
68. Bergers, G. and L.E. Benjamin, *Tumorigenesis and the angiogenic switch.* Nature reviews cancer, 2003. **3**(6): p. 401–410.

69. Chen, Y.-L., M.-C. Chang, and W.-F. Cheng, *Metronomic chemotherapy and immunotherapy in cancer treatment.* Cancer letters, 2017. **400**: p. 282–292.
70. Mansilla, S., L. Llovera, and J. Portugal, *Chemotherapeutic targeting of cell death pathways.* Anti-Cancer agents in medicinal chemistry, 2012. **12**(3): p. 226–238.
71. Zu, C., et al., *Low dose Emodin induces tumor senescence for boosting breast cancer chemotherapy via silencing NRARP.* Biochemical and biophysical research communications, 2018. **505**(4): p. 973–978.
72. Zsiros, E., et al., *Efficacy and safety of pembrolizumab in combination with bevacizumab and oral metronomic cyclophosphamide in the treatment of recurrent ovarian cancer: a phase 2 nonrandomized clinical trial.* JAMA oncology, 2021. **7**(1): p. 78–85.
73. Cazzaniga, M.E., et al., *Pan-European expert meeting on the use of metronomic chemotherapy in advanced breast cancer patients: the PENELOPE project.* Advances in therapy, 2019. **36**: p. 381–406.
74. Biziota, E., et al., *Metronomic chemotherapy: a potent macerator of cancer by inducing angiogenesis suppression and antitumor immune activation.* Cancer letters, 2017. **400**: p. 243–251.
75. Martin, S.C., *The experience and communicative management of identity threats among people with Parkinson's disease: Implications for health communication theory and practice.* Communication monographs, 2016. **83**(3): p. 303–325.

6 Metronomic Chemotherapy in Breast Cancer

Manzoor Ahmad Mir, Aisha Shabir, and Nusrat Jan

6.1 INTRODUCTION

Breast cancer (BC) is one of the leading public health concerns and the most common type of cancer, i.e., around 25% of all the cancers reported among women globally [1]. The cases of BC are at an all-time high and are projected to elevate for the next 20 years despite the significant advancements that have been made in the therapy of BC [2, 3].

It has been reported that at the early diagnostic stage, around 5–10% of the BC patients have metastasis, and after about 3 years of being diagnosed with BC, around 10–15% of patients develop metastatic disease [4]. Following the initial diagnosis, there is a possibility that some individuals may develop metastatic disease a decade or more later [5].

Given the broad spectrum of BC, it is necessary to find new techniques that address the process of metastatic progression and to investigate more specialized treatment options [6, 7] Although there have been indications of BC diversity, and several molecular subtypes have been identified (such as luminal A and B, triple negative, and HER2 positive), only a small number of attempts have been successful in establishing specific therapies for early detection of BC [8].

The term chemotherapy is attributed to Paul Ehrlich, a notable chemist from Germany, who extensively studied the application of medicines to combat infectious ailments. Based on historical documentation, the utilization of arsenicals was initially observed during the 1900s. By the 1960s, radiotherapy and surgical procedures formed the fundamental pillars of cancer therapy.

Cancer therapy, particularly oral derivatives like chlorambucil and finally cyclophosphamide (CTX), developed in response to the Lindskog article demonstrating nitrogen mustard's efficacy in the treatment of lymphoma [9]. The finding of actinomycin D sparked the search for additional chemotherapeutic agents, including anthracyclines, mitomycin, and bleomycin [10]. Rituximab was given the go-ahead to treat BC in the late 1990s [11].

Chemotherapy is becoming more crucial for the treatment of cancer, especially with its usage as an adjunct to local therapy. Additionally, chemotherapy plays a bigger part in advanced disease, when the tumor has spread from its original location, in attempts to reduce the symptoms of cancer and lengthen life.

Traditional chemotherapy is characterized by a 'maximum tolerated dose' (MTD) approach (i.e., giving the highest possible dose of the drug that a patient can tolerate), which has shown promising results for increased efficacy and reduced treatment duration, but this comes with a drawback concerning the long gaps between the therapy sessions. These long gaps (3–4 weeks) between the chemotherapy sessions lead to the gradual recovery of the tumor cells and development of resistance, followed by other side effects including hair loss, vomiting, nausea, bone marrow suppression, and even organ damage (Figure 6.1). Therefore, this led to the need to explore alternatives to increase the response to treatment and decrease the side effects, and metronomic chemotherapy (MCT) is one such alternative [12].

6.2 OVERVIEW OF METRONOMIC CHEMOTHERAPY

The difficulties associated with the MTD approach and the proposed side effects have led to the development of a negative attitude of patients toward receiving toxic chemotherapy for metastatic

 DOI: 10.1201/9781032632674-6

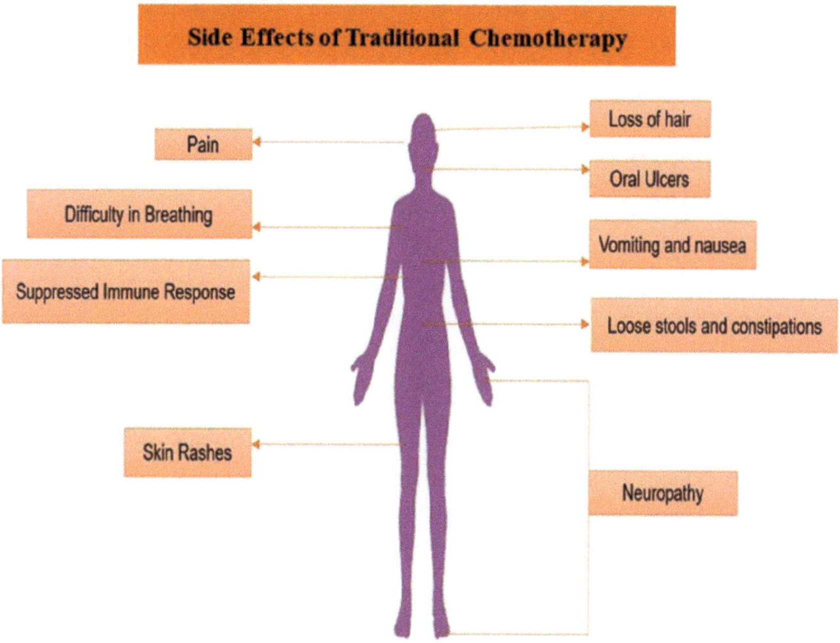

FIGURE 6.1 Side effects of traditional chemotherapy.

breast cancer (MBC). It is therefore desirable to introduce some new and more reliable techniques with enhanced or at least similar effectiveness and lowered toxic effects. These strategies might involve the use of less harmful medication and more practical (e.g., oral and outpatient) methods of drug delivery systems compared to traditional drug delivery systems [13]. In the context of MBC, it has been pointed out by a European expert committee that "Treatment should control disease in at least 20-30% of patients with an acceptance side effect profile" [14]. In this regard, MCT therefore proves to be the most promising alternative. MCT has been described as an alternative to dose-dense therapy, with a crucial distinction that it is not always dose-dense, which means that the overall dose received may be substantially less than the MTD approach, thereby diminishing or even eliminating the drug-induced toxic effects [15]. MCT refers to the low but frequent doses of chemotherapy drugs to patients without long breaks, as in the traditional MTD approach, and therefore MCT continuously targets the cancerous cells, giving it no time to recover or develop resistance [8]. The basic principle behind MCT involves continuous drug administration, low doses of the drugs, immune system modulation, and exploitation of anti-angiogenic effects of certain chemotherapy drugs [16]. Because it provides for a more satisfactory equilibrium between the activity and adverse effects of treatments, it may boost the therapeutic efficacy of medications and additionally enable prolonged therapy, thereby improving the chances of survival [17].

According to Kerbel and Kamen (2004), MCT treatment is typically characterized by the following:

- Regular, uninterrupted chemotherapeutic delivery.
- The dose does not increase to the MTD.
- Hematopoietic growth factor assistance is not required.
- The possibility of developing resistance is later than expected.
- Side effects are minimal to zero.
- Oral medication is preferred, which includes the use of drugs such as CTX.

The benefit of investigating MCT strategies is connected to the clinical pathway of later-stage BC. Several patients may experience a longer life expectancy with the diagnosis of advanced disease [18].

6.3 MECHANISMS OF ACTION

6.3.1 Mechanism of Action of Traditional Chemotherapy

The actual work of chemotherapy is to stop the growth of tumors and cell division, which ultimately prevents metastasis because it also has an impact on normal cells, which leads to hazardous side effects. Tumor growth can be slowed down at various degrees, both inside the cell and in its surrounding; some of the drugs act on cells from the outside.

However, the majority of cytotoxic medications interact with biological processes required for cell development and survival to exert their effects inside the cell. Drugs typically used in chemotherapy impact the proper functioning of the preformed molecule or the macromolecular synthesis and function of malignant cells by interfering with the production of DNA, RNA, or proteins. When there is enough disruption of macromolecular synthesis or function, it causes cell death either directly from the action of a chemotherapeutic agent or by inducing apoptosis [19]. The bloodstream carries chemotherapeutic drugs throughout the body. As a result, it can cure cancerous cells virtually anywhere in the body. This is referred to as systemic therapy. Chemotherapy destroys cells that are dividing to become two new cells by targeting the genes in the nucleus of the cell. The chemical ability of the big groups of alkylators and alkylator-related substances to attach to numerous cellular macromolecules, most significantly DNA, prevents DNA from performing as it should during cell division and gene expression. Certain antimetabolites also help in damaging the cancerous cells (Figure 6.2). In the creation of DNA and RNA, antimetabolites either substitute for the regular molecules or inhibit crucial stages, disrupting the normal function of the molecules and perhaps impairing cell development and survival. In addition to this certain topoisomerase inhibitors also

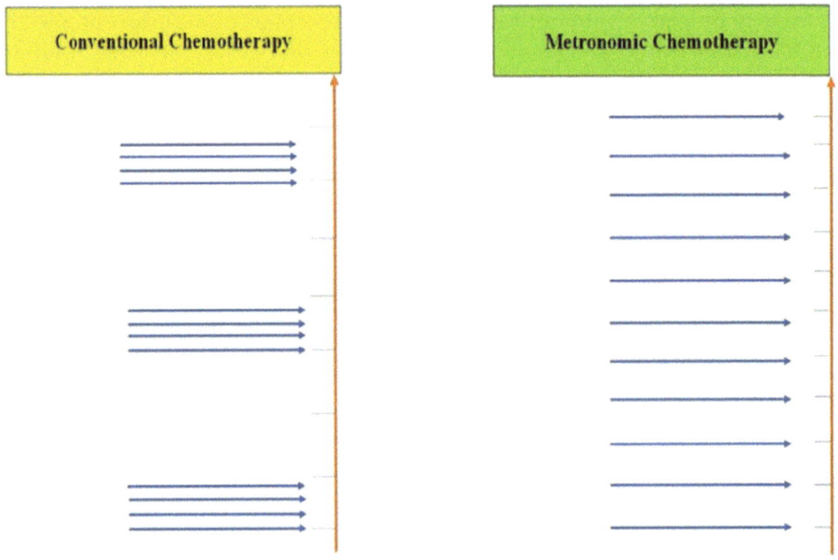

FIGURE 6.2 The two aspects of chemotherapy. Conventional chemotherapy is characterized by MTD approach followed by long gaps between the therapy sessions. Metronomic chemotherapy is characterized by the low but frequent dose of chemotherapy without long breaks.

halt the process of replication, transcription, and DNA repair and hence prevent cell division and multiplication [16].

6.3.2 Mechanism of Action of Metronomic Chemotherapy

MCT's mode of action was first thought to include the suppression of angiogenesis. The fact that MCT possesses several modes of action, including anti-angiogenic, anti-proliferative, and immunomodulatory properties, is now generally acknowledged [20, 21].

6.3.2.1 Angiogenesis and Its Effects

Folkman introduced the concept of fighting cancer by explicitly focusing on the process of tumor angiogenesis; he proposed the ground-breaking theory that tumor growth depends on angiogenesis [22]. Angiogenesis is a complex process in which certain biomolecules of the body are involved, which leads to the formation of new blood vessels from pre-existing ones and further facilitates the growth of tumor cells and metastasis [23]. The anti-angiogenic agents for the treatment of cancer block the process of angiogenesis and thereby suppress tumor growth [24].

Angiogenesis is intricately regulated by a diverse array of stimulatory and inhibitory factors working in precise coordination [25]. Angiogenesis is a key factor in the development and probable metastasis of BC development, according to preclinical research [26, 27]. Their high metabolic needs and frequent exposure to hypoxic conditions induce the expression of hypoxia-inducible factor 1-alpha (HIF1-alpha), which starts angiogenesis. Vascular endothelial growth factor (VEGF), which stimulates the growth and invasiveness of tumors, is in turn encouraged by HIF1 [28]. The transcriptional region of VEGF-A contains particular promoter regions that HIF1 directly interacts with, and HIF1 induces the expression of VEGF-A, an autocrine growth factor implicated in angiogenesis [29]. The VEGF family of ligands, which includes VEGF-A to -D, interacts with the VEGFR-1 and VEGFR-2 members of the VEGF tyrosine kinase receptor superfamily. The two ligands with the highest affinity for these receptors are VEGF-A and VEGF-B. Notably, the strongest angiogenic effects are produced by VEGF-A binding to VEGFR-2 [30]. As a result, anti-angiogenesis therapy would be a suitable method for treating cancer.

Regeneration of the anti-cancer impact of the immune system is the newly discovered pathway for MCT. Mounting evidence shows that the immune response against cancer can show promising results in the long-term management and control of cancer [31]. Some preclinical results have shown that chemotherapy medications, like CTX, may have a positive impact in enhancing anti-tumor immunity [32].

It has been demonstrated that metronomic treatment increases an immune response against cancer cells by reducing the quantity and suppressive activity of regulatory T cells (Tregs), which contributes to immunological tolerance and can inhibit the activity of other immune cells, including those engaged in anti-tumor responses [33].

The control of myeloid-derived suppressor cells [34] and the improvement of the cytotoxic activity of immune effector cells, such as tumor-specific T cells [35], are two additional immunological-based approaches (Figure 6.3).

Metronomic therapy can be used in conjunction with targeted therapies for a longer time due to minor side effects. Additionally compared to many MTD chemotherapies, drugs like CTX may prove to be much more affordable [13].

Recently other putative mechanisms of action of MCT, including the anti-cancer immune response and additional actors from the tumor microenvironment, have been proposed. Researchers have proposed several novel pathways that would be involved in anti-cancer activity based on the interaction between tumor cells and the microenvironment: Immune system activation and tumor dormancy induction.

According to several experts, tumor escape from immune tracking could be the most significant feature of cancer development. Additionally, the immune suppression brought about by

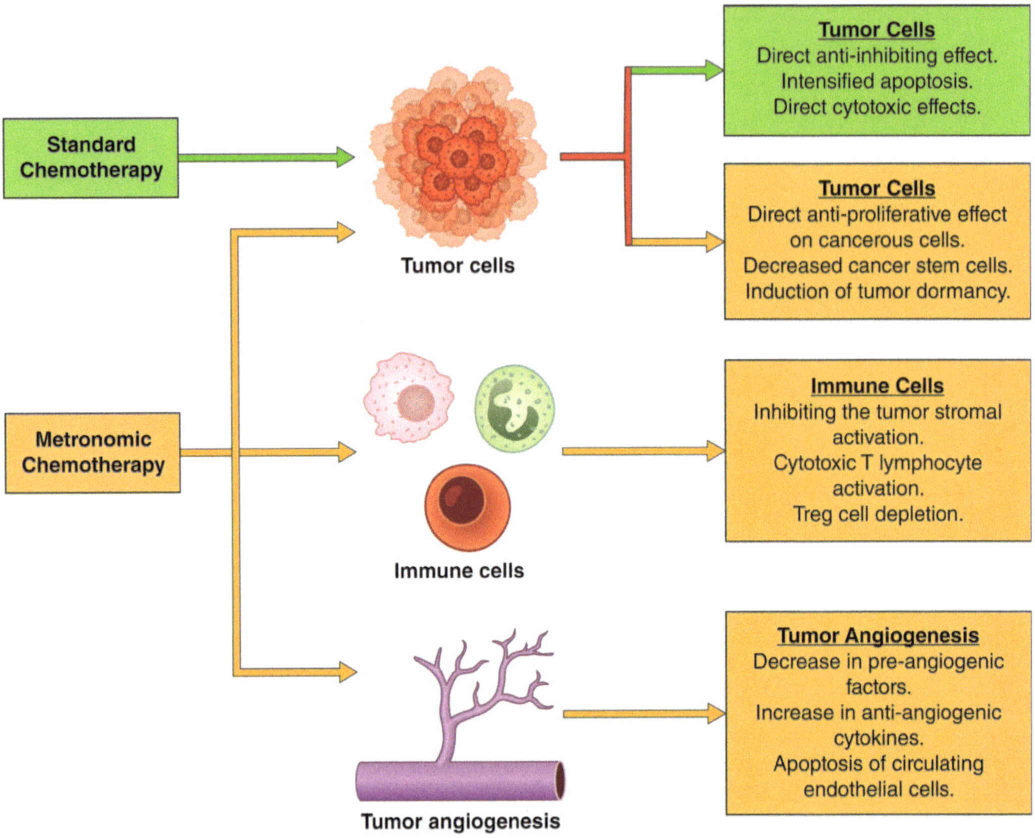

FIGURE 6.3 A comparison of the mechanism of action of standard chemotherapy and metronomic chemotherapy. Treg, regulatory T cells.

chemotherapy may promote the proliferation of tumor cells since it lowers the white blood cell (WBC) count [36, 37]. Several other researchers have also verified that many cytotoxic medications, including anthracyclines, taxanes, CTX, etc., may also have significant immunostimulatory effects. These medications may specifically affect Tregs, which appear to be important in the setting of metronomic schedules [38, 39].

6.4 CLINICAL APPLICATIONS OF METRONOMIC CHEMOTHERAPY IN BREAST CANCER

Numerous clinical trials have been carried out over the past 5 years to investigate the safety as well as effectiveness of MCT in various forms of human cancer. These novel therapeutic strategies have become persuasive substitutes for either primary systemic therapy or maintenance therapy, although the results have varied somewhat [40].

6.4.1 Adjuvant Therapy

In the last 40 years, extensive research has demonstrated that adjuvant endocrine therapy and chemotherapy have independently and consistently reduced the risk of BC recurrence and mortality. These treatment modalities have shown efficacy regardless of factors such as age, nodal status, and estrogen receptor (ER) status. In the adjuvant situation or as continued therapy, in which the tumor load is minimal, the implementation of an easily tolerated and practical metronomic treatment for patients is recommended [8].

BC that does not respond to endocrine therapy encompasses a diverse range of subtypes, making it especially important to explore personalized treatment approaches, particularly in the context of adjuvant chemotherapy. This is particularly relevant and notable in the case of triple-negative BC [41].

6.4.1.1 Clinical Trial Evidence

- In the CASA phase III trial, a comparison was conducted between adjuvant MCT and adjuvant pegylated liposomal doxorubicin (PLD) in frail and older patients who were not responsive to endocrine therapy and were not suitable candidates for standard chemotherapy. The PLD regimen involved intravenous administration of 20 mg/m^2 every 2 weeks. This dosing schedule was derived from studies in AIDS-associated Kaposi sarcoma and was considered safer for elderly patients with comorbidities. The MCT regimen consisted of continuous oral administration of CTX at 50 mg/day and methotrexate at 2.5 mg twice daily on days 1 and 4 of each week. However, the trial was prematurely closed due to slow and inadequate patient recruitment, with only 77 patients enrolled over 2 years. Nonetheless, initial results indicated that the metronomic regimen was associated with improved quality of life and cognitive function compared to the PLD regimen. At a median follow-up of 42 months, 81% of the patients remained free from BC recurrence. Despite the limited data, the authors concluded that both PLD and MCT could be considered viable options for further investigation in elderly and vulnerable patients with endocrine non-responsive BC [42].
- Nasr and colleagues [43] conducted a study that compared the outcomes of maintenance treatment using a metronomic approach versus no maintenance in 158 patients with triple-negative breast cancer (TNBC). The maintenance group received adjuvant chemotherapy followed by a year of MCT, while the no-maintenance group received chemotherapy alone. Results showed that the median disease-free survival (DFS) was 28 months in the maintenance group compared to 24 months in the no-maintenance group. The median overall survival (OS) was 37 months in the maintenance group and 29 months in the no-maintenance group. The 4-year DFS rates were 63% in the maintenance group and 42% in the no-maintenance group. The 4-year OS rates were 74% in the maintenance group and 55% in the no-maintenance group. Despite the study's limitations, the use of extended adjuvant metronomic CM (cyclophosphamide and methotrexate) regimen as maintenance treatment showed improved DFS and OS compared to no maintenance, and it was well tolerated by the patients.

6.4.2 Non-Adjuvant Therapy

For those patients for whom the primary surgical approach is technically impractical and also the patients with operable BC who want to preserve breast tissue, non-adjuvant chemotherapy comes into play in terms of enhancing surgical results [44, 45].

Neoadjuvant chemotherapy offers several advantages, including tumor downsizing, which increases the possibility of breast-conserving surgery and improves the feasibility of surgical removal. Moreover, it allows for an in vivo assessment of tumor response to chemotherapy, as achieving a pathologic complete response (pCR) following neoadjuvant treatment serves as a reliable prognostic factor [46, 47]. It also plays a crucial role in the early assessment of the efficacy of a systemic therapy, which may help doctors stop administering unsuccessful medication [48].

6.4.2.1 Clinical Trial Evidence

- The first-ever study to investigate the potential of MCT in the neoadjuvant setting for elderly women with ER-positive BC who were unfit for standard chemotherapy was conducted by

Bottini and colleagues [49]. The study was carried out on approximately 114 patients, who were randomly assigned to receive either primary letrozole therapy or a combination of letrozole and oral CTX. The overall response rates were 71.9% in the letrozole-only group and 87.7% in the letrozole plus cyclophosphamide group. Adverse events were comparable between the two treatment arms. Notably, the combination therapy group exhibited significantly greater suppression of Ki-67 and VEGF-A expression compared to the letrozole-only group. This resulted in lower levels of Ki-67 and VEGF expression in the residual histology after treatment. It was therefore concluded from the study that adding CTX to letrozole in an MCT approach led to higher overall response rates and more pronounced suppression of Ki-67 and VEGF-A expression in elderly patients with ER-positive BC.

- Another such study was carried out by [50] which aimed to assess the effectiveness and safety of using metronomic paclitaxel/cyclophosphamide/capecitabine (mPCX) followed by 5-fluorouracil/epirubicin/cyclophosphamide (FEC) as preoperative chemotherapy for patients diagnosed with TNBC. In the study, 40 primary TNBC patients were treated with four cycles of mPCX followed by four cycles of FEC. The primary measure of success was the rate of achieving pCR. It was found that the intent-to-treat population had a pCR rate of 47.5%, while the per-protocol population had a higher rate of 54.5%. Clinical response rates were 90.0% and 93.9% for the intent-to-treat and per-protocol populations, respectively. The study also revealed that 72.7% of patients were able to conserve their breasts, with some opting for partial resection instead of total mastectomy. Grade 3–4 adverse events, including neutropenia (35%), leukopenia (25%), and hand-foot syndrome (8%), were observed. The study concluded that the preoperative chemotherapy regimen of mPCX followed by FEC demonstrated significant efficacy with a high pCR rate and minimal toxicity in TNBC patients. Further research involving larger patient groups is necessary to validate these findings.

6.4.3 Combination Therapies

Combination therapy, which combines multiple therapeutic agents, is crucial in cancer treatment. The use of different drugs together enhances effectiveness by targeting pathways synergistically or additively. This approach reduces drug resistance and provides various anti-cancer benefits, such as inhibiting tumor growth, preventing metastasis, stopping cell division, reducing cancer stem cells, and inducing cell death [51]. Combining two or more treatments to specifically target pathways responsible for cancer development and cell survival is a fundamental aspect of cancer therapy [51, 52].

While utilizing chemotherapeutic drugs in combination therapy can be hazardous, the toxicity is significantly minimized because several pathways are being targeted. Lower dosages of each medicine are possible with this additive or synergistic approach, reducing toxicity while preserving therapeutic efficacy [53, 54]. While having cytotoxic effects on cancer cells, combination therapy can prevent harmful effects on healthy cells. This is possible if one drug in the combination regimen acts in a way that is cytotoxic to normal cells in opposition to another treatment. In essence, this protective association protects healthy cells from the effects of cytotoxicity, allowing impact on cancer cells with specificity [55].

6.4.3.1 Clinical Trial Evidence

- Bachelot and colleagues [56] performed a research experiment to assess the effectiveness and well-being of amalgamating everolimus, a blocker of the mammalian target of rapamycin (mTOR), with tamoxifen in postmenopausal women with MBC impervious to aromatase inhibitors (AIs). One hundred and eleven patients with AI-resistant, hormone receptor–positive, and human epidermal growth factor receptor 2–negative MBC were randomized to receive either tamoxifen + everolimus or tamoxifen alone in this open-label, phase II research study. The clinical benefit rate (CBR) at 6 months served as the main endpoint, with time to progression (TTP) and OS serving as supplementary goals. The findings of the study demonstrated that tamoxifen with everolimus led to a greater 6-month CBR (61%) than tamoxifen alone (42%).

The TTP was also significantly extended with the combined therapy, going from 4.5 months with tamoxifen alone to 8.6 months, resulting in a 46% lower risk of advancement. In addition, when tamoxifen was combined with everolimus as opposed to tamoxifen alone, the risk of death was decreased by 55%. Fatigue, stomatitis, rash, anorexia, and diarrhea were the primary toxicities related to tamoxifen and everolimus. The study concluded that the CBR, TTP, and OS of postmenopausal women with AI-resistant MBC were improved by the combination of tamoxifen and everolimus.

- The BOLERO-2 study evaluated the efficacy of combining the mTOR inhibitor everolimus with the AI exemestane in patients with hormone receptor–positive advanced BC who had encountered relapse or advancement while undergoing prior treatment with a nonsteroidal AI. The trial demonstrated enhanced DFS in this patient population. The investigation encompassed 724 patients, and the primary endpoint was DFS. The interim analysis showed that the median progression-free survival was 6.9 months with exemestane plus everolimus, compared to 2.8 months with exemestane plus placebo, based on evaluation by regional researchers. The central assessment also demonstrated a substantial improvement in progression-free survival (PFS) (10.6 months vs 4.1 months). The amalgamation treatment was predominantly well accepted, with the most frequent severe grade 3 and 4 unfavorable occurrences being oral inflammation, reduced red blood cell (RBC) count, difficulty in breathing, high blood sugar levels, exhaustion, and lung inflammation.

  The research indicated that incorporating everolimus with endocrine treatment can offer anti-tumor efficacy and potentially surmount resistance to endocrine therapy in advanced HR+ breast cancer [57]. Although the amalgamation of exemestane and everolimus did not show a statistically substantial advantage in OS compared to exemestane monotherapy in the study conducted by [58], there was a trend toward improved OS with combination therapy. The median OS was 31.0 months for the combination treatment group versus 26.6 months for the monotherapy group. However, the hazard ratio (HR) of 0.89 did not reach statistical significance (95% confidence interval [CI] 0.73–1.10; log-rank $P = 0.14$). Despite this, the study's findings on PFS led to approvals by the Food and Drugs Administration (FDA) and the European Medicines Agency (EMA) for the combination of everolimus and exemestane in hormone receptor–positive MBC patients who had previously been treated with nonsteroidal AIs. It is important to note that combination therapy has associated side effects, but careful clinical evaluation can help identify suitable candidates for this treatment approach [58]. Some recent clinical trials of MCT with results are given in Table 6.1.

6.5 ENHANCING WELL-BEING AND COST-EFFECTIVENESS: THE BENEFITS OF METRONOMIC CHEMOTHERAPY FOR PATIENTS AND HUMANITY

MCT offers a potential solution to the exorbitant costs of cancer treatment worldwide. By utilizing affordable oral medications and administering treatment at lower doses, this approach reduces expenses related to travel, hospital admissions, intravenous infusions, and the risks of infection. Moreover, MCT minimizes side effects and decreases the need for frequent monitoring exams, resulting in an improved quality of life for patients while simultaneously lowering the overall cost of cancer treatment. With estimates predicting a major increase in direct expenses, the rising costs of cancer treatment pose serious problems for economies and families, both in high-income and low- to middle-income countries [17].

TABLE 6.1
Recent Trials with MCT in Breast Cancer

Metronomic Therapy	Cancer Subtype	Phase	n	Response	Reference
Capecitabine 500 mg (3 times daily) + Vinorelbine 40 mg (day 1, 3, and 5)	HR+ TNBC	II; first line /≥second line	80	*ORR*: 35.5%/25.6% *CBR*: 45.7%/51.1% *PFS*: 6.7 months/ 7.2 months	[59]
Exemestane (25 mg daily) + Capecitabine (500 mg/m^2, twice every day)	HR+/HER2–	II	38	*ORR*: 68.4% *PFS*: 17 months *OS*: 20 months	[60]
Cyclophosphamide 50 mg daily + Vinorelbine 20–40 mg 3 times per week + Trastuzumab (every 3 weeks in HER2+ only) + Bevacizumab 15 mg/kg every 3 weeks	ER+ HER2+	I	15	*CBR*: 66.6% *PFS*: 6.9 months	[61]
Trastuzumab (4 mg/kg, once every 2 weeks) + Cyclophosphamide (50 mg/daily) + Capecitabine (1500 mg/daily)	HER2+	II	60	*ORR*: 56.7% *CBR*: 78.2% *PFS*: 11 months	[62]
Vinorelbine 40 mg (3 times a week) + Trastuzumab (6 mg/kg every 3 weeks)	HER2+	II	20	*ORR*: 20% *CBR*: 75% *PFS*: 7.4 months *OS*: 45.8 months	[63]
Capecitabine (1000 mg/m^2 twice a day for days 1–14) + Vinorelbine 50 mg (3 times a week)	HER2–	II	120	*ORR*: 29% *CBR*: 51.7% *PFS*: 6.3 months *OS*: 22.3 months	[64]
Vinorelbine (30 mg/day)	HR+/HER2–	II	9	*CBR*: 22.2% *PFS*: 12 weeks	[65]
Capecitabine 650 mg/m^2 twice daily between day 1 and day 21 every 28 days + Exemestane 25 mg daily or Anastrozole 1 mg every day + Letrozole 2.5 mg every day	HER2–/HR+	Retrospective	163	*ORR*: 29.5% *PFS*: 29.37 months	[66]

Abbreviations: CBR, clinical benefit rate; HER2, human epidermal growth factor receptor 2; HR, hormone receptor; n, number of patients; ORR, overall response rate; OS, overall survival; PFS, progression-free survival; TNBC, triple negative breast cancer.

6.6 TOXICITY AND SIDE EFFECTS ASSOCIATED WITH METRONOMIC CHEMOTHERAPY

The cumulative evidence from clinical studies consistently shows that low-dose chemotherapy, whether taken alone or combined with other drugs, has shown favorable tolerability, despite variances in efficacy observed across different study populations, medication combinations, and tumor types [40]. Furthermore, various studies have assessed its cost-effectiveness, further establishing it as a viable and economical treatment alternative [67]. Severe adverse reactions were infrequent or absent. The most prevalent side effects were mild tiredness, slight nausea, mild to moderate reduction in RBCs, and low levels of neutrophils, WBCs, and lymphocytes [68]. Certain potential dangers and precautions should be addressed while using continuous low-dose therapy for a long time, especially in youngsters.

Long-term MCT use may lead to an accumulation of anti-cancer medication toxicity, raising the risk of subsequent illnesses. For instance, secondary leukemia might arise from the cumulative dose of temozolomide or etoposide [69]. In children diagnosed with medulloblastoma, a recent study found a possible association between MCT and subdural hematoma [70]. Therefore, in general, MCT offers notable clinical breakthroughs that result in an enhanced quality of life with little harm. Nevertheless, the available data is insufficient to make definitive conclusions regarding the tolerance of different drug combinations.

6.7 RESISTANCE DEVELOPMENT TO METRONOMIC CHEMOTHERAPY AND OVERCOMING THE CHALLENGES

MCT has shown promising results with respect to the elimination of tumors in mouse models [15] and has also shown success in treating drug-resistant tumors in human studies [40]. However, despite these encouraging outcomes, clinical studies often reveal occurrences of both inherent and acquired resistance to MCT, resulting in relapse in a sizable number of patients [17]. Hence it becomes very crucial to analyze the resistance mechanism in treating the patient through MCT for better effectiveness and promising results along with reduced treatment-related toxicity.

The growth of resistance to the prescribed treatment regimen is a major problem in chemotherapy. Due to the presumed genetic stability of endothelial cells, the possibility of developing acquired drug resistance is improbable when employing an anti-angiogenic medication regimen specifically aimed at these cells. Nevertheless, some clinical studies have shown how escape mechanisms develop in response to anti-angiogenic treatment regimens. These results provide evidence for the possibility of two novel resistance mechanisms: Evasive resistance and intrinsic resistance. Evasive resistance is a modification to stop a particular anti-angiogenic action. It includes elevated metastasis and tumor growth in lymph nodes and other organs, revascularization due to higher levels of additional pro-angiogenic signals, and enhanced tumor cell spread into nearby tissues to utilize normal vasculature. Intrinsic resistance is the in-built tolerance to drugs [71].

The anti-neoplastic effects of MCT are believed to be influenced significantly by reduced oxygen levels. Consequently, the increased ability of tumor cells to tolerate hypoxia promotes resistance to these therapies. In this regard, studies have reported a synergistic effect between the hypoxia-activated prodrug tirapazamine and metronomic CTX in breast, colon, and prostate cancer models [72]. Our existing understanding of the mechanisms governing inherent and acquired resistance to continuous low-dose therapy is still restricted. Emphasizing the need for further research, dedicated efforts should be directed toward unraveling the resistance mechanism to low-dose MCT. Such investigations have the potential to improve patient selection, treatment surveillance, and biomarker identification in addition to maximizing therapeutic effectiveness.

6.8 CONCLUSION

In the past few years, MCT has garnered growing attention from researchers. In the form of low doses, decreased toxicity, or increased quality of life, MCT has been to some extent successful in filling the loopholes left by the MTD regimen. The recent revelation of the multi-targeted nature of MCT, coupled with its integration with combination therapies, drug repurposing, and adjuvant and non-adjuvant therapy, has opened up new avenues for the advancement of personalized targeted chemotherapy. Furthermore, the cost of cancer treatment is at an all-time high, which makes MCT a viable solution. The ultimate superiority of MCT over MTD-based therapies in treating metastatic disease is still yet to be established. However, ongoing research seems to suggest that MCT may hold promise in this regard.

The exploration of low-dose chemotherapy as a new approach in the management of BC has shown encouraging potential and opened new avenues in cancer therapeutics. This alternative strategy, which involves the continuous administration of low doses of chemotherapeutic agents, has demonstrated several advantages over traditional MTD regimens. An important benefit of continuous low-dose chemotherapy is its ability to inhibit tumor angiogenesis actively. By continuously administering low doses of certain chemotherapeutic agents, this approach can normalize the tumor vasculature, leading to reduced tumor hypoxia and creating an unfavorable environment for tumor growth. Additionally, the anti-angiogenic effects limit the tumor's access to nutrients, further impeding its progression and metastasis. Moreover, it has been demonstrated that MCT directly kills cancer cells while also stimulating the body's immune system. By promoting anti-tumor immune responses, this treatment modality effectively enhances the ability of the body to identify and target metastatic cells. This immunomodulatory effect not only targets the primary tumor but also provides systemic protection against metastatic cells, which is a crucial factor in BC management.

Furthermore, the lower and more frequent doses of chemotherapy in metronomic regimens contribute to reduced toxicity and fewer adverse side effects compared to traditional chemotherapy. Patients undergoing MCT experience improved tolerability and can sustain treatment over more extended periods, which is particularly beneficial in managing metastatic BC as a chronic condition. While MCT holds tremendous promise, its integration into standard treatment protocols requires further investigation and optimization. Large-scale clinical trials are necessary to establish its efficacy, identify the most effective drugs and dosing schedules, and explore its potential in combination with other therapeutic approaches.

In conclusion, MCT represents a paradigm shift in BC treatment, focusing on continuous low-dose therapy that targets cancerous cells directly along with modification of tumor microecology and boosts the body's immune defenses. By mitigating the drawbacks of traditional chemotherapy, such as toxicity and drug resistance, MCT offers a more sustainable and patient-friendly treatment option. As we continue to unravel its full potential, MCT promises to be an indispensable component of the evolving BC treatment landscape, ultimately leading to improved patient outcomes, prolonged life expectancy, and enhanced life quality.

For further reading, some textbooks mentioned here also give a further understanding of MCT:

- *Progress in Anti-cancer Chemotherapy* by Gabriel N. Hortobagyi and David Khayat.
- *Triple-Negative Breast Cancer: Biomarkers, Emerging Therapeutic Strategies and Clinical Challenges* by Marion Montgomery.
- *Current Cancer Treatment: Novel Beyond Conventional Approaches* by Oner Ozdemir.
- *Breast Cancer: Current and Alternative Therapeutic Modalities* by Esra Gunduz and Mehmet Gunduz.

The following links for video lectures may also be helpful:

https://youtu.be/TAqO0Mq19JQ
https://youtu.be/6wEnYfvoBqk
https://youtu.be/A9fRRCUlMms
https://youtu.be/zIWvmCsKr34

For more insight about the topic, we would suggest detailed findings from the books of (Mir MA, 2022) https://doi.org/10.1016/C2021-0-02565-7, https://doi.org/10.1016/C2022-0-00074-X, (Mir MA, 2021) https://doi.org/10.52305/WXJL6770, and (Mir MA, 2015) https://doi.org/10.1016/C2014-0-02898-5 and from the cancer.net website on the following mentioned links:

https://www.cancer.net/cancer-types/breast-cancer/types-treatment
https://discovery.ucl.ac.uk/id/eprint/1472740/
https://www.jmedsciences.com/doi/JMEDS/pdf/10.5005/jp-journals-10045-00138

For diagrammatic illustrations and descriptive tables (Lazzeroni, 2012): http://www.eurekaselect.com/article/49928

See video links on the overall status of cancer, its various types, and current and new possible treatment options available: https://www.sciencedirect.com/science/article/pii/S205970292032278X

REFERENCES

1. Torre, L.A., et al., *Global cancer statistics, 2012.* CA: a cancer journal for clinicians, 2015. **65**(2): p. 87–108.
2. Rahib, L., et al., *Projecting cancer incidence and deaths to 2030: the unexpected burden of thyroid, liver, and pancreas cancers in the United States.* Cancer research, 2014. **74**(11): p. 2913–2921.
3. Colditz, G.A. and K. Bohlke, *Priorities for the primary prevention of breast cancer.* CA: a cancer journal for clinicians, 2014. **64**(3): p. 186–194.
4. Bloom, H.J.G., W.W. Richardson, and E.J. Harries, *Natural history of untreated breast cancer (1805–1933).* British medical journal, 1962. **2**(5299): p. 213.
5. Harris, J.R., et al., Diseases of the Breast. 2012: Lippincott Williams & Wilkins.
6. Perou, C.M., et al., *Molecular portraits of human breast tumours.* Nature, 2000. **406**(6797): p. 747–752.
7. Stephens, P.J., et al., *The landscape of cancer genes and mutational processes in breast cancer.* Nature, 2012. **486**(7403): p. 400–404.
8. Munzone, E. and M. Colleoni, *Clinical overview of metronomic chemotherapy in breast cancer.* Nature reviews. Clinical oncology, 2015. **12**(11): p. 631–644.
9. DeVita Jr, V.T. and E. Chu, *A history of cancer chemotherapy.* Cancer research, 2008. **68**(21): p. 8643–8653.
10. Pinkel, D., *Actinomycin D in childhood cancer: A preliminary report.* Pediatrics, 1959. **23**(2): p. 342–347.
11. McLaughlin, P., et al., *Rituximab chimeric anti-CD20 monoclonal antibody therapy for relapsed indolent lymphoma: half of patients respond to a four-dose treatment program.* Journal of clinical oncology, 1998. **16**(8): p. 2825–2833.
12. Banys-Paluchowski, M., et al., *Metronomic chemotherapy for metastatic breast cancer–a systematic review of the literature.* Senologie-Zeitschrift für Mammadiagnostik und-therapie, 2016. **13**(04): p. 180–188.
13. Bocci, G., et al., *Cyclophosphamide-methotrexate 'metronomic'chemotherapy for the palliative treatment of metastatic breast cancer. A comparative pharmacoeconomic evaluation.* Annals of oncology, 2005. **16**(8): p. 1243–1252.
14. Crown, J., et al., *Chemotherapy for metastatic breast cancer–report of a European expert panel.* The Lancet oncology, 2002. **3**(12): p. 719–727.
15. Kerbel, R.S. and B.A. Kamen, *The anti-angiogenic basis of metronomic chemotherapy.* Nature reviews cancer, 2004. **4**(6): p. 423–436.
16. Nygren, P., *What is cancer chemotherapy?* Acta oncologica, 2001. **40**(2-3): p. 166–174.
17. André, N., et al., *Has the time come for metronomics in low-income and middle-income countries?* The Lancet oncology, 2013. **14**(6): p. e239–e248.
18. Kontani, K., et al., *Metronomic chemotherapy for metastatic breast cancer to prolong time to treatment failure to 12 months or more.* Molecular and clinical oncology, 2013. **1**(2): p. 225–230.
19. Amjad, M.T., A. Chidharla, and A. Kasi, *Cancer chemotherapy.* 2020.
20. Cazzaniga, M.E., et al., *Metronomic chemotherapy with oral vinorelbine (mVNR) and capecitabine (mCAPE) in advanced HER2-negative breast cancer patients: is it a way to optimize disease control? Final results of the VICTOR-2 study.* Breast cancer research and treatment, 2016. **160**: p. 501–509.
21. Kerbel, R.S. and Y. Shaked, *The potential clinical promise of 'multimodality' metronomic chemotherapy revealed by preclinical studies of metastatic disease.* Cancer letters, 2017. **400**: p. 293–304.
22. Folkman, J., *Tumor angiogenesis: therapeutic implications.* New England journal of medicine, 1971. **285**(21): p. 1182–1186.
23. Rajabi, M. and S.A. Mousa, *The role of angiogenesis in cancer treatment.* Biomedicines, 2017. **5**(2): p. 34.
24. Mackey, J.R., et al., *Controlling angiogenesis in breast cancer: a systematic review of anti-angiogenic trials.* Cancer treatment reviews, 2012. **38**(6): p. 673–688.

25. Bergers, G. and L.E. Benjamin, *Tumorigenesis and the angiogenic switch.* Nature reviews cancer, 2003. **3**(6): p. 401–410.
26. Otrock, Z.K., et al., *Understanding the biology of angiogenesis: review of the most important molecular mechanisms.* Blood cells, molecules, and diseases, 2007. **39**(2): p. 212–220.
27. Cao, Y. and Q. Liu, *Therapeutic targets of multiple angiogenic factors for the treatment of cancer and metastasis.* Advances in cancer research, 2007. **97**: p. 203–224.
28. Holash, J., S.J. Wiegand, and G.D. Yancopoulos, *New model of tumor angiogenesis: dynamic balance between vessel regression and growth mediated by angiopoietins and VEGF.* Oncogene, 1999. **18**(38): p. 5356–5362.
29. Ferrara, N., H.-P. Gerber, and J. LeCouter, *The biology of VEGF and its receptors.* Nature medicine, 2003. **9**(6): p. 669–676.
30. Ahmed, S.I., A.L. Thomas, and W.P. Steward, *Vascular endothelial growth factor (VEGF) inhibition by small molecules.* Journal of chemotherapy, 2004. **16**(Suppl 4): p. 59–63.
31. André, N., M. Carré, and E. Pasquier, *Metronomics: towards personalized chemotherapy?* Nature reviews. Clinical oncology, 2014. **11**(7): p. 413–431.
32. Radojcic, V., et al., *Cyclophosphamide resets dendritic cell homeostasis and enhances antitumor immunity through effects that extend beyond regulatory T cell elimination.* Cancer immunology, immunotherapy, 2010. **59**: p. 137–148.
33. Chaudhary, B. and E. Elkord, *Regulatory T cells in the tumor microenvironment and cancer progression: role and therapeutic targeting.* Vaccines, 2016. **4**(3): p. 28.
34. Michels, T., et al., *Paclitaxel promotes differentiation of myeloid-derived suppressor cells into dendritic cells in vitro in a TLR4-independent manner.* Journal of immunotoxicology, 2012. **9**(3): p. 292–300.
35. Sierro, S.R., et al., *Combination of lentivector immunization and low-dose chemotherapy or PD-1/PD-L1 blocking primes self-reactive T cells and induces anti-tumor immunity.* European journal of immunology, 2011. **41**(8): p. 2217–2228.
36. Dunn, G.P., et al., Cancer immunoediting: from immunosurveillance to tumor escape. Nature immunology, 2002. 3(11): p. 991–998.
37. Hoenicke, L. and L.J.C. Zender, *Immune surveillance of senescent cells—biological significance in cancer-and non-cancer pathologies.* Carcinogenesis, 2012. 33(6): p. 1123–1126.
38. Kosmaczewska, A., et al., The significance of Treg cells in defective tumor immunity. Archivum immunologiae et therapiae experimentalis, 2008. 56(3): p. 181–191.
39. Kono, K., et al., *CD4 (+) CD25 high regulatory T cells increase with tumor stage in patients with gastric and esophageal cancers.* Cancer immunology, immunotherapy, 2006. 55: p. 1064–1071.
40. Pasquier, E., M. Kavallaris, and N. André, *Metronomic chemotherapy: new rationale for new directions.* Nature reviews. Clinical oncology, 2010. **7**(8): p. 455–465.
41. Metzger-Filho, O., et al., *Dissecting the heterogeneity of triple-negative breast cancer.* Journal of clinical oncology, 2012. **30**(15): p. 1879–1887.
42. Crivellari, D., et al., *Adjuvant pegylated liposomal doxorubicin for older women with endocrine non-responsive breast cancer who are NOT suitable for a "standard chemotherapy regimen": the CASA randomized trial.* The breast, 2013. **22**(2): p. 130–137.
43. Nasr, K.E., et al., *Metronomic methotrexate and cyclophosphamide after carboplatin included adjuvant chemotherapy in triple negative breast cancer: a phase III study.* Annals of translational medicine, 2015. **3**(19): p. 284.
44. Gralow, J.R., et al., *Preoperative therapy in invasive breast cancer: pathologic assessment and systemic therapy issues in operable disease.* Journal of clinical oncology, 2008. **26**(5): p. 814–819.
45. Kaufmann, M., et al., *Recommendations from an international expert panel on the use of neoadjuvant (primary) systemic treatment of operable breast cancer: an update.* Journal of clinical oncology, 2006. **24**(12): p. 1940–1949.
46. Fisher, B., et al., *Effect of preoperative chemotherapy on the outcome of women with operable breast cancer.* Journal of clinical oncology, 1998. **16**(8): p. 2672–2685.
47. Kuerer, H.M., et al., *Clinical course of breast cancer patients with complete pathologic primary tumor and axillary lymph node response to doxorubicin-based neoadjuvant chemotherapy.* Journal of clinical oncology, 1999. **17**(2): p. 460–460.
48. Schwartz, G.F., G.N. Hortobagyi, and C. Consensus Conference, *Proceedings of the consensus conference on neoadjuvant chemotherapy in carcinoma of the breast, April 26–28, 2003, Philadelphia, Pennsylvania.* Cancer, 2004. **100**(12): p. 2512–2532.

49. Bottini, A., et al., *Randomized phase II trial of letrozole and letrozole plus low-dose metronomic oral cyclophosphamide as primary systemic treatment in elderly breast cancer patients.* Journal of clinical oncology, 2006. **24**: p. 3623–3628.
50. Masuda, N., et al., *A phase II study of metronomic paclitaxel/cyclophosphamide/capecitabine followed by 5-fluorouracil/epirubicin/cyclophosphamide as preoperative chemotherapy for triple-negative or low hormone receptor expressing/HER2-negative primary breast cancer.* Cancer chemotherapy and pharmacology, 2014. **74**: p. 229–238.
51. Mokhtari, R.B., et al., *Combination therapy in combating cancer.* Oncotarget, 2017. **8**(23): p. 38022.
52. Blagosklonny, M.V., *Analysis of FDA approved anticancer drugs reveals the future of cancer therapy.* Cell cycle, 2004. **3**(8): p. 1033–1040.
53. Mokhtari, R.B., et al., *Combination of carbonic anhydrase inhibitor, acetazolamide, and sulforaphane, reduces the viability and growth of bronchial carcinoid cell lines.* BMC cancer, 2013. **13**(1): p. 1–18.
54. Albain, K.S., et al., *Gemcitabine plus paclitaxel versus paclitaxel monotherapy in patients with metastatic breast cancer and prior anthracycline treatment.* Journal of clinical oncology, 2008. **26**(24): p. 3950–3957.
55. Blagosklonny, M.V., *Overcoming limitations of natural anticancer drugs by combining with artificial agents.* Trends in pharmacological sciences, 2005. **26**(2): p. 77–81.
56. Bachelot, T., et al., *Randomized phase II trial of everolimus in combination with tamoxifen in patients with hormone receptor–positive, human epidermal growth factor receptor 2–negative metastatic breast cancer with prior exposure to aromatase inhibitors: A GINECO study.* Journal of clinical oncology, 2012. **30**(22): p. 2718–2724.
57. Baselga, J., et al., *Everolimus in postmenopausal hormone-receptor–positive advanced breast cancer.* New England journal of medicine, 2012. **366**(6): p. 520–529.
58. Piccart, M., et al., *Everolimus plus exemestane for hormone-receptor-positive, human epidermal growth factor receptor-2-negative advanced breast cancer: overall survival results from BOLERO-2.* Annals of oncology, 2014. **25**(12): p. 2357–2362.
59. Cazzaniga, M.E., et al., *Metronomic oral vinorelbine in advanced breast cancer and non-small-cell lung cancer: current status and future development.* Future oncology, 2016. **12**(3): p. 373–387.
60. Ghannam, A. and R.A. Khedr, *Preliminary results of capecitabine metronomic chemotherapy combined with exemestane in advanced breast cancer–a single-arm Phase II study.* Cancer biology, 2019. **9**(1): p. 43–50.
61. Sanna, G., et al., *A dose-finding study of metronomic oral vinorelbine in combination with oral cyclophosphamide and bevacizumab in patients with advanced breast cancer.* Clinical breast cancer, 2021. **21**(4): p. e332–e339.
62. Orlando, L., et al., *Metronomic oral chemotherapy with cyclophosphamide plus capecitabine combined with trastuzumab (HEX) as first line therapy of HER-2 positive advanced breast cancer: a phase II trial of the Gruppo Oncologico Italia Meridionale (GOIM).* The breast, 2020. **53**: p. 18–22.
63. Wang, Z., et al., *Safety and efficacy study of oral metronomic vinorelbine combined with trastuzumab (mNH) in HER2-positive metastatic breast cancer: a phase II trial.* Breast cancer research and treatment, 2021. **188**: p. 441–447.
64. Brems-Eskildsen, A.S., et al., *Metronomic treatment of vinorelbine with oral capecitabine is tolerable in the randomized Phase 2 study XeNa including patients with HER2 non-amplified metastatic breast cancer.* Acta oncologica, 2021. **60**(2): p. 157–164.
65. Krajnak, S., et al., *Phase II study of metronomic treatment with daily oral vinorelbine as first-line chemotherapy in patients with advanced/metastatic HR+/HER2– breast cancer resistant to endocrine therapy: VinoMetro—AGO-B-046.* Journal of cancer research and clinical oncology, 2021. **147**(11): p. 3391–3400.
66. Alvarado-Miranda, A., et al., *Capecitabine Plus Aromatase Inhibitor as First Line Therapy for Hormone Receptor Positive, HER2 Negative Metastatic Breast Cancer.* Current oncology, 2023. **30**(7): p. 6097–6110.
67. Kieran, M.W., et al., *A feasibility trial of antiangiogenic (metronomic) chemotherapy in pediatric patients with recurrent or progressive cancer.* Journal of pediatric hematology/oncology, 2005. **27**(11): p. 573–581.
68. Mross, K. and S. Steinbild, *Metronomic anti-cancer therapy–an ongoing treatment option for advanced cancer patients.* Journal of cancer therapeutics and research, 2012. **1**(1): p. 32.
69. Le Deley, M.C., et al., *High cumulative rate of secondary leukemia after continuous etoposide treatment for solid tumors in children and young adults.* Pediatric blood & cancer, 2005. **45**(1): p. 25–31.

70. Rome, A., et al., *Metronomic chemotherapy-induced bilateral subdural hematoma in a child with meningeal carcinomatosis.* Pediatric blood & cancer, 2009. **53**(2): p. 246–247.
71. Maiti, R., *Metronomic chemotherapy.* Journal of pharmacology and pharmacotherapeutics, 2014. **5**(3): p. 186–192.
72. Emmenegger, U., et al., *Low-dose metronomic daily cyclophosphamide and weekly tirapazamine: a well-tolerated combination regimen with enhanced efficacy that exploits tumor hypoxia.* Cancer research, 2006. **66**(3): p. 1664–1674.

7 Metronomic Chemotherapy in Combination with an Anti-Cancer Vaccine

Manzoor Ahmad Mir, Tabish Javed, and Burhan Ul Haq

7.1 INTRODUCTION

Conventional chemotherapy is now the cornerstone of accepted practice for treating cancer in humans. This treatment strategy, also known as the maximum tolerated dose (MTD) regimen, involves administering the largest dosage of a chemotherapeutic agent with comparatively long drug-free periods. This strategy has been shown effective in the experimental therapy of numerous kinds of cancer in humans [1, 2] and attempts to eliminate cancerous cells by affecting the cell cycle. Due to the genetic heterogeneity of most tumors and the cytotoxicity of conventional chemotherapy, which primarily kills the therapeutically sensitive tumor cell fraction, this treatment may promote the development of therapeutically unaffected tumor cell clones in between the treatment cycles. Due to this reason, some cancer patients have tumor resistance and/or disease recurrence after standard chemotherapeutic therapy [3].

Metronomic chemotherapy (MCT), which is defined as the continuous administration of a low dosage of chemotherapeutic drugs with few drug-free periods, may provide an alluring substitute for traditional chemotherapy treatment [4]. Although MCT was primarily intended to overcome tumor resistance by focusing on tumor endothelial cells, recent research has demonstrated that MCT may induce tumor cytotoxicity pathways, adaptive immunity as well and anti-tumor innate immunity [5]. Based on this, MCT has lately been combined with various pharmacological substances, particularly immune checkpoint inhibitors.

Intriguingly, cancer patients who had failed conventional chemotherapy reported therapeutic benefits from MCT. An excellent outcome was seen in the treatment of non–small cell lung cancer (NSCLC), metastatic breast cancer, and colorectal carcinoma, especially following the injection of MCT [6]. Additionally, the therapeutic arrangement of MCT with unconventional cytotoxic drugs, such as anti-angiogenic drugs, may help overcome a multitude of toxicities while preserving therapeutic efficiency [7]. Some naturally occurring substances, such as cordycepin, which prompts cell death and prevents the transformation of oral squamous cell carcinoma cells from epithelial to mesenchymal, and, cephalomannine, which in hypoxic lung cancer cells limits cell viability and reactive oxygen species generation, also had a metronomic anti-tumor effect [8]. Despite these findings, MCT treatment currently serves mainly as an analgesic-appropriate maintenance instrument and as an initial therapy. At present, more research is required to understand the core mechanisms of MCT activity and to determine the most efficient chemotherapeutic agents explicitly formulated for a given tumor type along with the right biological dose (Figure 7.1).

The anti-tumor MCT routes are examined in this chapter along with recent investigations, with a special emphasis on the possible treatment benefit of MCT in combination with inhibitors of immune checkpoints.

DOI: 10.1201/9781032632674-7

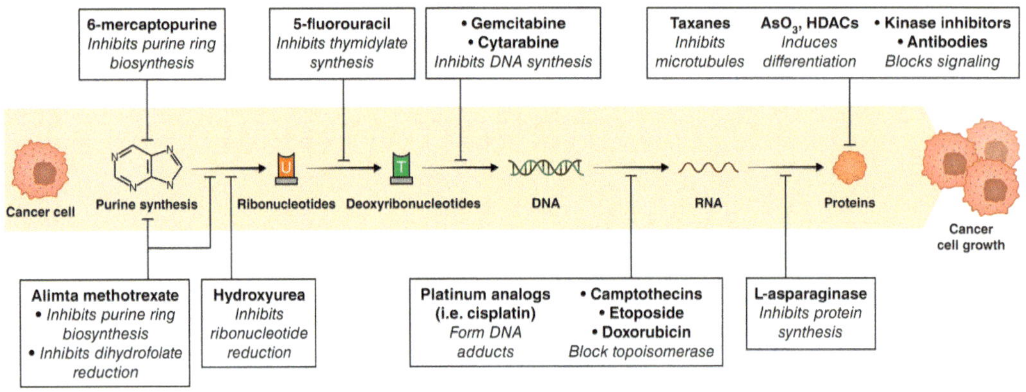

FIGURE 7.1 Basic overview of metronomic chemotherapy and the mechanisms involved.

7.2 METRONOMIC CHEMOTHERAPY AND ITS MECHANISM OF ACTION

There is growing evidence that several chemotherapy drugs may restrict angiogenesis, activate the immune system, and stop tumor development when they are given in modest doses and on a regular schedule.

7.2.1 Anti-Angiogenic Impacts

Angiogenesis, often known as the creation of new blood vessels from the vasculature, may take place regionally when differentiated endothelium cells are involved or systemically when bone marrow endothelial cells are recruited [9]. Because tumor-associated endothelium cells have a lower rate of reproduction than other tumor cells, conventional chemotherapy has been shown to only cause minor or transient endothelial damage [10]. In comparison to the standard treatment regimen, MCT cyclophosphamide therapy suppressed angiogenesis in resistant endothelial cell lines three-fold more effectively. Likely, the endothelial cell's repair activation mechanisms during the therapeutic rest interval prevented the therapeutic effect from being seen following the standard treatment with cyclophosphamide [11]. Studies have revealed endothelial cell replication in vitro might be impacted by even very low doses of MCT (picomolar and/or nanomolar scale) [5]. For this reason, Bocci et al. observed that endothelial cells exposed to several low-dosage anti-cancer drugs in vitro enhanced the production of thrombospondin-1 [12], a matricellular protein that directly affects endothelial cell motility to prevent angiogenesis [13]. These results were further supported in vivo as mouse models with various tumor types showed lower manifestations of thrombospondin-1 and other angiogenesis inhibitors after receiving MCT [14]. Circulating endothelial cells stimulate neovascularization, which is essential for the development of tumors [15]. The exposure to traditional chemotherapy after the cycle treatment considerably mobilized circulatory endothelial cells, whereas MCT decreased the number of such cells in mice tumor models [16]. Both hypoxia inducible factor (HIF-1α) and vascular endothelial growth factor (VEGF) are well acknowledged as significant pro-angiogenic factors [17]. In general, a hypoxic microenvironment increases the transcription of HIF-1, which in turn boosts the production of VEGF and the VEGF receptors 1–2 in both normal and malignant cells [18]. In this setting, lower HIF-1 levels were shown in a variety of tumor types after MCT, and comparable outcomes were seen in ovarian cell lines after MCT topotecan therapy [19]. Latent cells are cancer cells with the capacity to live in a non-proliferative state for longer periods of time before recurring from the site of metastasis. It has been demonstrated that MCT may successfully target these cells [20]. On the other hand, this latent cell phenotype might be seen subsequently at the end of anti-cancer therapy [21]. So, after the accomplishment of conventional chemotherapy, MCT might be utilized to induce angiogenic dormancy [22].

7.2.2 Metronomic Chemotherapy and Immunomodulation Effects

It has been shown that various chemotherapy regimens have varying degrees of immunostimulatory effects. Particularly, the immunogenicity of MCT may be caused by on-target effects including immune cells with anti-tumor-promoting activities [22]. Regarding the on-target effects, MCT might promote immunogenic cell death (ICD) and/or make tumor cells more susceptible to immune effectors. An immunocompetent syngeneic host must experience an adaptive immune response in response to ICD [23]. The two primary factors of adjuvanticity and antigenicity determine how the adaptive immune system is stimulated. While the expulsion or exposure of warning signs called damage-related molecular patterns (DAMPs) enables the adjuvanticity of ICD and draws in and triggers the development of specialized antigen-presenting cells (APCs), such as dendritic cells (DCs), antigenicity depends on mechanisms that enable a beneficial identification of tumor-associated antigens (TAAs) by the adaptive immunity of the host [24]. In this situation, it was discovered that several chemotherapeutics increased the antigenicity and adjuvanticity of dying tumor cells. For illustration purposes, by altering tumor cell metabolism to favor oxidative phosphorylation, mitomycin C enhanced ICD. As a consequence of the enhanced permeabilization of the mitochondrial membranes, mitochondrial DNA was released into the cytoplasm, favoring DC activation [25]. Other DNA-damaging chemotherapy drugs, for example, anthracyclines (doxorubicin and idarubicin), can expose or release DAMPs that cause dying tumor cells to translocate calreticulin to the surface or release high-mobility group box 1 (HMGB1), both of which cause DCs to be attracted to the site and activated. Another DAMP capable of activating APC in the tumor microenvironment, cyclophosphamide, may facilitate the discharge of uric acid following tumor cell death [26]. Choi et al. studied a novel metronomic oral oxaliplatin formulation and found that it boosted ICD markers in vivo and in vitro, increasing the tumor antigen absorption and activating DCs in lymph nodes that are draining tumors [27]. By activating the mannose-6-phosphate receptors and increasing the granzyme B permeability generated by cytotoxic T lymphocytes (CTLs), chemotherapy drugs like doxorubicin, paclitaxel, and cisplatin can also make tumor cells more vulnerable to immune effectors [28]. As a result, these not only recognize and destroy tumor cells that express TAA but also impact nearby tumor cells that do not express TAA [28]. Pemetrexed, in a similar manner, sensitizes NSCLC to natural killer (NK) cells as well as CTL activity [29]. Finally, modest doses of gemcitabine in pancreatic cancer cell lines may boost MICA/B expression on the cell surface, improving innate immune action as compared to tumor cells [30]. Immunological effectors can be stimulated to produce more anti-tumor immunity, whereas immunosuppressive cells like myeloid-derived suppressor cells (MDSCs), M2-like macrophages, and regulatory T (Treg) cells can be eliminated. DCs are undoubtedly the main aim of chemo-immunomodulation. By transporting DC precursors from the bone marrow to the tumor site, low doses of cyclophosphamide have been found to re-establish the reversing tumor-induced paralysis of the DCs and myelopoiesis seen both systemically and in situ [26]. Low doses of paclitaxel, in contrast to large doses of the same medication, which induced DC death, promoted the digestion of tumor antigens by DCs and improved their development at the tumor site [31]. Furthermore, the capability of DCs to successfully transport tumor antigens to T cells in vitro was directly improved by non-cytotoxic doses of methotrexate, mitomycin C, and doxorubicin. On the cell surfaces of DCs, many markers, including CD86, CD80, CD40, and major histocompatibility class (MHC) class II, were up-regulated, favoring this capability [32]. The activation of CD40 prevented tumor-induced dendritic cells from death, hence counteracting the DCs suppression caused by tumor growth [33]. Shortly after receiving the treatment of relatively low doses of 0.1–1 mol/L of vinorelbine (VNR) and vinblastine, the activation of the previously characterized molecules on the surface of DCs amplified production of interleukin (IL)-6, IL-1, and IL-12 [34]. Some chemotherapy drugs, such as oxaliplatin and 5-fluorouracil, decreased immunosuppressive ligand expression on the surface of DCs [35]. On the other hand, by improving the DC development and antigen presentation, a number of chemotherapeutic drugs, including methotrexate, when taken at extremely low doses, might significantly boost the potential of DCs to promote T-cell proliferation [36]. A small number of chemotherapeutic medicines, as well as certain metronomic regimens or treatment combinations, have been shown to directly affect CTLs and NK cells. In vitro studies demonstrated that T-cell activation was dependent on mitochondrial respiratory capacity, which pemetrexed therapy increased, promoting the cytotoxic effects of CTLs and the synthesis of co-stimulatory receptors [37]. In a lung cancer xenograft model, combination treatment with modest dosages of paclitaxel with DC immunization led to improved tumor infiltration by CD4+ and CD8+ T cells [38]. In in vivo tumor models, low dosages of cyclophosphamide in NK cells enhanced anti-tumor cytotoxic activity [38]. The tumor microenvironment was more favorable to the initiation of anti-tumor M1-like macrophages when cyclophosphamide was given in metronomic dosages [39]. A critical problem in the area of oncology is the capacity to reverse the immunosuppressive environment characteristic of a tumor microenvironment and anti-tumor immunity activation. Targeting immune-suppressive cells and repolarizing them into immune-stimulating effectors may be how MCT achieves this function. In detail, Michels et al. showed via the use of cancer cell models that ultra-low dosages of paclitaxel might trigger MDSC differentiation into DCs without relying on TLR4 [40]. When gemcitabine was paired with genetic immunization using an adenovirus-producing interferon (IFN), this impact was further increased [41]. It is interesting to note that, compared to higher levels, the modulation of MDSC-mediated immunosuppression was more potent at low doses than at higher doses [42]. Similar outcomes were shown when metronomic dosages of capecitabine were administered to glioblastoma patients in a clinical setting. These patients' tumor microenvironments had better cytotoxic immune infiltration as a consequence of lower levels of circulating MDSCs [43]. Currently, there is a consensus in the scientific community that metronomic cyclophosphamide has been widely recognized to have a substantial impact on the number of Tregs at both local and systemic levels, as shown by studies conducted on mouse models as well as cancer patients. In vivo cancer models demonstrated a disruption in the regeneration of Tregs after the administration of low-dose cyclophosphamide, specifically at intervals of 45 days [44].The capacity of cyclophosphamide to damage the transforming growth factor (TGF)-beta pathway [45], an important controller of Tregs formation, may also play a role in the selective depletion of Tregs. In advanced cancer patients receiving metronomic cyclophosphamide therapy, many clinical studies indicated Treg depletion [46, 47]. Metronomic therapy, in particular, reduced CTLA-4 expression in Foxp3+ Tregs, specifically suppressing this important fraction of Tregs [46]. The use of chemotherapeutic agents, such as cyclophosphamide, has been seen to potentially induce the depletion of Tregs when provided at lower dosages. In tumor-bearing animal models, it was shown that the alkylating medication temozolomide caused a reduction in the ratios of intratumor Tregs/CD4+ cells. Similarly, gemcitabine was found to limit the formation of Tregs in mice with pancreatic adenocarcinoma [48].

7.2.3 Metronomic Chemotherapy Targeted at Tumor Stem Cells

Although MCT has traditionally been considered a palliative treatment option, scientific investigations have shown that this therapy has a direct cytotoxic effect on tumor cells [49]. Orlandi et al. conducted an investigation whereby they observed that the administration of metronomic VNR resulted in a reduced rate of proliferation in two human NSCLC cell lines. These cell lines were treated with both wild-type and mutant EGFR tyrosine kinase inhibitors (TKIs). In contrast, it was shown that the EGFR wild-type NSCLC cell line exhibited an effective response to the same VNR dosage as the conventional MTD chemotherapeutic treatment [49]. The important discoveries show that MCT has cytotoxic effects on carcinoma stem cells and stem-like tumor-initiating cells. Despite constituting a small fraction of the tumor mass, these cellular subtypes play a pivotal role in tumor dissemination and recurrence. An investigation into xenografts of human pancreatic tumors demonstrated a reduction in the number of cancer stem cells subsequent to metronomic cyclophosphamide treatment [50]. Similar to this, Chan and associates reported that in vivo administration of MCT inhibited carcinoma cells in tumor stem cells from changing their phenotype and increased mouse survival. On the other hand, the same medication as traditional chemotherapy with MTD induced the phenotype – a transformation of surviving cancer cells into tumor cells that resemble stem cells [51].

In a recent study, Bodarenko et al. examined the effects of MCT and traditional MTD treatment on tumor heterogeneity. The rate of proliferation of both types of NSCLC cells, as well as the total development of 2D and 3D co-cultures, was revealed to be constrained by NSCLC cell lines receiving in vitro MCT therapy, whether they were sensitive to patriline and cisplatin or resistant to them. In contrast, it was shown that growth suppression occurred after administering the MTD in drug-sensitive cells, but no such suppression was seen in resistant cells. Significantly, the observed restricted viability of chemically sensitive cell lines after metronomic treatment was effective in impeding the rate of proliferation of resilient cells [52]. Together, these results demonstrate the therapeutic potential of MCT in the treatment of tumor resistance.

By using a variety of methods, including cell death, caspase-independent cell death, senescence, immunogenic cell death, non-apoptotic cell death, and immunogenic cell death, MCT can target tumor cells in a specific manner. In vivo and in vitro studies have revealed that the application of pan-caspase inhibitors, either alone or in conjunction with low doses of metronomic actinomycin D, led to cell death in neuroblastoma tumors that was independent of apoptosis [53]. The ability of MCT to induce tumor senescence in prostatic cells following treatment with a modest dosage (25 nM) of doxorubicin has been also documented [54]. Low-dosage topotecan was also used to further induce senescence both in neuroblastoma xenografts and the neuroblastoma cell line [55].

MCT may indirectly influence tumor development by changing the metabolism of cancer cells. In a similar vein, Fares et al.'s work showed that the combination of metronomic paclitaxel and the AKT inhibitor perifosine enhanced the overall survival of NSCLC mice harboring tumors. The suppression of the two major metabolic processes, glycolysis and/or oxidative phosphorylation, is likely what caused the observed effectiveness [55]. The long-term use of chemotherapy might make tumor cells dependent on the medication, and the cessation of chemotherapy could result in the death of the tumor cells, according to many preclinical studies [56]. As a result, adding long-term exposure to MCT may improve the treatment's ability to destroy cancer cells of all types [56].

7.3 DOSE AND TIME EXPOSURE

There is mounting data that suggests medication dosage and timing of delivery may have non-linear effects on the anti-tumor MCT pathways. Therefore, even with the same total quantity of the delivered medicine, the dosage and exposure period of a specific anti-cancer mediator might influence the overall result of cancer treatment [57]. Numerous preclinical investigations have been conducted to assess the influence of dose and time exposure of MCT on tumor regression and the stimulation of immunological anti-tumor responses. In a seminal study conducted by Wu et al., it was shown that human cancer cells exhibited a more pronounced response when subjected to prolonged treatment (lasting 14 days) with a concentration of 0.29 M paclitaxel, as compared to a brief exposure (lasting 1 hour) to the same dose. The study demonstrates the significance of time duration in influencing the toxicity of paclitaxel in human cancer cells. Moreover, it argues that extended treatment regimens have the potential to augment the effectiveness of paclitaxel as an anti-cancer agent [57]. In contrast to short-term exposure, the results of in vitro topotecan exhibited a greater response rate in a clonogenic test at long-term exposure, suggesting that topotecan had increased efficacy under prolonged and continuous therapy [58].

More recent research indicates that dosage and time exposure have a significant impact on the immunological anti-tumor response and that this route may be the primary driver of tumor regression. Chen et al. conducted a study whereby they observed that the administration of cyclophosphamide at a dosage of 140 mg/kg for 6 days, following an intermittent schedule, led to a significant up-regulation of innate immune cells such as macrophages, NK cells, and DCs. This activation subsequently resulted in the regression of tumors. It is noteworthy that the administration of cyclophosphamide at the same dosage showed a significant decrease in the quantity of infiltrating NK cells and subsequent tumor regrowth following 24 days when administered using an extended intermittent regimen (at intervals of either 9 or 12 days). Hence, metronomic drug treatment must be administered at an adequate dose and with appropriate temporal intervals to effectively stimulate sustained recruitment of immune cells with anti-cancer properties [59]. The breast cancer model consistently showed an immunomodulatory impact of low-dosage metronomic cyclophosphamide [60]. A metronomic dosage of chemotherapeutic medications is often computed as a sub-toxic dose of the MTD. Notably, the therapy failed to stop tumor development when the metronomic schedule's daily dosage was too low, highlighting the need for a minimum effective dose [61]. However, many studies suggested that a unique measure for MCT would be the optimum biological dosage (OBD) with the maximum effectiveness and the absence of or the lowest toxicity [62]. In tumor-bearing animal models, Shaked et al. investigated the OBD of four metronomic chemotherapy regimens to establish viable circulating endothelial progenitor cells (CEPs) in blood samples [62]. Another cutting-edge method that could determine the best medication, dosage, and schedule combination while accounting for all aspects of tumor biology is the use of mathematical models [63]. For example, an in silico model developed by Faivre et al. indicated that a temozolomide metronomic regimen would be the most effective course of therapy to cause tumor regression due to its anti-angiogenic properties; importantly, clinical data further supported this prediction [64]. When compared to other metronomic regimens that were explored, the outcomes of in silico simulations showed that the administration of an alternative VNR dosage regimen (60/30/60 mg) was both well-tolerated and more efficacious. These observations were seen when the identical model was employed to include metronomic VNR treatment for patients with lung cancer [65]. It has been shown elsewhere that metronomic gemcitabine, as opposed to normal MTD treatment, significantly slowed the development of the tumor in a human neuroblastoma xenograft model that was resistant to the drug [66]. Overall, these results show that an effective anti-tumor treatment requires a combination of chemotherapeutic dosage and time exposure. Furthermore, it is difficult to predict the ideal biological dosage and the exact schedule since they may differ among tumor types.

7.4 METRONOMIC CHEMOTHERAPY AND IMMUNE CHECKPOINT INHIBITORS

The pleiotropic immunomodulatory actions of MCT make it an ideal adjunct to immunological checkpoint inhibitors (ICIs). Immunotherapy using ICIs may enhance the natural immune mechanism that will fight against the tumor cells by activating pronounced cytotoxic efficacy. The T-cell surface receptors, together with their corresponding ligands, have garnered significant interest as targets for ICIs so far. Outstanding outcomes were seen for many tumors in the clinical setting, including melanoma [67], lung cancer [68], and hematological malignancies [69]. Despite the encouraging advancements in cancer immunotherapy, only a small percentage of patients have success using immune checkpoint-blocking antibodies to treat different forms of cancer. Hence, the investigation of the amalgamation of immunotherapy and chemotherapy has been undertaken across several cancer types, yielding favorable outcomes that justify the need for further clinical investigation [70].

By comparing the efficacy of three various cyclophosphamide regimens, Khan et al. evaluated triple-negative breast cancer (TNBC) in terms of anti-tumor response. The results of the study suggest that cyclophosphamide at a dosage of 140 mg/kg every 6 days led to a more pronounced inhibition of tumor growth in mice orthotopic models, in comparison with the conventional MTD or daily metronomic dose. The study revealed that both in vivo and in vitro studies with cyclophosphamide treatment led to an improvement in PD-L1 expression in breast tumors. However, the combined treatment of cyclophosphamide and an anti–PD-1 antibody did not provide a more pronounced immunomodulatory effect. Despite this sudden outcome, the study demonstrated a

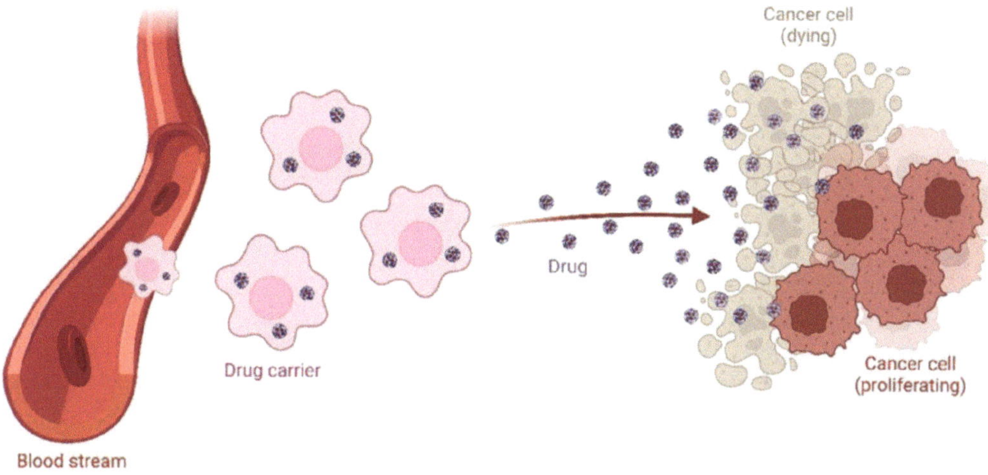

FIGURE 7.2 A detailed mechanism of drug delivery through blood and its role in cancer cell treatment.

metronomic potential system of cyclophosphamide for use in the treatment of breast cancer [71]. Intriguingly, a study by He and colleagues using syngeneic murine models of squamous cell lung cancer revealed that MCT had a strong anti-tumor impact when paired using monoclonal antibody (mAb) treatment against PD-1. The combined effects of anti–PD-1 mAb and MTD chemotherapy, on the other hand, seemed to be more additive than synergistic. The research also suggested that the gut microbiota, CD8+ T cells, and macrophages may be crucial players in mediating the anti-cancer synergistic effects. In total, the research supported a unique approach for small-cell lung cancer (SCLC) patients' chemotherapy dosage optimization [72]. The anti-tumor impact of PD-1 inhibitors was reportedly amplified without side effects by a modest dosage of carboplatin in the lung cancer mice model by increasing CD8+ T-cell infiltration and PD-L1 expression [73] (Figure 7.2).

A significant rise in cytotoxic CD8+ anti-tumorigenic T cells, DCs, and M1 macrophage levels was observed in the model of SCLC following the administration of metronomic doses of gemcitabine combined with an inhibitor of checkpoint kinase 1 (CHK1), anti–PD-1/PD-L1, and a CHK1 antagonist. As a result of the combined therapy, however, immunosuppressive M2 macrophage and MDSC levels were dramatically lowered [68]. By reducing tumor formation of the mice with melanoma by 66.6%, Petrizzo et al. found that a triple combination method consisting of a multi-peptide vaccination, MCT with many drugs, and an anti–PD-1 treatment successfully increased the anti-tumor efficacy of the vaccine [74]. In their study, Skavatsou et al. demonstrated that the administration of a combination of gemcitabine and anti–PD-1 monoclonal antibody resulted in the upregulation of TSP-1 and VEGF-A expression. This upregulation effectively restricted tumor angiogenesis, as seen in both in vitro and in vivo models of NSCLC. By boosting the number of CD8+ T cells, encouraging the production of pro-inflammatory cytokines, and decreasing the proportion of Treg cells, the group therapy showed increased anti-tumor immunity [75].

The use of nivolumab, an anti–PD-1 monoclonal antibody, in the treatment of patients with metastatic or incurable, locally advanced TNBC was examined in the TONIC study conducted by Voorwerk et al. This examination included patients who had undergone short-term induction with MCT (cyclophosphamide, cisplatin, or doxorubicin), radiation therapy, or no induction [76]. The study found a 20% overall objective response rate (ORR), with doxorubicin and cisplatin cohorts showing the highest rates of responses (35% and 23%, respectively). The elevated response rate in these cohorts was associated with a marked rise in tumor-infiltrating T cells and activation of PD-L1/PD-1–related genes [76]. An analogous increase in the expression of CD8+ T cells has been seen in SCLC patients who had treatment with a combination of low-dose albumin-paclitaxel and

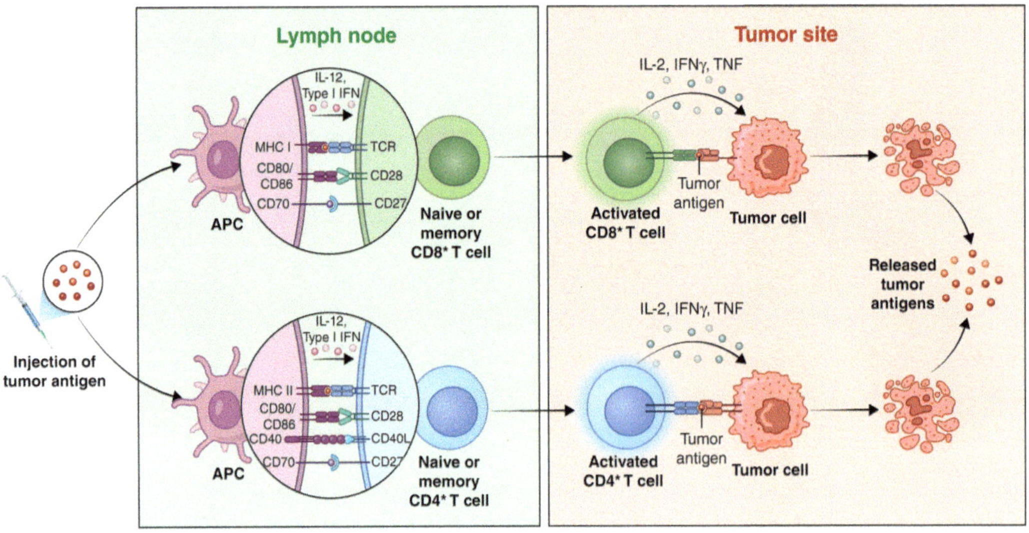

FIGURE 7.3 A possible mechanism of how a vaccine works and the underlying principles for the treatment of cancer.

nivolumab. Furthermore, the study has shown evidence suggesting that the optimal timing for achieving successful treatment outcomes in SCLC may include administering immunotherapy by injection within a 24-hour window after low-dose chemotherapy [72]. In the investigation of the combined impact of metronomic VNR and atezolizumab, an anti–PD-L1 monoclonal antibody, on patients with advanced-stage NSCLC, the clinical study conducted by Vergnenegre et al. did not yield any significant therapeutic advantages in terms of effectiveness and safety [77] (Figure 7.3).

7.5 CONCLUSION AND FUTURE CONTEMPLATIONS

MCT has not yet been used as an initial treatment, but rather as a palliative standard care tool. However, mounting data suggests that MCT might be used as a therapeutic tool to treat a variety of tumor forms. To target tumor heterogeneity and effectively block anti-angiogenesis, MCT is effective. Additionally, MCT has the potential to be used as a tool in the treatment of cancer due to its capacity to turn an immunological "cold" tumor into a "hot" tumor. MCT was created to address the medication resistance associated with traditional chemotherapy. After receiving MCT therapy for a long time, certain tumor cells may likely become resistant to the treatment. However, the many modes of action of MCT may be able to successfully counteract this impact by inhibiting angiogenesis, maintaining the therapeutic effectiveness.

Further investigation is necessary to fully comprehend the inherent capabilities of MCT, due to the extremely low likelihood that a single metronomic regimen will consistently demonstrate efficacy across all distinct tumor types. The optimal biological dose (pharmacokinetics/pharmacodynamics [PK/PD]) of each drug, whether administered alone or in combination, as well as the most appropriate timing for medicine administration, remain areas of limited comprehension. This is especially true for the selection of the most successful chemotherapeutic agents based on tumor type. Despite the existing studies indicating the potential efficacy of MCT at different medication dosages, there is a significant knowledge gap regarding the PK and PD of MCT. Due to the lack of direct proportionality between the micro-dose (MCT) PK/PD and those seen at large doses, the MCT PK/PD may exhibit varied kinetics, leading to distinct plasma concentrations of free drug. The purpose of this study is to determine whether PK/PD measurements may be used as a biomarker to predict the clinical outcome of MCT. The existing biomarkers employed for identifying patients who may derive advantages from MCT predominantly consist of clinical parameters associated with advanced cancer stage and inability to take high-dose chemotherapy due to severe toxicity or lack of response to conventional treatment (relapsed or refractory tumor). It is hypothesized that certain biomarkers found in the bloodstream may have predictive and prognostic significance in the perspective of immunotherapy using ICIs. These biomarkers might be used to guide the selection of patients who would benefit from a combined treatment approach including ICIs and MCT. When used in conjunction with standard MTD chemotherapy, there exists a vast array of potential combinations including MCT, targeted treatment, and/or radiation. Utilizing mathematical models might be seen as a viable approach for selecting optimal combinations before their use in preclinical and clinical studies. The identification of easily measurable MCT-related biomarkers, both pre- and post-medication, remains an essential need for the effective implementation of metronomic therapy. The identification of biomarkers capable of predicting the sustained advantage of MCT in cancer patients who exhibit resistance to current treatment or experience illness recurrence is a persistent problem. To enhance the efficacy of MCT, it is advisable, wherever possible, to examine the causative processes in these patients.

For further reading, some textbooks mentioned here also give a further understanding of MCT:

- *Progress in Anti-cancer Chemotherapy* by Gabriel N. Hortobagyi and David Khayat.
- *Triple-Negative Breast Cancer: Biomarkers, Emerging Therapeutic Strategies and Clinical Challenges* by Marion Montgomery.
- *Current Cancer Treatment: Novel Beyond Conventional Approaches* by Oner Ozdemir.
- *Breast Cancer: Current and Alternative Therapeutic Modalities* by Esra Gunduz and Mehmet Gunduz.

The following links for video lectures may also be helpful:

https://youtu.be/TAqO0Mq19JQ
https://youtu.be/6wEnYfvoBqk
https://youtu.be/A9fRRCUlMms
https://youtu.be/zIWvmCsKr34

For more insight about the topic, we would suggest detailed findings from the books of (Mir MA, 2022) https://doi.org/10.1016/C2021-0-02565-7, https://doi.org/10.1016/C2022-0-00074-X, (Mir MA, 2021) https://doi.org/10.52305/WXJL6770, and (Mir MA, 2015) https://doi.org/10.1016/C2014-0-02898-5 and from the cancer.net website on the following mentioned links:

https://www.cancer.net/cancer-types/breast-cancer/types-treatment
https://discovery.ucl.ac.uk/id/eprint/1472740/
https://www.jmedsciences.com/doi/JMEDS/pdf/10.5005/jp-journals-10045-00138

For diagrammatic illustrations and descriptive tables (Lazzeroni, 2012): http://www.eurekaselect.com/article/49928

See video links on the overall status of cancer, its various types, and current and new possible treatment options available: https://www.sciencedirect.com/science/article/pii/S205970292032278X

REFERENCES

1. Cazzaniga, M.E., et al., *Metronomic chemotherapy.* Cancers, 2021. **13**(9): p. 2236.
2. Wysocki, P.J., et al., *Metronomic Chemotherapy Based on Topotecan or Topotecan and Cyclophosphamide Combination (CyTo) in Advanced, Pretreated Ovarian Cancer.* Cancers, 2023. **15**(4): p. 1067.

3. Tang, Y., et al., *Prognosis and dissection of immunosuppressive microenvironment in breast cancer based on fatty acid metabolism-related signature.* Frontiers in immunology, 2022. **13**: p. 843515.
4. Krajnak, S., et al., *Metronomic chemotherapy for metastatic breast cancer.* Oncology research and treatment, 2022. **45**(1-2): p. 12–17.
5. Muraro, E., et al., *Metronomic Chemotherapy: Anti-Tumor Pathways and Combination with Immune Checkpoint Inhibitors.* Cancers, 2023. **15**(9): p. 2471.
6. Rampioni Vinciguerra, G.L., et al., *CDK4/6 inhibitors in combination therapies: better in company than alone: a mini review.* Frontiers in oncology, 2022. **12**: p. 891580.
7. Bocci, G., et al., *Remarkable Remission Rate and Long-Term Efficacy of Upfront Metronomic Chemotherapy in Elderly and Frail Patients, with Diffuse Large B-Cell Lymphoma.* Journal of clinical medicine, 2022. **11**(23): p. 7162.
8. Ullah, A., et al., *Molecular mechanisms of Sanguinarine in cancer prevention and treatment.* Anti-cancer agents in medicinal chemistry, 2023. **23**(7): p. 765–778.
9. Dudley, A.C. and A.W. Griffioen, *Pathological angiogenesis: mechanisms and therapeutic strategies.* Angiogenesis, 2023. **26**(3): p. 313–347.
10. Li, Z., et al., *Smart nanotherapeutic targeting of tumor vasculature.* Accounts of chemical research, 2019. **52**(9): p. 2703–2712.
11. Chen, Y.-L., M.-C. Chang, and W.-F. Cheng, *Metronomic chemotherapy and immunotherapy in cancer treatment.* Cancer letters, 2017. **400**: p. 282–292.
12. Bocci, G., K. Nicolaou, and R.S. Kerbel, *Protracted low-dose effects on human endothelial cell proliferation and survival in vitro reveal a selective antiangiogenic window for various chemotherapeutic drugs.* Cancer research, 2002. **62**(23): p. 6938–6943.
13. Lawler, J., *Thrombospondin-1 as an endogenous inhibitor of angiogenesis and tumor growth.* Journal of cellular and molecular medicine, 2002. **6**(1): p. 1–12.
14. Beerepoot, L., et al., *Increased levels of viable circulating endothelial cells are an indicator of progressive disease in cancer patients.* Annals of oncology, 2004. **15**(1): p. 139–145.
15. Weskamp, G., et al., *Pathological neovascularization is reduced by inactivation of ADAM17 in endothelial cells but not in pericytes.* Circulation research, 2010. **106**(5): p. 932–940.
16. Kim, J.Y. and Y.-M. Kim, *Tumor endothelial cells as a potential target of metronomic chemotherapy.* Archives of pharmacal research, 2019. **42**: p. 1–13.
17. Albertsson, P., B. Lennernäs, and K. Norrby, *On metronomic chemotherapy: modulation of angiogenesis mediated by VEGE-A.* Acta oncologica, 2006. **45**(2): p. 144–155.
18. Schito, L. and G.L. Semenza, *Hypoxia-inducible factors: master regulators of cancer progression.* Trends in cancer, 2016. **2**(12): p. 758–770.
19. Merritt, W.M., et al., *Anti-angiogenic properties of metronomic topotecan in ovarian carcinoma.* Cancer biology & therapy, 2009. **8**(16): p. 1596–1603.
20. Chen, Z., et al., *Surgical stress and cancer progression: the twisted tango.* Molecular cancer, 2019. **18**(1): p. 1–11.
21. Udagawa, T., *Tumor dormancy of primary and secondary cancers.* APMIS, 2008. **116**(7-8): p. 615–628.
22. Galluzzi, L., et al., *Immunostimulation with chemotherapy in the era of immune checkpoint inhibitors.* Nature reviews. Clinical oncology, 2020. **17**(12): p. 725–741.
23. Galluzzi, L., et al., *Molecular mechanisms of cell death: recommendations of the Nomenclature Committee on Cell Death 2018.* Cell death & differentiation, 2018. **25**(3): p. 486–541.
24. Bedognetti, D., et al., *The biology of immune-active cancers and their regulatory mechanisms.* Cancer treatment and research, 2020. **180**: p. 149–172.
25. Oresta, B., et al., *Mitochondrial metabolic reprogramming controls the induction of immunogenic cell death and efficacy of chemotherapy in bladder cancer.* Science translational medicine, 2021. **13**(575): p. eaba6110.
26. Schiavoni, G., et al., *Cyclophosphamide synergizes with type I interferons through systemic dendritic cell reactivation and induction of immunogenic tumor apoptosis.* Cancer research, 2011. **71**(3): p. 768–778.
27. Choi, J.U., et al., *Modulating tumor immunity by metronomic dosing of oxaliplatin incorporated in multiple oral nanoemulsion.* Journal of controlled release, 2020. **322**: p. 13–30.
28. Ramakrishnan, R., et al., *Chemotherapy enhances tumor cell susceptibility to CTL-mediated killing during cancer immunotherapy in mice.* The journal of clinical investigation, 2010. **120**(4): p. 1111–1124.
29. Okimoto, T., et al., *Pemetrexed sensitizes human lung cancer cells to cytotoxic immune cells.* Cancer science, 2020. **111**(6): p. 1910–1920.

30. Miyashita, T., et al., *Low-dose gemcitabine induces major histocompatibility complex class I-related chain A/B expression and enhances an antitumor innate immune response in pancreatic cancer.* Clinical and experimental medicine, 2017. **17**: p. 19–31.
31. He, Q., et al., *Low-dose paclitaxel enhances the anti-tumor efficacy of GM-CSF surface-modified whole-tumor-cell vaccine in mouse model of prostate cancer.* Cancer immunology, immunotherapy, 2011. **60**: p. 715–730.
32. Shurin, G.V., et al., *Chemotherapeutic agents in noncytotoxic concentrations increase antigen presentation by dendritic cells via an IL-12-dependent mechanism.* The journal of immunology, 2009. **183**(1): p. 137–144.
33. Pinzon-Charry, A., C.W. Schmidt, and J.A. López, *The key role of CD40 ligand in overcoming tumor-induced dendritic cell dysfunction.* Breast cancer research, 2006. **8**: p. 1–2.
34. Cazzaniga, M.E., et al., *Metronomic oral vinorelbine in advanced breast cancer and non-small-cell lung cancer: current status and future development.* Future oncology, 2016. **12**(3): p. 373–387.
35. Hong, X., et al., *Impact of 5-Fu/oxaliplatin on mouse dendritic cells and synergetic effect with a colon cancer vaccine.* Chinese journal of cancer research, 2018. **30**(2): p. 197.
36. Kaneno, R., et al., *Chemomodulation of human dendritic cell function by antineoplastic agents in low noncytotoxic concentrations.* Journal of translational medicine, 2009. **7**: p. 1–10.
37. Schaer, D.A., et al., *The folate pathway inhibitor pemetrexed pleiotropically enhances effects of cancer immunotherapy.* Clinical cancer research, 2019. **25**(23): p. 7175–7188.
38. Moschella, F., et al., *Combination strategies for enhancing the efficacy of immunotherapy in cancer patients.* Annals of the New York Academy of Sciences, 2010. **1194**(1): p. 169–178.
39. Doloff, J.C. and D.J. Waxman, *VEGF receptor inhibitors block the ability of metronomically dosed cyclophosphamide to activate innate immunity–induced tumor regression.* Cancer research, 2012. **72**(5): p. 1103–1115.
40. Michels, T., et al., *Paclitaxel promotes differentiation of myeloid-derived suppressor cells into dendritic cells in vitro in a TLR4-independent manner.* Journal of immunotoxicology, 2012. **9**(3): p. 292–300.
41. Suzuki, E., et al., *Gemcitabine selectively eliminates splenic Gr-1+/CD11b+ myeloid suppressor cells in tumor-bearing animals and enhances antitumor immune activity.* Clinical cancer research, 2005. **11**(18): p. 6713–6721.
42. Tongu, M., et al., *Metronomic chemotherapy with low-dose cyclophosphamide plus gemcitabine can induce anti-tumor T cell immunity in vivo.* Cancer immunology, immunotherapy, 2013. **62**: p. 383–391.
43. Peereboom, D.M., et al., *Metronomic capecitabine as an immune modulator in glioblastoma patients reduces myeloid-derived suppressor cells.* JCI insight, 2019. **4**(22), p. e130748.
44. Lutsiak, M.C., et al., *Inhibition of CD4+ 25+ T regulatory cell function implicated in enhanced immune response by low-dose cyclophosphamide.* Blood, 2005. **105**(7): p. 2862–2868.
45. Zhong, H., et al., *Low dose cyclophosphamide modulates tumor microenvironment by TGF-β signaling pathway.* International journal of molecular sciences, 2020. **21**(3): p. 957.
46. Noordam, L., et al., *Low-dose cyclophosphamide depletes circulating naïve and activated regulatory T cells in malignant pleural mesothelioma patients synergistically treated with dendritic cell-based immunotherapy.* Oncoimmunology, 2018. **7**(12): p. e1474318.
47. Banissi, C., et al., *Treg depletion with a low-dose metronomic temozolomide regimen in a rat glioma model.* Cancer immunology, immunotherapy, 2009. **58**: p. 1627–1634.
48. Shevchenko, I., et al., *Low-dose gemcitabine depletes regulatory T cells and improves survival in the orthotopic Panc02 model of pancreatic cancer.* International journal of cancer, 2013. **133**(1): p. 98–107.
49. Orlandi, P., et al., *Metronomic vinorelbine is directly active on non small cell lung cancer cells and sensitizes the $EGFR^{L858R/T790M}$ cells to reversible EGFR tyrosine kinase inhibitors.* Biochemical pharmacology, 2018. **152**: p. 327–337.
50. Vives, M., et al., *Metronomic chemotherapy following the maximum tolerated dose is an effective anti-tumour therapy affecting angiogenesis, tumour dissemination and cancer stem cells.* International journal of cancer, 2013. **133**(10): p. 2464–2472.
51. Chan, T.-S., et al., *Metronomic chemotherapy prevents therapy-induced stromal activation and induction of tumor-initiating cells.* Journal of experimental medicine, 2016. **213**(13): p. 2967–2988.
52. Ayipo, Y.O., et al., *Epigenetic oncogenesis, biomarkers and emerging chemotherapeutics for breast cancer.* Biochimica et biophysica acta. Gene regulatory mechanisms, 2022. **1865**(7): p. 194873.
53. Cortes, C.L., et al., *Effect of low doses of actinomycin D on neuroblastoma cell lines.* Molecular cancer, 2016. **15**(1): p. 1–13.
54. Schwarze, S.R., et al., *The identification of senescence-specific genes during the induction of senescence in prostate cancer cells.* Neoplasia, 2005. **7**(9): p. 816–823.

55. Fares, C.M., et al., *Mechanisms of resistance to immune checkpoint blockade: why does checkpoint inhibitor immunotherapy not work for all patients?* American Society of Clinical Oncology educational book, 2019. **39**: p. 147–164.
56. André, N. and E. Pasquier, *Response to 'Intermittent androgen blockade should be regarded as standard therapy in prostate cancer'.* Nature clinical practice. Oncology, 2009. **6**(2): p. E1–E1.
57. Wu, J. and D.J. Waxman, *Immunogenic chemotherapy: dose and schedule dependence and combination with immunotherapy.* Cancer letters, 2018. **419**: p. 210–221.
58. Gerrits, C., et al., *Topoisomerase I inhibitors: the relevance of prolonged exposure for present clinical development.* British journal of cancer, 1997. **76**(7): p. 952–962.
59. Chen, C.-S., J.C. Doloff, and D.J. Waxman, *Intermittent metronomic drug schedule is essential for activating antitumor innate immunity and tumor xenograft regression.* Neoplasia, 2014. **16**(1): p. 84–96.
60. Parra, K., et al., *Impact of CTLA-4 blockade in conjunction with metronomic chemotherapy on preclinical breast cancer growth.* British journal of cancer, 2017. **116**(3): p. 324–334.
61. Scharovsky, O.G., L.E. Mainetti, and V.R. Rozados, *Metronomic chemotherapy: changing the paradigm that more is better.* Current oncology, 2009. **16**(2): p. 7–15.
62. Shaked, Y., et al., *Optimal biologic dose of metronomic chemotherapy regimens is associated with maximum antiangiogenic activity.* Blood, 2005. **106**(9): p. 3058–3061.
63. Bhattacharjee, A. and V.M. Patil, *Determining an optimum biological dose of a metronomic chemotherapy.* Journal of data science, 2017. **15**(1): p. 77–94.
64. Faivre, C., et al., *A mathematical model for the administration of temozolomide: comparative analysis of conventional and metronomic chemotherapy regimens.* Cancer chemotherapy and pharmacology, 2013. **71**: p. 1013–1019.
65. Barbolosi, D., et al., *Metronomics chemotherapy: time for computational decision support.* Cancer chemotherapy and pharmacology, 2014. **74**: p. 647–652.
66. Ciccolini, J., et al., *Computational-driven metronomics: application to gemcitabine in neuroblastoma-bearing mice.* Cancer research, 2015. **75**(15_Supplement): p. 4506–4506.
67. Hodi, F.S., et al., *Improved survival with ipilimumab in patients with metastatic melanoma.* New England journal of medicine, 2010. **363**(8): p. 711–723.
68. Sen, T., et al., *Combination treatment of the oral CHK1 inhibitor, SRA737, and low-dose gemcitabine enhances the effect of programmed death ligand 1 blockade by modulating the immune microenvironment in SCLC.* Journal of thoracic oncology, 2019. **14**(12): p. 2152–2163.
69. Brahmer, J.R., et al., *Safety and activity of anti–PD-L1 antibody in patients with advanced cancer.* New England journal of medicine, 2012. **366**(26): p. 2455–2465.
70. Liikanen, I., et al., *Oncolytic adenovirus with temozolomide induces autophagy and antitumor immune responses in cancer patients.* Molecular therapy, 2013. **21**(6): p. 1212–1223.
71. Khan, K.A., et al., *Immunostimulatory and anti-tumor metronomic cyclophosphamide regimens assessed in primary orthotopic and metastatic murine breast cancer.* NPJ breast cancer, 2020. **6**(1): p. 29.
72. He, X., et al., *Upfront dose-reduced chemotherapy synergizes with immunotherapy to optimize chemoimmunotherapy in squamous cell lung carcinoma.* Journal for immunotherapy of cancer, 2020. **8**(2): p. e000807.
73. Zhou, L., et al., *Low-dose carboplatin reprograms tumor immune microenvironment through STING signaling pathway and synergizes with PD-1 inhibitors in lung cancer.* Cancer letters, 2021. **500**: p. 163–171.
74. Petrizzo, A., et al., *Inhibition of tumor growth by cancer vaccine combined with metronomic chemotherapy and anti-PD-1 in a pre-clinical setting.* Oncotarget, 2018. **9**(3): p. 3576.
75. Skavatsou, E., et al., *Immunotherapy combined with metronomic dosing: an effective approach for the treatment of NSCLC.* Cancers, 2021. **13**(8): p. 1901.
76. Voorwerk, L., et al., *Immune induction strategies in metastatic triple-negative breast cancer to enhance the sensitivity to PD-1 blockade: the TONIC trial.* Nature medicine, 2019. **25**(6): p. 920–928.
77. Kareva, I., D.J. Waxman, and G.L. Klement, *Metronomic chemotherapy: an attractive alternative to maximum tolerated dose therapy that can activate anti-tumor immunity and minimize therapeutic resistance.* Cancer letters, 2015. **358**(2): p. 100–106.

8 Metronomics

Toward Personalized Chemotherapy

Manzoor Ahmad Mir, Asma Jan, Aisha Shabir, and Nusrat Jan

8.1 INTRODUCTION

The idea of inhibiting tumor angiogenesis as a cancer therapy was first proposed by Judah Folkman in 1971 [1], and it took another 30 years before "pure" antiangiogenic medicines were approved in the management of colon and breast malignancy [2, 3]. Significantly, several standard cytotoxic medicines also have significant anti-angiogenic properties [4], particularly when administered in lower dosages, on frequently scheduled basis, and with no extended drug-free pauses [5] The term "metronomic chemotherapy," which refers to the aforementioned modes of delivery, was originally described in 2000 in two important works [6, 7]. The maximum toxic dose (MTD), or extremely near to it, is used for administering conventional chemotherapy, which is done every 3 weeks on average. During this time, the patient can recover from the side effects of chemotherapy, but the cancer cells can restart their growth [8] by recruiting circulating endothelial progenitor cells, a process known as tumor neovascularization [9, 10]. Metronomic chemotherapy, which uses lower doses administered more frequently to minimize vascular rebound (the fast rebuilding of vascular tissue that can contribute to tumor development during therapy interruptions), is effective in reducing adverse effects and prolonging survival for cancer patients [6, 9, 11]. The mechanism of metronomic chemotherapy is shown in Figure 8.1. The advent of metronomic chemotherapy allowed for the creation of an orally administered, low-cost, and well-tolerated treatment that may delay the progression of cancer for a long time [12–14]. Earlier theories held that metronomic chemotherapy's anti-tumor action was achieved only by anti-angiogenic mechanisms were proven incorrect. However, metronomic chemotherapy is now understood to be a type of multi-targeted chemotherapy since it has direct anti-cancer effects and also stimulates an anti-tumor immune response [15]. Metronomic chemotherapy's effects on drug resistance are discussed, as are the anti-tumor modes of action it employs. We then explore the place of metronomic chemotherapy in this new era of individualized anti-cancer therapies, after reviewing the acquired clinical experience.

8.2 AN EVOLVING TERMINOLOGY

The word "metronomic" was first used to characterize the anti-angiogenic chemotherapeutic approach developed by the two pioneering groups led by Judah Folkman and Robert Kerbel and was coined by Hanahan and coworkers [6, 7, 16]. The term "metronomic chemotherapy" was first coined to describe a continuous dosing regimen of a chemotherapeutic agent at or below MTD with no drug-free intervals longer than a few days. More recently, "metronomic chemotherapy" has come to mean the lowest efficacious dosage of a chemotherapeutic substance administered similarly [17]. Despite some differences, these definitions are consistent in highlighting the three main characteristics of metronomic chemotherapy: Regular dosing, a relatively modest dosage, and few interdose intervals. Other names for this approach to cancer treatment include metronomic scheduling [18, 19] low-dose maintenance chemotherapy [20], and low-dose anti-angiogenic chemotherapy [21]. Although metronomic chemotherapy is based on the low-dose administration of chemotherapeutics,

DOI: 10.1201/9781032632674-8

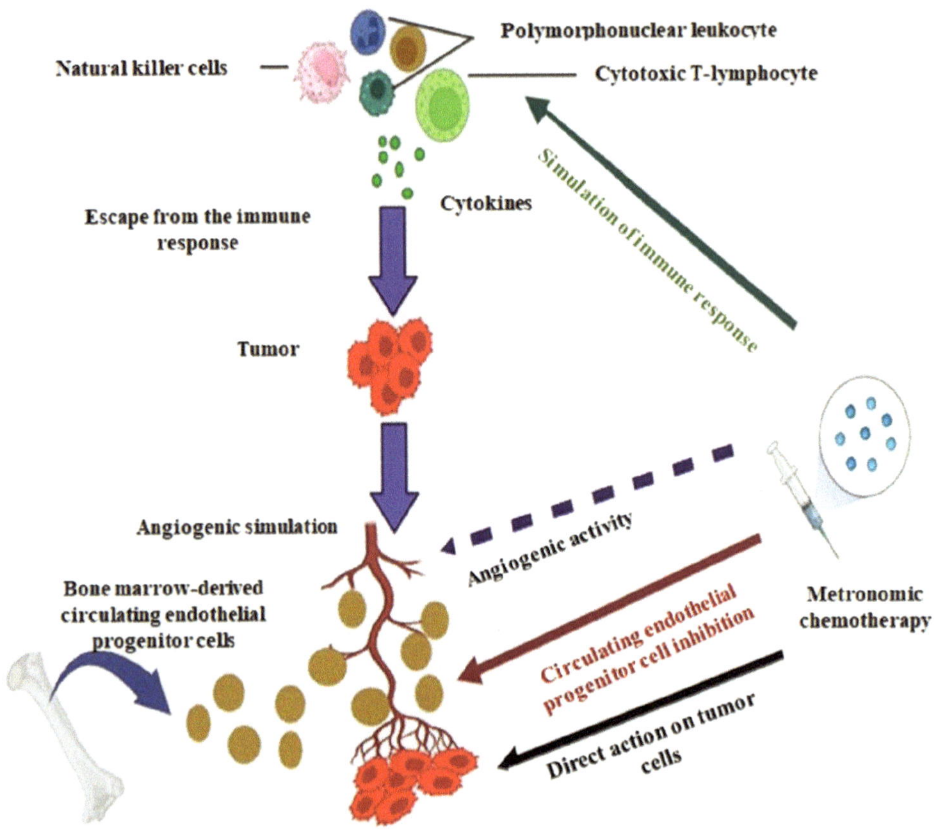

FIGURE 8.1 Mechanism of metronomic chemotherapy.

it is important to note that compared to standard MTD regimens, the overall dosages given for long-term metronomic therapy can be the same or considerably greater. The difference between conventional and non-conventional chemotherapy is shown in Table 8.1.

Many medications that were not designed to treat cancer are now used in conjunction with metronomic chemotherapy in clinical use [29, 30]. Repurposing or repositioning a medicine is using it for a different medical use than it was originally intended for; this strategy has many benefits over the standard drug development process, including reduced risk and reduced development time and expense [29–31]. In several therapeutic trials, for instance, COX-2 inhibitors were administered in conjunction with metronomic chemotherapy. Drugs including rapamycin, thalidomide, valproic acid, statins, metformin, and retinoic acid are among the 15 utilized in metronomic regimens. Metronomic chemotherapy and drug repositioning have a lot in common. Both are relatively inexpensive and safe even when used for extended periods of time. For these reasons, we suggest the more general term "metronomics" to describe any anti-cancer therapy regimens that utilize both medication repositioning and metronomic chemotherapy [32].

8.3 A MULTI-TARGETED THERAPY

Metronomic chemotherapy was originally thought of as an anti-angiogenic treatment, but it has now expanded to include other targets. The clinical use of metronomic chemotherapy might benefit from a deeper understanding of its complicated mechanisms of action. Figure 8.2 depicts a summary of metronomic chemotherapy's many anticancer effects.

TABLE 8.1
Conventional and Non-Conventional Chemotherapy in Cancer

Features	Conventional	Non-Conventional	References
Based on	Maximum tolerated dose	Metronomic	[16, 22, 23]
Schedule	High doses/relative free long intervals between doses	Frequent administration of low doses/ short drug free intervals	[6, 11, 24]
Therapeutic effect	Mainly tumor cell death	Tumor cell and cancer stem cell death/ tumor dormancy/anti-angiogenic effects/host immunostimulation	[25, 26]
Adverse effects	Normal cell death/systemic toxicity/resistance	No or low incidence of toxicity/ resistance	[23, 27]
Chemotherapeutic	Cytotoxic drugs/hormones/ antibodies	Cytotoxic drugs (cyclophosphamide, capecitabine, etoposide, and vinorelbine)/repositioning drugs (propranolol, thalidomide)	[28]

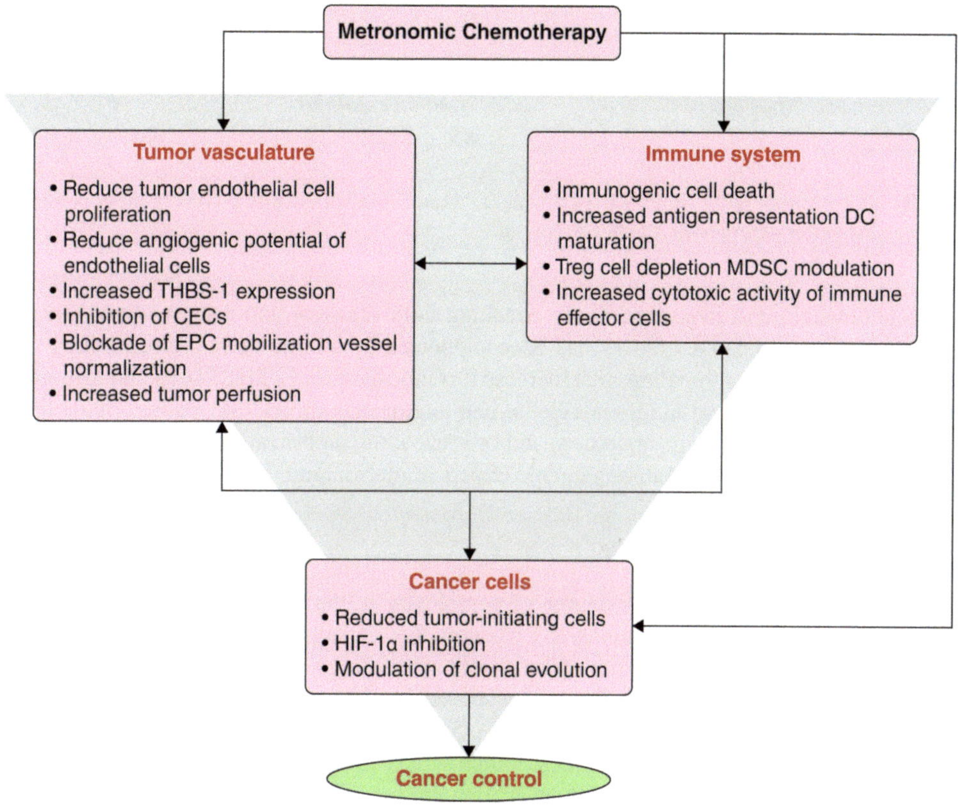

FIGURE 8.2 Metronomic chemotherapy as a multi-targeted therapy.

8.4 ANTI-ANGIOGENIC EFFECT

The initial idea of Folkman and Kerbel underpins the traditional anti-angiogenic paradigm of metronomic chemotherapy [6, 7]. Metronomic chemotherapy may be able to damage existing arteries and inhibit neo-angiogenesis, resulting in hypoxia and food deprivation, by targeting fast-proliferating

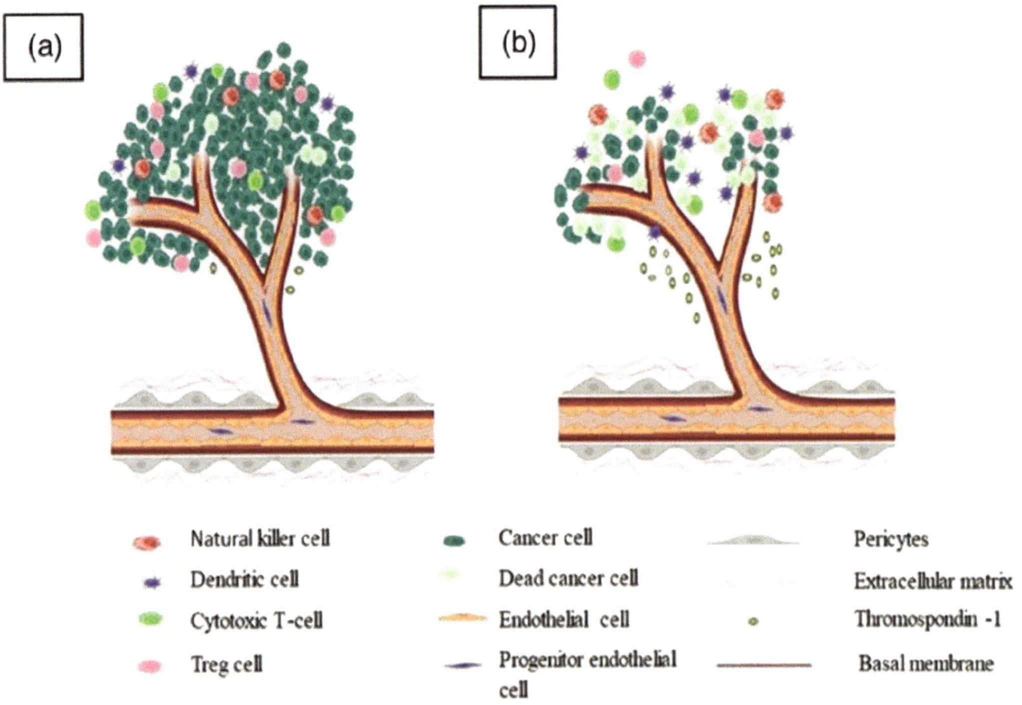

FIGURE 8.3 A tumor and its microenvironment are depicted schematically (a) before metronomic chemotherapy and (b) after metronomic chemotherapy.

tumor endothelial cells, hence indirectly attacking both sensitive and drug-resistant cancer cells, as shown in Figure 8.3. This theory has been supported by a large body of preliminary research [5, 33]. Drugs or drug combinations that increase thrombospondin 1 (THBS1), an endogenous angiogenesis inhibitor, impair vascular endothelial cell angiogenic ability [34], cause circulating endothelial cells (CECs) to undergo apoptosis, and/or prevent the mobilization of endothelial progenitor cells. It can also mediate the anti-angiogenic effects of metronomic chemotherapy [9, 10, 35].

8.5 IMMUNOMODULATION

There is a wide range and a high degree of complexity in the immunological effects of chemotherapy medicines [36, 37]. Chemotherapy has complex impacts on the immune system that vary in terms of factors such as medication, dosage, and scheduling [37, 38]. Growing research suggests that immunological responses that fight against cancer are important for keeping cancer under control after chemotherapy has been administered [15]. Shifting the immunological scales away from immunosuppression and toward immunostimulation is a primary goal of metronomic chemotherapy. First, metronomic chemotherapy induces immune-mediated apoptosis of tumor cells [39]; second, metronomic chemotherapy regulates dendritic cells (DCs) to improve antigen presentation [40] and increases the ability of cancer cells to elicit an immune response [41]; third, metronomic chemotherapy selectively eliminates T_{REG} cells [42–45]; fourth, metronomic chemotherapy improves the function of myeloid-derived suppressor cells (MDSCs) [46]; and finally, metronomic chemotherapy improves the cytotoxic activity of effector immune cells like $\gamma\delta$ T cells [47] and tumor-specific T cells [48, 49]. Metronomic chemotherapy methods can be optimized to reduce immunosuppressive effects while increasing therapeutic efficacy if more is known about the underlying processes. These results also lend credence to the idea that immunotherapy and metronomic chemotherapy approaches like $\gamma\delta$ T-cell–based treatment and cancer vaccines should be used together. Breast cancer, melanoma, and lung cancer pre-clinical models have shown synergy between metronomic chemotherapy and immunotherapy, including the usage of HER2 and vascular endothelial growth factor (VEGF) peptide mimics as well as DNA vaccinations [50–52].

To keep the immune system stimulated with metronomic cyclophosphamide, Chen and his coworkers demonstrated that compared to once-a-day dosing, a pulsed regimen showed much higher efficacy [53]. Similarly, in a mouse model of colorectal cancer, a single treatment of low-dose cyclophosphamide was superior to metronomic cyclophosphamide (thrice weekly) in improving the effectiveness of interleukin-12 gene therapy [54]. This data suggests that slightly varied schedules or many medications with various dose regimens may be required to optimize the anti-angiogenic and immunological benefits of metronomic regimens. The cyclophosphamide and methotrexate regimen for breast cancer patients is one example of a therapeutic plan that combines daily and less frequent medication delivery [55]. The total response rate was 19%, while the clinical benefit was 32% due to this method.

8.6 IMPACT ON CANCER-CAUSING CELLS

There is mounting proof suggesting metronomic chemotherapy can affect cancer cells directly. It was found that glioma cells isolated from drug-treated tumors generated fewer primary and secondary tumor spheres when treated with metronomic cyclophosphamide [56]. In addition, metronomic cyclophosphamide treatment of human pancreatic tumor xenografts decreased the concentration of CD133+ progenitor cells and triple-positive CD24+/CD44+/CD133+ cancer stem cells [57]. When cancerous cells of the colon and prostate were exposed to paclitaxel and etoposide in vitro for extended periods, drug-resistant cells developed as a result of epigenetic processes [58]. Interestingly, fewer CD44+ cells were present in the drug-tolerant population in vivo compared to the original cell population, suggesting that this subset of cells was less tumorigenic. These results show that metronomic chemotherapy can effectively kill cancer stem cells and the cells responsible for starting tumors (CD44+). It has been established time and time again that sensitizing leukemia cells in vivo to minimal doses of cisplatin requires dual blockage of CD44 and CD49d [59].

8.7 ADDITIONAL MECHANISMS

Metronomic chemotherapy has been found in several investigations to improve tumor perfusion and normalize blood vessel function. Using metronomic gemcitabine or LY2334737, a prodrug of gemcitabine for the treatment of breast, pancreatic, and ovarian cancer models [60, 61], and metronomic cyclophosphamide for the treatment of xenografted glioma [62, 63] showed improved circulation and/or oxygenation in the tumor. As a result, the concentration of oxygen in the tumor's blood has been suggested as a surrogate sign for finding the best metronomic dosage. Targeted suppression of hypoxia-inducible factor 1 (HIF-1) is another strategy of metronomic treatment. Melillo and his colleagues found that camptothecin and topotecan, two topoisomerase I inhibitors, significantly reduced the transcriptional activation of HIF-1α [64]. As a result, a novel pilot clinical study with specific objectives of metronomic topotecan for individuals with metastatic solid tumors was developed [65]. Researchers showed that metronomic topotecan induced alterations in the expression of HIF-1α and genes regulated by HIF-1 in a subset of patient tumor samples. In addition, doxorubicin and daunorubicin were shown to be effective inhibitors of HIF-1 after being examined alongside Food and Drug Administration (FDA)–approved and clinically tested medicines [66]. Treatment with these medicines, even at modest doses and regularly, significantly decreased tumor vascularization and transcription of HIF1 target genes in animal models. In addition, resistance to VEGF suppression in sarcoma xenografts was reversed by metronomic doxorubicin by reducing HIF-1 binding to DNA [67], offering another reason for the combination of antiangiogenic medicines with metronomic chemotherapy.

Chemotherapy administered at regular intervals may also impede cancer clonal evolution. Each tumor is made up of a diverse population of cells that share some genetic material but are otherwise unique. These duplicates vie with one another for access to the same materials. Cancer cells that exhibit phenotypic resistance to chemotherapy tend to proliferate less when they are surrounded by "fitter" chemosensitive cells. To prevent the spread of the resistant population, Gatenby and his coworkers advise using chemotherapy at dosages below the MTD to preserve a critical mass of susceptible cells within the tumor [12]. Due to its use of sub-cytotoxic doses, metronomic chemotherapy has the potential to slow the growth of the most drug-resistant clones, opening the door to chemo-switch approaches that alternate metronomic and increased chemotherapeutic dosages [68]. Recent research by Li and his colleagues suggests that metronomic chemotherapy may also inhibit subclonal switching signals, thus preventing quiescent subclones (like cancerous stem cells) from becoming dominant and causing tumor growth and post-treatment recurrence [69].

Potentially effective against even the most advanced forms of metastatic illness, metronomic chemotherapy works by disrupting the metastatic cascade. Hepatocellular carcinoma; neuroblastoma; pancreatic, ovarian, and colorectal cancer xenografts; and spontaneous metastatic models have all indicated that metronomic chemotherapy, either separately or in conjunction with inhibitors of tyrosine kinase, can prevent metastatic dissemination [70–74]. This leads us to hypothesize that this strategy can interfere with a number of the biological activities essential to metastasis, including resistance to anoikis, cell movement, extravasation, intravasation, invasion, and adaption to a new milieu. Metronomic chemotherapy may be beneficial in the adjuvant situation, as suggested by these preclinical investigations, even in the presence of preexisting metastases.

Figure 8.2 shows how metronomic treatment can have an immediate impact on the tumor microvasculature, the immune system, and cancerous cells. Moreover, the anti-cancer effects of this treatment method can be amplified by the interaction between these three compartments. Inhibiting angiogenesis, for instance, may disturb the stem cell niche, hence indirectly hindering the maintenance and expansion of cancer stem cells [75]. Numerous factors, including VEGF, interleukin-17, leukemia inhibitory factor, and transforming growth factor (TGF)β, mediate the mutual interactions directly and indirectly between cancer stem cells and T_{REG} and may be influenced by metronomic treatment [76]. It is widely established that inflammation promotes tumor angiogenesis. Stimulated by HIF1-α, tumor-associated macrophages (TAMs) secrete angiogenic mediators like interleukin 8 (IL-8) and VEGF [77]. These in vitro investigations reveal the intricate workings of metronomic chemotherapy and provide light on its multifaceted character. Because of this intricacy, several agents will be deployed at various times to enhance various action mechanisms [53].

8.8 RESISTANCE MECHANISM

Although metronomic chemotherapy is effective in drug-refractory tumors in humans [15] and in eradicating tumors in mouse models [5], it is frequently met with resistance from both intrinsic and acquired mechanisms, resulting in the development of the illness in a large number of individuals [32]. To increase the effectiveness of treatment and decrease toxic effects, it is essential to investigate the causes of resistance or cancer growth during metronomic therapy.

New research suggests that different pathways contribute to metronomic chemotherapy resistance vs. resistance to standard chemotherapy. Initially, it is a common clinical observation [78] that tumor cells that develop resistance to a particular chemotherapy treatment when it is supplied at the MTD typically retain sensitivity to a similar medication when it is provided in a metronomic form [79, 80]. In addition, tumor cells in the prostate that acquired in vivo resistance to the chemotherapy drug metronomic cyclophosphamide remained sensitive to MTD cyclophosphamide [79] and using metronomic cyclophosphamide in conjunction with the MTD chemotherapeutics [81]. Third, it was shown that administering docetaxel intermittently (one or three times in a week) upregulated many genes associated with drug resistance, but constant docetaxel therapy did not [82]. These results are consistent with those observed by Pasquier and colleagues [34]. They studied the effects of prolonged low-dose or MTD etoposide and vinblastine exposure on the expansion of vascular endothelial cells. Metronomic chemotherapy, in contrast to MTD chemotherapy, greatly reduced endothelial cells' angiogenic capacity and, unexpectedly, improved their sensitivity to MTD chemotherapy, suggesting that the two treatments may have additive effects.

Resistance to metronomic treatment may be facilitated by cancer stem cells. After administration of cyclophosphamide to xenografts of orthotopic hepatocellular cancer, Martin-Padura and coworkers showed that recurrence in tumors was caused by latent CD13+ cancer stem cells [83]. Bestatin, an inhibitor of CD13, combined with metronomic cyclophosphamide dramatically enhanced cancer shrinkage and decreased the amount of remaining cancer cells in the liver [83]. Similarly to its well-characterized involvement in evading resistance to anti-angiogenic treatment, hypoxia may contribute to metronomic chemotherapeutic resistance, as shown by indirect evidence. Breast, colon, gliosarcoma, and prostate cancer models [84] showed synergism between tirapazamine, a hypoxia-activated prodrug, and metronomic cyclophosphamide.

Acquired resistance to metronomic treatment is beginning to reveal its molecular pathways. The initial proteomics study of cyclophosphamide-resistant tumor cells of the prostate was conducted by Thoenes and coworkers [85]. They found that the expression of cathepsin B, thioredoxin domain-containing protein 5 (TXNDC5), and annexin A3 (ANXA3) was higher in resistant cells compared to parental cells. It is interesting to note that ANXA3 was elevated in drug-resistant cells according to a thorough gene expression study [86] that was investigating the molecular basis of tumor cells of the prostate resistant to metronomic administration of cyclophosphamide. Coagulation signaling pathways, complement system, steroid biosynthesis, and axon guidance were also identified to have differently expressed genes. TXNDC5 has recently been linked to the development of several different tumors, including those of the cervix, stomach, colon, lungs, breast, and liver [87–90]. Cathepsin B release and consequent inflammasome activation in MDSCs have been shown to reduce the efficacy of MTD chemotherapy [91]. Collectively, these trials' findings imply that metronomic chemotherapy could benefit from targeting cathepsin B, TXNDC5, and ANXA3.

8.9 CLINICAL EXPERIENCE IN ADULTS

Multiple studies have corroborated the promising results of metronomic chemotherapy in treating prostate, breast, and ovarian carcinoma and have expanded the scope of its prospective use. The use of metronomic chemotherapy separately or combined with more traditional chemotherapeutic agents, immunotherapeutic methods, radiotherapeutic techniques, targeted medications, and repurposed pharmaceuticals has been used to treat over 2,500 individuals with various kinds of resistant, advanced, relapsed, or metastatic cancers in the previous 4 years.

8.10 METRONOMIC CHEMOTHERAPY ALONE

Metronomic chemotherapy regimens have been the subject of more than 15 published research studies in the last 4 years [6, 92–108]. As interest in metronomic treatment grows, medicines that target microtubules – like the orally administered vinca alkaloid vinorelbine – are of special interest because of their powerful anti-angiogenic and pro-immune characteristics at lower dosages [109, 110]. Previously, vinorelbine has been administered in weekly doses to patients with non-small cell lung cancer (NSCLC) and breast cancer [110]. Vinorelbine has a half-life in the body of 49 hours; thus, it may be best to administer it more frequently at regular intervals to get the most out of it. The optimal dosage of 50 mg vinorelbine for use at thrice-weekly intervals was determined by two separate phase I studies. Age-related activity was later documented in individuals with metastatic breast cancer [95] and relapsed/refractory NSCLC. Individuals with advanced breast cancer have also benefited from Navelbine when combined with capecitabine [96]. The combination was well tolerated in this phase I study and demonstrated significant effectiveness with a response rate of 33% [96].

Capecitabine's orally bioavailable form makes it an attractive choice for metronomic chemotherapy, and the wide spectrum of action of fluoropyrimidines is well recognized [111]. Capecitabine, when given daily, was active and well tolerated in two phase II studies including individuals with breast cancer [98, 100]. As has been proven with etoposide, cyclophosphamide, and temozolomide [98], metronomic treatment may overcome resistance to MTD chemotherapy, and its effectiveness has been seen even in patients who had received the same medication at conventional dosages in the past [112, 113]. El-Arab and coworkers showed the efficacy of a metronomic administration of cyclophosphamide and capecitabine to individuals with metastatic breast carcinoma in a developing nation, achieving a 57% clinical benefit and 22% overall response rate. This further confirms the potential of this approach in resource-limited countries.

Several metronomic temozolomide regimens have been tried because of the agent's potential anti-angiogenic and pro-immune effects, as shown by preclinical research. Metronomic temozolomide has been studied for the first time as a single drug in two recent studies as a means of preventing recurring high-grade gliomas. Patients with glioblastoma who did not react to adjuvant temozolomide were shown to tolerate metronomic temozolomide well; however, it did not improve survival. Omuro and his coworkers, in contrast, found substantial therapeutic action in a comparable patient cohort, with 36% of patients reporting clinical benefit. While the latter trial employed temozolomide constantly every day at a dose of 50 mg/m^2, the former followed a procedure that involved taking temozolomide every day for 7 days at a dose of 130 mg/m^2, with 7-day drug-free intervals every week; these findings further emphasize the requirement for optimizing metronomic chemotherapy dosages and administration times for maximum effectiveness. Metronomic chemotherapy has recently evolved to emphasize the use of several drugs to improve activity and response rate. Using a metronomic schedule of cyclophosphamide and methotrexate, Gebbia and coauthors found promising outcomes in individuals with recurrent breast carcinoma or metastatic prostate carcinoma. Response rates in colorectal carcinoma and in recurrent metastatic breast cancers were from 47% to 49%, while clinical benefits ranged from 84% to 97% when irinotecan was combined with oral fluoropyrimidines S1 or tegafur-gimeracil-oteracil potassium every week. A randomized experiment found that individuals who received uracil-tegafur as maintenance for 2 years had a considerably higher 2-year survival rate than those who did not get the drug (88% against 85%) after having stage I lung adenocarcinoma surgically removed [114].

8.11 METRONOMIC CHEMOTHERAPY COMBINATION REGIMENS

More effective and less toxic chemotherapy and/or radiation therapy regimens have been developed by combining metronomic and MTD approaches. A phase II study looked at the effectiveness of combining metronomic chemotherapy and radiation in treating 36 patients with breast cancer with brain metastasis and found the findings to be encouraging [115]. Treatment began with temozolomide and whole-brain radiation and was then followed with vinorelbine three times weekly and temozolomide once a day for 4 weeks. They saw a response rate of 52% (with three full responses) and overall clinical benefit of 77% [115], which was higher than the expected goal activity. Overall, patients lived for 11 months on average, whereas progression-free survival (PFS) was 8 months. This group of patients, who typically had substantial morbidity and a 1-year life expectancy of 20%, also had a notable increase in quality of life, while side effects were minor overall [116]. With surgery, metronomic temozolomide, and radiation treatment, individuals with newly diagnosed glioblastoma had a 50% 1-year PFS with carmustine wafers placed after chemotherapy [117]. Immunosuppression caused by temozolomide was the leading cause of treatment termination in 20% of patients, while 63% of patients experienced side effects of some kind. Temozolomide's anti-angiogenic and immunological properties can be preserved at a lower dosage, which may help reduce side effects. In a current phase III research study, individuals with squamous cell lung cancer are being randomly assigned to either a conventional MTD etoposide and cisplatin regimen or the same regimen combined with metronomic cyclophosphamide [118].

Even though pegylated liposomal doxorubicin and metronomic treatment showed significant clinical effectiveness and a good security record in breast cancer [119, 120], administering MTD chemotherapy along with metronomic chemotherapy concurrently does not appear to be a viable approach due to higher toxic effects [121]. As an alternative, a so-called chemo-switch regimen might mix metronomic chemotherapy with MTD treatment in a sequential fashion [68]. Patients with advanced renal cell carcinoma were included in a phase II trial to examine the effectiveness of this strategy [122]. Patients treated with sorafenib, metronomic capecitabine, and MTD gemcitabine had a median PFS of 11 months, a 92.5% clinical benefit, and a 50% response rate. Metronomic chemotherapy is used in conjunction with radiotherapy to lessen the overall radiation exposure a patient receives while still providing a systemic benefit [123]. There have been some promising outcomes from using this method, and it merits more investigation in future comparison research.

8.12 METRONOMIC CHEMOTHERAPY AND TARGETED THERAPY

As was previously mentioned, anti-tumor effectiveness can be boosted by combining metronomic chemotherapy with anti-angiogenic drugs [6, 7]. Several ongoing clinical investigations (Table 8.2)

TABLE 8.2
Targeted and Metronomic Chemotherapy in Adults

Medication Therapy	Disease Treated	No. of Individuals	Study Type	Best response	Efficiency	References
Bevacizumab + etoposide + cisplatin	Advanced non-small-cell lung cancer (NSCLC)	45	Phase II	PR: 31/45	CB: 87% RR: 69%	[133]
Cyclophosphamide + imatinib + bevacizumab	Advanced colorectal cancer	35	Phase I	SD: 7/35	CB: 20% RR: 0%	[132]
Cyclophosphamide + bevacizumab	Metastatic breast cancer	13	Phase II	PR: 1/13	CB: 20% RR: 0%	[131]
Erlotinib + capecitabine + bevacizumab	Metastatic HER2 breast cancer	24	Phase I/II	PR:14/24 CR:1/24	CB: 83% RR: 62%	[130]
Cyclophosphamide + methotrexate + vandetanib	Metastatic breast cancer	23	Phase I	PR: 2/20	CB: 25% RR: 10%	[129]
Paclitaxel + sorafenib	Advanced refractory adrenocortical carcinoma	10	Phase II	PD: 9/9	CB: 0% RR: 0%	[126]
Bevacizumab + temozolomide or etoposide	Recurrent glioblastoma	23	Phase II	SD: 12/23	CB: 52% RR: 0%	[125]
Sorafenib + temozolomide	Relapsed glioblastoma	43	Phase II	PR: 5/43	CB: 53% RR: 12%	[124]
Gefitinib vs gefitinib + tegafur/uracil	Refractory/relapsing pulmonary adenocarcinoma	Arm G: 58 Arm GU: 57	Randomized phase II	PR: 20/58 PR: 21/57	CB: 66% RR: 34% CB: 75% RR: 37%	[134]
Cyclophosphamide + bevacizumab	Recurrent ovarian cancer	38	Retrospective	PR: 12/37 CR: 3/37	CB: 49% RR: 40%	[128]
Cyclophosphamide + bevacizumab	Platinum-resistant recurrent ovarian cancer	66	Retrospective	PR: 21/66 CR: 7/66	CB: 65% RR: 42%	[127]

aim to verify this result and expand the application of metronomic chemotherapy in tandem with various forms of targeted treatment [124–138]. Bevacizumab is typically used in conjunction with metronomic chemotherapy. Bevacizumab in conjunction with metronomic cyclophosphamide has been studied in two retrospective trials for the treatment of recurrent ovarian cancer that has not responded to previous therapies [127, 128]. There was a significant improvement in overall survival and PFS for responders (median survival was 20 months compared with 9 months for responders and non-responders, respectively; P = 0.004) in this heavily pretreated patient population: 49–65% in terms of overall clinical benefit and 10–12 months of median overall survival.

Patients with HER2-negative metastatic breast cancer were evaluated for their response to metronomic cyclophosphamide and capecitabine in concert with erlotinib and bevacizumab [130]. Overall, 75% of patients experienced clinical improvement from this treatment plan, and the median time to progression was 11 months. These outcomes appear to be better than those seen with bevacizumab and erlotinib combined in individuals with relapsed metastatic breast cancer, although there was little difference in the population of patients (this means individuals affected had metastatic HER2-negative breast cancer and were treatment-naïve compared to those with recurrent metastatic breast cancer) [139].

Although the mixture of high-dose cisplatin, metronomic etoposide, and bevacizumab was clinically effective in 87% of patients with advanced-stage NSCLC, it was associated with severe toxic effects that necessitated careful patient selection and thus diminished the benefits of metronomic chemotherapy [133]. Individuals with recurrent glioblastoma were recruited for a phase II trial comparing metronomic temozolomide with sorafenib [124]. Clinical action was modest, with 7.4 months of median overall survival and 3.2 months of median PFS; similar outcomes have been observed using metronomic temozolomide separately in the identical clinical environment.

Patients with lung adenocarcinoma were evaluated in a randomized phase II study that assessed the effectiveness of gefitinib separately and when combined with metronomic tegafur-uracil [134]. Overall, the response rate was not improved with the use of metronomic chemotherapy; however, the 1-year PFS was increased from 18% to 36% (P = 0.03). Kelley and his coworkers demonstrated the MTD of metronomic cyclophosphamide, imatinib, and bevacizumab in individuals with late-stage colon cancer, and 7 of 35 individuals experienced stable illness after more than 6 months [132]. An increase in imatinib exposure was seen, albeit only modestly, when cyclophosphamide was used. Patients with lymphomas and resistant solid tumors responded well to a regimen of metronomic cyclophosphamide combined with the veliparib inhibitor of poly(ADP-ribose) polymerase (PARP) [136]. In light of the encouraging findings from this phase I investigation, two phase II studies comparing the effectiveness of metronomic cyclophosphamide separately and in conjunction with veliparib in individuals with lymphoma, breast cancer, and ovarian carcinoma have begun. The combination of vandetanib plus a standard chemotherapy schedule (cyclophosphamide every day and methotrexate every week) for metastatic breast cancer was well tolerated, but had low activity (overall clinical benefit of 25% and response rate of 10%) [129]. It has been demonstrated that several metronomic chemotherapy regimens, like administration of paclitaxel weekly and sorafenib in adrenocortical cancer [126], recurrent glioblastoma treated with bevacizumab with metronomic temozolomide or etoposide [128], and bevacizumab with metronomic vinorelbine in the treatment of breast carcinoma [131], are ineffective. The most recent research study's disappointing findings might be explained by the fact that the response rate was employed as an outcome measure instead of the more conventional clinical benefit rate (53%) used in metronomic studies. Clinical data shows that combining metronomic chemotherapy with targeted therapy does not significantly enhance the toxicity of targeted medicines and may even improve their efficacy. To further verify these early results, randomized clinical investigations are required.

8.13 METRONOMIC CHEMOTHERAPY AND IMMUNOTHERAPY

Researchers in the field of cancer immunology have been using metronomic scheduling of anti-cancer drugs to modify immunity for decades [140]. However, it is well acknowledged that there is

TABLE 8.3
Immunotherapy and Metronomic Chemotherapy in Adults

Disease Treated	Medication Therapy	No. of Individuals	Study Type	Best Response	Efficiency	References
Unresectable pancreas/lung/ prostate cancer	Standard chemotherapy	Arm G1: 30	Randomized phase I/II	NA	NA	[146]
	Standard chemotherapy + Cyclophosphamide + Celecoxib + G-CSF+ n-acetyl cysteine + Tumor antigens	Arm G2: 30		NA	NA	
Refractory tumors	Oncolytic virus + MTD and MC cyclophosphamide	7	Pilot	NA	CB: 77% RR: NA	[144]
	Oncolytic virus + MTD cyclophosphamide	7		NA	CB: 55% RR: NA	
	Oncolytic virus + MC cyclophosphamide	21		NA	CB: 61% RR: NA	
	Oncolytic virus	8		NA	CB: 22% RR: NA	
Refractory tumors	Temozolomide +Oncolytic virus	17	Retrospective	PR:1/17	CB: 76% RR: 6%	[143]
Metastatic renal cell carcinoma	Pioglitazone + Etoricoxib + IFN + Capecitabine	45	Phase II	PR:12/45 CR:4/45	CB: 76% RR:36%;	[149]
Metastatic melanoma	DC vaccine + IL2 + cyclophosphamide + celecoxib	28	Phase II	SD:16/28	CB: 57% RR: 0%	[147]

Abbreviations: RR, response rate; CB, clinical benefit; MTD, maximum tolerated dose.

a communication gap among oncology specialists such as immunologists, biologists, pharmacologists, and clinicians [141]. The infrequent usage of the term "metronomic" in studies and review publications concentrating on the immunological consequences of chemotherapy [15] in the context of metronomic chemotherapy serves as a demonstration of this, and the immune system's significance in metronomic chemotherapy's mode of action was only recently discovered [142]. Metronomic chemotherapy and its combination with immunotherapy have not progressed as quickly as they could due to a lack of communication across different areas of study. However, new clinical research suggests that this strategy has great potential to enhance and maintain the anti-tumor effectiveness of metronomic chemotherapy (Table 8.3) [143–149].

In order to determine whether "an induction regimen of anti-angiogenesis and antitumor immunity" improves the outcomes of individuals with cancers with a poor prognosis and late stages, Lasalvai-Prisco and his colleagues devised a randomized trial to accomplish just that. The induction protocol included autologous tumor antigens, granulocyte colony-stimulating factor, celecoxib, metronomic cyclophosphamide, and N-acetylcysteine [146]. Overall survival was improved in all three patient populations after the integration of immunotherapy and metronomic chemotherapy ($P = 0.036$ for pancreatic cancer, $P = 0.042$ for NSCLC, and $P = 0.048$ for hormone-refractory prostate cancer). There was no significant difference in quality of life or negative impacts between the two groups of therapy, and the longer life expectancy was linked to changes including lower T_{REG} numbers and VEGF expression, higher angiostatin levels, and activated DC numbers. Since PFS and response rates were not disclosed, it was not possible to draw any firm conclusions on the success of this approach.

Patients with metastatic melanoma who received DC immunization and IL-2 with metronomic cyclophosphamide and celecoxib significantly improved total life expectancy at 6 months than in a prior experiment that lacked cyclophosphamide and celecoxib [147]. Despite using metronomic cyclophosphamide, this combination did not reduce T_{REG}, which would have been predicted to happen as a result of its improved effectiveness. Surprisingly low response rates (14% in patients with breast cancer) were accompanied by a remarkable event-free survival (EFS) of 10 months and a total life expectancy of 13 months when an anti-idiotype immunization was added to standard dosing of cyclophosphamide and methotrexate [148]. In individuals with metastatic breast cancer, the prognosis was poorer for those treated with an oncolytic virus alone compared to those treated with metronomic cyclophosphamide, MTD cyclophosphamide, or both [144]. At 1 year, a significant 77% clinical benefit rate was seen, and the chemo-switch schedule emerged as the most effective treatment plan in this varied patient group. However, the mechanisms behind these changes are still a mystery. The reduction in T_{REG} attained without affecting cytotoxic T-cell responses in the groups administered metronomic cyclophosphamide may be connected to this. The direct anti-cancer impact of high-dose cyclophosphamide may have helped the "chemo-switch" group as well.

In individuals with metastatic renal cell cancer that is progressing or recurring, it was found that the combination of the metronomic capecitabine, interferon (IFN)-α (an immunomodulator), etoricoxib (an anti-inflammatory agent), and peroxisome proliferator-activated receptor gamma agonist (pioglitazone) resulted in an extremely significant rate of response of 36% and 76% of clinical benefit [149].

Although clinical evidence is limited, it appears that combining immunotherapy with metronomic chemotherapy is an effective strategy that helps to boost the efficacy of immunotherapy, and the therapeutic impact of metronomic chemotherapy is extended.

8.14 PERSONALIZED MEDICINE AND METRONOMIC ANALYSIS

The rise of personalized therapy is a very encouraging trend in cancer. This model proposes tailoring cancer care to each patient by determining which medications to use depending on the presence of molecular targets in the tumor. In the future era of precision medicine, metronomics is expected to perform a significant role.

Metronomic chemotherapy is a suitable method in elderly or fragile individuals or people who are intensively treated when adjusting treatment to the overall condition of the patient. In older breast cancer individuals, the combination of letrozole plus metronomic cyclophosphamide showed an 88% response rate in a randomized phase II study [150]. In unresectable sarcoma among older individuals, metronomic cyclophosphamide exhibited signs of potential success, achieving a 69% clinical benefit rate and a 27% response rate [151]. Thirty individuals who had late-stage hepatocellular cancer and severe thrombosis of the portal vein were evaluated for the security and effectiveness of a 5-fluorouracil (5-FU) regimen and metronomic cisplatin, and a clinical benefit of 40% was noted [152]. The results of this research suggest that metronomic treatment may be of great benefit to those with low-performance status and/or many medical conditions.

In oncology, the notion of personalized medicine refers to individualizing therapies for each patient diagnosed with a specific malignancy. The phrase "at a given time" is a vital qualifier that is often left out. The epigenetic and genomic landscape of a particular cancer can be diagnostically distinct, depending on post-treatment and recurrence status, as shown by the intricate and multi-faceted pathways of resistance to therapy in melanoma [153]. This landscape undergoes continuous change throughout therapy, consistent with principles of evolutionary adaptation to selection pressure. Thus, the therapeutic environment is crucial for the success of metronomic treatment. For instance, vinblastine is a successful therapy for patients with recurrent anaplastic large cell lymphoma [154], with a full response rate of 80% and a 50% long-term survival rate. Contrary to expectations, a randomized study in a comparable patient group found that vinblastine as a maintenance drug did not enhance survival [155], demonstrating the critical relevance of the clinical environment for the application of metronomic chemotherapy [156].

According to the results of certain clinical trials, metronomic regimens may not be able to be modified for all varieties of tumors. Over 50% of disease control has been reported with the use of either capecitabine separately or in combination or the combination of cyclophosphamide-methotrexate [55] in the management of breast cancer. Metronomic etoposide orally every day [157] or weekly Navelbine separately [158] or when combined with cyclophosphamide [159] or the more sophisticated Combat II regimen, where etoposide and Navelbine were also used in the rhabdomyosarcoma arm, appears to be effective in treating children with rhabdomyosarcoma [160]. However, when temozolomide was left out of the treatment plan for relapsed osteosarcoma [158, 160] and glioblastoma, metronomic chemotherapy failed to improve survival rates [161, 162].

It's worth noting that the phrase "personalized" isn't always applicable because precision medicine rarely considers the patient's individual, family, cultural, and socioeconomic background [163]. Since metronomics is low-cost, well-tolerated, and simple to administer, it can be adapted to meet the needs of patients who prefer oral medication or who live in areas where social and/or economic constraints prevent them from accessing care meeting the standards set by those in high-income countries [32].

8.15 CONCLUSION

Metronomic chemotherapy is still relatively new, having been developed about 15 years ago, but it is gaining popularity among cancer biologists, pharmacologists, and doctors. Drug repositioning, targeted treatments, and immunotherapy, in addition to this method, have recently been discovered to have a multi-targeted character, which has opened the way for the development of personalized precision chemotherapy. There is no need to pit metronomic chemotherapy against targeted treatments, with the former representing the more "traditional" and "toxic" method of cancer treatment and the latter the more "modern" and "less toxic" one. When used together, however, these strategies have the potential to increase therapy efficacy and expand how anti-cancer medications can be used. However, it is still up for discussion whether metronomic treatments need to be personalized or kept as a straightforward, user-friendly, all-purpose therapeutic approach. There is growing concern, both domestically and internationally, over the rising expense of cancer treatment. Medicines in both high- and low-income nations may be produced at a lower cost thanks to metronomics, which makes use of off-patent medications through drug repositioning. Accumulating clinical and pre-clinical data backs up metronomics' potential to revolutionize medicine with individualized precision. Now more than ever, worldwide cooperation is required to fully realize the promise of metronomics by investigating the numerous possible medication combinations, the wide variety of illnesses, and the many treatment contexts.

For further reading, some textbooks mentioned here also give a further understanding of metronomic chemotherapy:

- *Progress in Anti-cancer Chemotherapy* by Gabriel N. Hortobagyi and David Khayat.
- *Triple-Negative Breast Cancer: Biomarkers, Emerging Therapeutic Strategies and Clinical Challenges* by Marion Montgomery.
- *Current Cancer Treatment: Novel Beyond Conventional Approaches* by Oner Ozdemir.
- *Breast Cancer: Current and Alternative Therapeutic Modalities* by Esra Gunduz and Mehmet Gunduz.

The following links for video lectures may also be helpful:

https://youtu.be/TAqO0Mq19JQ
https://youtu.be/6wEnYfvoBqk
https://youtu.be/A9fRRCUlMms
https://youtu.be/zIWvmCsKr34

For more insight about the topic, we would suggest detailed findings from the books of (Mir MA, 2022) https://doi.org/10.1016/C2021-0-02565-7, https://doi.org/10.1016/C2022-0-00074-X, (Mir MA, 2021) https://doi.org/10.52305/WXJL6770, and (Mir MA, 2015) https://doi.org/10.1016/C2014-0-02898-5 and from the cancer.net website on the following mentioned links:

https://www.cancer.net/cancer-types/breast-cancer/types-treatment
https://discovery.ucl.ac.uk/id/eprint/1472740/
https://www.jmedsciences.com/doi/JMEDS/pdf/10.5005/jp-journals-10045-00138

For diagrammatic illustrations and descriptive tables (Lazzeroni, 2012): http://www.eurekaselect.com/article/49928

REFERENCES

1. Folkman, J., *Tumor angiogenesis: therapeutic implications.* New England journal of medicine, 1971. **285**(21): p. 1182–1186.
2. Miller, K., et al., *Paclitaxel plus bevacizumab versus paclitaxel alone for metastatic breast cancer.* New England journal of medicine, 2007. **357**(26): p. 2666–2676.
3. Hurwitz, H., et al., *Bevacizumab plus irinotecan, fluorouracil, and leucovorin for metastatic colorectal cancer.* New England journal of medicine, 2004. **350**(23): p. 2335–2342.
4. Miller, K.D., C.J. Sweeney, and G.W. Sledge Jr, *Redefining the target: chemotherapeutics as antiangiogenics.* Journal of clinical oncology, 2001. **19**(4): p. 1195–1206.
5. Kerbel, R.S. and B.A. Kamen, *The anti-angiogenic basis of metronomic chemotherapy.* Nature Reviews. Cancer, 2004. **4**(6): p. 423–436.
6. Browder, T., et al., *Antiangiogenic scheduling of chemotherapy improves efficacy against experimental drug-resistant cancer.* Cancer research, 2000. **60**(7): p. 1878–1886.
7. Klement, G., et al., *Continuous low-dose therapy with vinblastine and VEGF receptor-2 antibody induces sustained tumor regression without overt toxicity.* The Journal of clinical investigation, 2000. **105**(8): p. R15–R24.
8. Kim, J.J. and I.F. Tannock, *Repopulation of cancer cells during therapy: an important cause of treatment failure.* Nature Reviews. Cancer, 2005. **5**(7): p. 516–525.
9. Bertolini, F., et al., *Maximum tolerable dose and low-dose metronomic chemotherapy have opposite effects on the mobilization and viability of circulating endothelial progenitor cells.* Cancer research, 2003. **63**(15): p. 4342–4346.
10. Shaked, Y., et al., *Therapy-induced acute recruitment of circulating endothelial progenitor cells to tumors.* Science, 2006. **313**(5794): p. 1785–1787.
11. Shaked, Y., et al., *Optimal biologic dose of metronomic chemotherapy regimens is associated with maximum antiangiogenic activity.* Blood, 2005. **106**(9): p. 3058–3061.
12. Gatenby, R.A., *A change of strategy in the war on cancer.* Nature, 2009. **459**(7246): p. 508–509.
13. André, N. and E. Pasquier, *For cancer, seek and destroy or live and let live?* Nature, 2009. **460**(7253): p. 324–324.
14. Fidler, I.J. and L.M. Ellis, *Chemotherapeutic drugs—more really is not better.* Nature medicine, 2000. **6**(5): p. 500–502.
15. Pasquier, E., M. Kavallaris, and N. André, *Metronomic chemotherapy: new rationale for new directions.* Nature reviews. Clinical oncology, 2010. **7**(8): p. 455–465.
16. Hanahan, D., G. Bergers, and E. Bergsland, *Less is more, regularly: metronomic dosing of cytotoxic drugs can target tumor angiogenesis in mice.* The journal of clinical investigation, 2000. **105**(8): p. 1045–1047.
17. Klement, G.L. and B.A. Kamen, *Nontoxic, fiscally responsible, future of oncology: could it be beginning in the Third World?* Journal of pediatric hematology/oncology, 2011. **33**(1): p. 1.
18. Andre, N., L. Padovani, and E. Pasquier, *Metronomic scheduling of anticancer treatment: the next generation of multitarget therapy?* Future oncology, 2011. **7**(3): p. 385–394.
19. Pai, P.S., et al., *Oral metronomic scheduling of anticancer therapy-based treatment compared to existing standard of care in locally advanced oral squamous cell cancers: A matched-pair analysis.* Indian journal of cancer, 2013. **50**(2): p. 135–141.

20. Mannel, R.S., et al., *A randomized phase III trial of IV carboplatin and paclitaxel× 3 courses followed by observation versus weekly maintenance low-dose paclitaxel in patients with early-stage ovarian carcinoma: a Gynecologic Oncology Group Study.* Gynecologic oncology, 2011. **122**(1): p. 89–94.
21. Kerbel, R.S., et al., *Continuous low-dose anti-angiogenic/metronomic chemotherapy: from the research laboratory into the oncologyclinic.* Annals of oncology, 2002. **13**(1): p. 12–15.
22. Gasparini, G., *Metronomic scheduling: the future of chemotherapy?* The Lancet oncology, 2001. **2**(12): p. 733–740.
23. Skipper H.E., F.M. Schabel, and W.S. Wilcox, *Experimental evaluation of potential anticancer agents. XIII. On the criteria and kinetics associated with" curability" of experimental leukemia.* Cancer chemotherapy reports, 1964. **61**: p. 1307–1317.
24. Loven, D., et al., *Low-dose metronomic chemotherapy: from past experience to new paradigms in the treatment of cancer.* Drug discovery today, 2013. **18**(3-4): p. 193–201.
25. Correale, P., et al., *Dose/dense metronomic chemotherapy with fractioned cisplatin and oral daily etoposide enhances the anti-angiogenic effects of bevacizumab in advanced non-small-cell-lung cancer patients.* Cancer biology & therapy, 2010. **9**(9): p. 685–693.
26. Ge, Y., et al., *Metronomic cyclophosphamide treatment in metastasized breast cancer patients: immunological effects and clinical outcome.* Cancer immunology, immunotherapy, 2012. **61**: p. 353–362.
27. Thomas, E.S., et al., *Ixabepilone plus capecitabine for metastatic breast cancer progressing after anthracycline and taxane treatment.* Journal of clinical oncology, 2007. **25**(33): p. 5210–7.
28. Lien, K., et al., *Low-dose metronomic chemotherapy: a systematic literature analysis.* European journal of cancer, 2013. **49**(16): p. 3387–3395.
29. Dueñas-González, A., et al., *The prince and the pauper. A tale of anticancer targeted agents.* Molecular cancer, 2008. **7**: p. 1–33.
30. Ashburn, T.T. and K.B. Thor, *Drug repositioning: identifying and developing new uses for existing drugs.* Nature reviews. Drug discovery, 2004. **3**(8): p. 673–683.
31. Blatt, J. and S.J. Corey, *Drug repurposing in pediatrics and pediatric hematology oncology.* Drug discovery today, 2013. **18**(1-2): p. 4–10.
32. André, N., et al., *Has the time come for metronomics in low-income and middle-income countries?* The Lancet oncology, 2013. **14**(6): p. e239–e248.
33. Laquente, B., F. Vinals, and J.R. Germa, *Metronomic chemotherapy: an antiangiogenic scheduling.* Clinical and translational oncology, 2007. **9**: p. 93–98.
34. Pasquier, E., et al., *Concentration-and schedule-dependent effects of chemotherapy on the angiogenic potential and drug sensitivity of vascular endothelial cells.* Angiogenesis, 2013. **16**: p. 373–386.
35. Mancuso, P., et al., *Circulating endothelial-cell kinetics and viability predict survival in breast cancer patients receiving metronomic chemotherapy.* Blood, 2006. **108**(2): p. 452–459.
36. Zitvogel, L., et al., *The anticancer immune response: indispensable for therapeutic success?* The journal of clinical investigation, 2008. **118**(6): p. 1991–2001.
37. Galluzzi, L., et al., *The secret ally: immunostimulation by anticancer drugs.* Nature reviews. Drug discovery, 2012. **11**(3): p. 215–233.
38. Landreneau, J.P., et al., *Immunological mechanisms of low and ultra-low dose cancer chemotherapy.* Cancer microenvironment, 2015. **8**: p. 57–64.
39. Tesniere, A., et al., *Immunogenic cancer cell death: a key-lock paradigm.* Current opinion in immunology, 2008. **20**(5): p. 504–511.
40. Kaneno, R., et al., *Chemomodulation of human dendritic cell function by antineoplastic agents in low noncytotoxic concentrations.* Journal of translational medicine, 2009. **7**: p. 1–10.
41. Kaneno, R., et al., *Chemotherapeutic agents in low noncytotoxic concentrations increase immunogenicity of human colon cancer cells.* Cellular oncology, 2011. **34**: p. 97–106.
42. Ghiringhelli, F., et al., *Metronomic cyclophosphamide regimen selectively depletes CD4+ CD25+ regulatory T cells and restores T and NK effector functions in end stage cancer patients.* Cancer immunology, immunotherapy, 2007. **56**: p. 641–648.
43. Banissi, C., et al., *Treg depletion with a low-dose metronomic temozolomide regimen in a rat glioma model.* Cancer immunology, immunotherapy, 2009. **58**: p. 1627–1634.
44. Zhao, J., et al., *Selective depletion of CD4+ CD25+ Foxp3+ regulatory T cells by low-dose cyclophosphamide is explained by reduced intracellular ATP levels.* Cancer research, 2010. **70**(12): p. 4850–4858.
45. Kan, S., et al., *Suppressive effects of cyclophosphamide and gemcitabine on regulatory T-cell induction in vitro.* Anticancer research, 2012. **32**(12): p. 5363–5369.
46. Michels, T., et al., *Paclitaxel promotes differentiation of myeloid-derived suppressor cells into dendritic cells in vitro in a TLR4-independent manner.* Journal of immunotoxicology, 2012. **9**(3): p. 292–300.

47. Todaro, M., et al., *Combining conventional chemotherapy and γδ T cell-based immunotherapy to target cancer-initiating cells.* Oncoimmunology, 2013. **2**(9): p. e25821.
48. Sierro, S.R., et al., *Combination of lentivector immunization and low-dose chemotherapy or PD-1/PD-L1 blocking primes self-reactive T cells and induces anti-tumor immunity.* European journal of immunology, 2011. **41**(8): p. 2217–2228.
49. Geary, S.M., et al., *The combination of a low-dose chemotherapeutic agent, 5-fluorouracil, and an adenoviral tumor vaccine has a synergistic benefit on survival in a tumor model system.* PloS one, 2013. **8**(6): p. e67904.
50. Hermans, I.F., et al., *Synergistic effect of metronomic dosing of cyclophosphamide combined with specific antitumor immunotherapy in a murine melanoma model.* Cancer research, 2003. **63**(23): p. 8408–8413.
51. Chen, C.-A., et al., *Metronomic chemotherapy enhances antitumor effects of cancer vaccine by depleting regulatory T lymphocytes and inhibiting tumor angiogenesis.* Molecular therapy, 2010. **18**(6): p. 1233–1243.
52. Foy, K.C., et al., *Immunotherapy with HER-2 and VEGF peptide mimics plus metronomic paclitaxel causes superior antineoplastic effects in transplantable and transgenic mouse models of human breast cancer.* Oncoimmunology, 2012. **1**(7): p. 1004–1016.
53. Chen, C.-S., J.C. Doloff, and D.J. Waxman, *Intermittent metronomic drug schedule is essential for activating antitumor innate immunity and tumor xenograft regression.* Neoplasia, 2014. **16**(1): p. 84–96.
54. Malvicini, M., et al., *Single low-dose cyclophosphamide combined with interleukin-12 gene therapy is superior to a metronomic schedule in inducing immunity against colorectal carcinoma in mice.* Oncoimmunology, 2012. **1**(7): p. 1038–1047.
55. Colleoni, M., et al., *Low-dose oral methotrexate and cyclophosphamide in metastatic breast cancer: antitumor activity and correlation with vascular endothelial growth factor levels.* Annals of oncology, 2002. **13**(1): p. 73–80.
56. Folkins, C., et al., *Anticancer therapies combining antiangiogenic and tumor cell cytotoxic effects reduce the tumor stem-like cell fraction in glioma xenograft tumors.* Cancer research, 2007. **67**(8): p. 3560–3564.
57. Vives, M., et al., *Metronomic chemotherapy following the maximum tolerated dose is an effective antitumour therapy affecting angiogenesis, tumour dissemination and cancer stem cells.* International journal of cancer, 2013. **133**(10): p. 2464–2472.
58. Yan, H., et al., *Drug-tolerant cancer cells show reduced tumor-initiating capacity: depletion of CD44+ cells and evidence for epigenetic mechanisms.* PloS one, 2011. **6**(9): p. e24397.
59. Singh, V., U. Erb, and M. Zöller, *Cooperativity of CD44 and CD49d in leukemia cell homing, migration, and survival offers a means for therapeutic attack.* The journal of immunology, 2013. **191**(10): p. 5304–5316.
60. Cham, K.K.Y., et al., *Metronomic gemcitabine suppresses tumour growth, improves perfusion, and reduces hypoxia in human pancreatic ductal adenocarcinoma.* British journal of cancer, 2010. **103**(1): p. 52–60.
61. Francia, G., et al., *Low-dose metronomic oral dosing of a prodrug of gemcitabine (LY2334737) causes antitumor effects in the absence of inhibition of systemic vasculogenesis.* Molecular cancer therapeutics, 2012. **11**(3): p. 680–689.
62. Doloff, J.C., et al., *Increased Tumor Oxygenation and Drug Uptake During Anti-Angiogenic Weekly Low Dose Cyclophosphamide Enhances the Anti-Tumor Effect of Weekly Tirapazamine (Supplementry Material).* Current cancer drug targets, 2009. **9**(6): p. 777–788.
63. Mupparaju, S., et al., *Repeated tumor oximetry to identify therapeutic window during metronomic cyclophosphamide treatment of 9L gliomas.* Oncology reports, 2011. **26**(1): p. 281–286.
64. Rapisarda, A., et al., *Identification of small molecule inhibitors of hypoxia-inducible factor 1 transcriptional activation pathway.* Cancer research, 2002. **62**(15): p. 4316–4324.
65. Kummar, S., et al., *Multihistology, target-driven pilot trial of oral topotecan as an inhibitor of hypoxia-inducible factor-1α in advanced solid tumors.* Clinical cancer research, 2011. **17**(15): p. 5123–5131.
66. Lee, K., et al., *Anthracycline chemotherapy inhibits HIF-1 transcriptional activity and tumor-induced mobilization of circulating angiogenic cells.* Proceedings of the National Academy of Sciences, 2009. **106**(7): p. 2353–2358.
67. Kim, Y.J., et al., *Overcoming evasive resistance from vascular endothelial growth factor a inhibition in sarcomas by genetic or pharmacologic targeting of hypoxia-inducible factor 1α.* International journal of cancer, 2013. **132**(1): p. 29–41.

68. Pietras, K. and D. Hanahan, *A multitargeted, metronomic, and maximum-tolerated dose "chemo-switch" regimen is antiangiogenic, producing objective responses and survival benefit in a mouse model of cancer.* Journal of clinical oncology, 2005. **23**(5): p. 939–952.
69. Li, S.C., K.L. Lee, and J. Luo, *Control dominating subclones for managing cancer progression and posttreatment recurrence by subclonal switchboard signal: implication for new therapies.* Stem cells and development, 2012. **21**(4): p. 503–506.
70. Tran Cao, H.S., et al., *Metronomic gemcitabine in combination with sunitinib inhibits multisite metastasis and increases survival in an orthotopic model of pancreatic cancer.* Molecular cancer therapeutics, 2010. **9**(7): p. 2068–2078.
71. Hashimoto, K., et al., *Potent preclinical impact of metronomic low-dose oral topotecan combined with the antiangiogenic drug pazopanib for the treatment of ovarian cancer.* Molecular cancer therapeutics, 2010. **9**(4): p. 996–1006.
72. Jang, J.W., et al., *Suppression of hepatic tumor growth and metastasis by metronomic therapy in a rat model of hepatocellular carcinoma.* Experimental & molecular medicine, 2011. **43**(5): p. 305–312.
73. Kumar, S., et al., *Metronomic oral topotecan with pazopanib is an active antiangiogenic regimen in mouse models of aggressive pediatric solid tumor.* Clinical cancer research, 2011. **17**(17): p. 5656–5667.
74. Hackl, C., et al., *Metronomic oral topotecan prolongs survival and reduces liver metastasis in improved preclinical orthotopic and adjuvant therapy colon cancer models.* Gut, 2013. **62**(2): p. 259–271.
75. Butler, J.M., H. Kobayashi, and S. Rafii, *Instructive role of the vascular niche in promoting tumour growth and tissue repair by angiocrine factors.* Nature Reviews. Cancer, 2010. **10**(2): p. 138–146.
76. Yu, X., H. Li, and X. Ren, *Interaction between regulatory T cells and cancer stem cells.* International journal of cancer, 2012. **131**(7): p. 1491–1498.
77. Werno, C., et al., *Knockout of HIF-1α in tumor-associated macrophages enhances M2 polarization and attenuates their pro-angiogenic responses.* Carcinogenesis, 2010. **31**(10): p. 1863–1872.
78. Ashley, D.M., et al., *Response of recurrent medulloblastoma to low-dose oral etoposide.* Journal of clinical oncology, 1996. **14**(6): p. 1922–1927.
79. Emmenegger, U., et al., *Tumors that acquire resistance to low-dose metronomic cyclophosphamide retain sensitivity to maximum tolerated dose cyclophosphamide.* Neoplasia, 2011. **13**(1): p. 40–48.
80. Pan, Q., et al., *Chemoresistance to temozolomide in human glioma cell line U251 is associated with increased activity of O 6-methylguanine-DNA methyltransferase and can be overcome by metronomic temozolomide regimen.* Cell biochemistry and biophysics, 2012. **62**: p. 185–191.
81. Chow, A., et al., *Preclinical analysis of resistance and cross-resistance to low-dose metronomic chemotherapy.* Investigational new drugs, 2014. **32**: p. 47–59.
82. De Souza, R., et al., *Chemotherapy dosing schedule influences drug resistance development in ovarian cancer.* Molecular cancer therapeutics, 2011. **10**(7): p. 1289–1299.
83. Martin-Padura, I., et al., *Residual dormant cancer stem-cell foci are responsible for tumor relapse after antiangiogenic metronomic therapy in hepatocellular carcinoma xenografts.* Laboratory investigation, 2012. **92**(7): p. 952–966.
84. Emmenegger, U., et al., *Low-dose metronomic daily cyclophosphamide and weekly tirapazamine: a well-tolerated combination regimen with enhanced efficacy that exploits tumor hypoxia.* Cancer research, 2006. **66**(3): p. 1664–1674.
85. Thoenes, L., et al., *In vivo chemoresistance of prostate cancer in metronomic cyclophosphamide therapy.* Journal of proteomics, 2010. **73**(7): p. 1342–1354.
86. Kubisch, R., et al., *A comprehensive gene expression analysis of resistance formation upon metronomic cyclophosphamide therapy.* Translational oncology, 2013. **6**(1): p. 1–9.
87. Wang, Y., et al., *Differential expression of mimecan and thioredoxin domain–containing protein 5 in colorectal adenoma and cancer: A proteomic study.* Experimental biology and medicine, 2007. **232**(9): p. 1152–1159.
88. Zhang, L., et al., *The influence of TXNDC5 gene on gastric cancer cell.* Journal of cancer research and clinical oncology, 2010. **136**: p. 1497–1505.
89. Vincent, E.E., et al., *Overexpression of the TXNDC5 protein in non-small cell lung carcinoma.* Anticancer research, 2011. **31**(5): p. 1577–1582.
90. Chang, X., et al., *Investigating a pathogenic role for TXNDC5 in tumors.* International journal of oncology, 2013. **43**(6): p. 1871–1884.
91. Bruchard, M., et al., *Chemotherapy-triggered cathepsin B release in myeloid-derived suppressor cells activates the Nlrp3 inflammasome and promotes tumor growth.* Nature medicine, 2013. **19**(1): p. 57–64.

92. Briasoulis, E., et al., *Dose selection trial of metronomic oral vinorelbine monotherapy in patients with metastatic cancer: a hellenic cooperative oncology group clinical translational study.* BMC cancer, 2013. **13**: p. 1–12.
93. Kontopodis, E., et al., *A phase II study of metronomic oral vinorelbine administered in the second line and beyond in non-small cell lung cancer (NSCLC): a phase II study of the Hellenic Oncology Research Group.* Journal of chemotherapy, 2013. **25**(1): p. 49–55.
94. Rajdev, L., et al., *Phase I trial of metronomic oral vinorelbine in patients with advanced cancer.* Cancer chemotherapy and pharmacology, 2011. **68**: p. 1119–1124.
95. Addeo, R., et al., *Low-dose metronomic oral administration of vinorelbine in the first-line treatment of elderly patients with metastatic breast cancer.* Clinical breast cancer, 2010. **10**(4): p. 301–306.
96. Saridaki, Z., et al., *A phase I trial of oral metronomic vinorelbine plus capecitabine in patients with metastatic breast cancer.* Cancer chemotherapy and pharmacology, 2012. **69**: p. 35–42.
97. He, S., et al., *Capecitabine "metronomic" chemotherapy for palliative treatment of elderly patients with advanced gastric cancer after fluoropyrimidine-based chemotherapy.* Medical Oncology, 2012. **29**: p. 100–106.
98. Fedele, P., et al., *Efficacy and safety of low-dose metronomic chemotherapy with capecitabine in heavily pretreated patients with metastatic breast cancer.* European journal of cancer, 2012. **48**(1): p. 24–29.
99. Yoshimoto, M., et al., *Metronomic oral combination chemotherapy with capecitabine and cyclophosphamide: a phase II study in patients with HER2-negative metastatic breast cancer.* Cancer chemotherapy and pharmacology, 2012. **70**: p. 331–338.
100. Wang, Z., et al., *An all-oral combination of metronomic cyclophosphamide plus capecitabine in patients with anthracycline-and taxane-pretreated metastatic breast cancer: a phase II study.* Cancer chemotherapy and pharmacology, 2012. **69**: p. 515–522.
101. Omuro, A., et al., *Phase II trial of continuous low-dose temozolomide for patients with recurrent malignant glioma.* Neuro-oncology, 2013. **15**(2): p. 242–250.
102. Santoni, M., et al., *Protracted low doses of temozolomide for the treatment of patients with recurrent glioblastoma: a phase II study.* Oncology letters, 2012. **4**(4): p. 799–801.
103. Bojko, P., et al., *Metronomic oral cyclophosphamide in patients with advanced solid tumors.* Oncology research and treatment, 2012. **35**(1-2): p. 35–38.
104. El-Arab, L.R.E., M. Swellam, and M.M. El Mahdy, *Metronomic chemotherapy in metastatic breast cancer: impact on VEGF.* Journal of the Egyptian National Cancer Institute, 2012. **24**(1): p. 15–22.
105. Gebbia, V., et al., *Salvage therapy with oral metronomic cyclophosphamide and methotrexate for castration-refractory metastatic adenocarcinoma of the prostate resistant to docetaxel.* Urology, 2011. **78**(5): p. 1125–1130.
106. Gebbia, V., H. Boussen, and M.R. Valerio, *Oral metronomic cyclophosphamide with and without methotrexate as palliative treatment for patients with metastatic breast carcinoma.* Anticancer research, 2012. **32**(2): p. 529–536.
107. Otsuka, H., et al., *Phase II clinical trial of metronomic chemotherapy with combined irinotecan and tegafur–gimeracil–oteracil potassium in metastatic and recurrent breast cancer.* Breast cancer, 2015. **22**: p. 335–342.
108. Kurume, S. and G. Clinical Oncology, *Multicenter phase II study of a new effective S-1 and irinotecan combination schedule in patients with unresectable metastatic or recurrent colorectal cancer.* Clinical medicine insights: Oncology, 2013. **7**: p. 21–30.
109. Pasquier, E., N. Andre, and D. Braguer, *Targeting microtubules to inhibit angiogenesis and disrupt tumour vasculature: implications for cancer treatment.* Current cancer drug targets, 2007. **7**(6): p. 566–581.
110. Galano, G., et al., *Efficacy and tolerability of vinorelbine in the cancer therapy.* Current drug safety, 2011. **6**(3): p. 185–193.
111. Zielinski, C., J. Gralow, and M. Martin, *Optimising the dose of capecitabine in metastatic breast cancer: confused, clarified or confirmed?* Annals of oncology, 2010. **21**(11): p. 2145–2152.
112. Kushner, B.H., K. Kramer, and N.-K.V. Cheung, *Oral etoposide for refractory and relapsed neuroblastoma.* Journal of clinical oncology, 1999. **17**(10): p. 3221–3225.
113. Baruchel, S., et al., *Safety and pharmacokinetics of temozolomide using a dose-escalation, metronomic schedule in recurrent paediatric brain tumours.* European journal of cancer, 2006. **42**(14): p. 2335–2342.
114. Kato, H., et al., *A randomized trial of adjuvant chemotherapy with uracil–tegafur for adenocarcinoma of the lung.* New England journal of medicine, 2004. **350**(17): p. 1713–1721.

115. Addeo, R., et al., *Protracted low dose of oral vinorelbine and temozolomide with whole-brain radiotherapy in the treatment for breast cancer patients with brain metastases.* Cancer chemotherapy and pharmacology, 2012. **70**: p. 603–609.
116. Lin, N.U., J.R. Bellon, and E.P. Winer, *CNS metastases in breast cancer.* Journal of clinical oncology, 2004. **22**(17): p. 3608–3617.
117. Salmaggi, A., et al., *Prospective study of carmustine wafers in combination with 6-month metronomic temozolomide and radiation therapy in newly diagnosed glioblastoma: preliminary results.* Journal of Neurosurgery, 2013. **118**(4): p. 821–829.
118. U.S. National Library of Medicine, *ClinicalTrials.gov [online].* 2013.
119. Crivellari, D., et al., *Adjuvant pegylated liposomal doxorubicin for older women with endocrine nonresponsive breast cancer who are NOT suitable for a "standard chemotherapy regimen": the CASA randomized trial.* The breast, 2013. **22**(2): p. 130–137.
120. Dellapasqua, S., et al., *Pegylated liposomal doxorubicin in combination with low-dose metronomic cyclophosphamide as preoperative treatment for patients with locally advanced breast cancer.* The breast, 2011. **20**(4): p. 319–323.
121. Pallis, A.G., et al., *A multicenter phase I trial of metronomic oral vinorelbine plus cisplatin in patients with NSCLC.* Cancer chemotherapy and pharmacology, 2011. **67**: p. 1239–1245.
122. Bellmunt, J., et al., *Activity of a multitargeted chemo-switch regimen (sorafenib, gemcitabine, and metronomic capecitabine) in metastatic renal-cell carcinoma: a phase 2 study (SOGUG-02-06).* The Lancet oncology, 2010. **11**(4): p. 350–357.
123. Padovani, L., et al., *Neurocognitive function after radiotherapy for paediatric brain tumours.* Nature reviews. Neurology, 2012. **8**(10): p. 578–588.
124. Zustovich, F., et al., *Sorafenib plus daily low-dose temozolomide for relapsed glioblastoma: a phase II study.* Anticancer research, 2013. **33**(8): p. 3487–3494.
125. Reardon, D.A., et al., *Phase II study of metronomic chemotherapy with bevacizumab for recurrent glioblastoma after progression on bevacizumab therapy.* Journal of neuro-oncology, 2011. **103**: p. 371–379.
126. Berruti, A., et al., *Phase II study of weekly paclitaxel and sorafenib as second/third-line therapy in patients with adrenocortical carcinoma.* European journal of endocrinology, 2012. **166**(3): p. 451–458.
127. Barber, E.L., et al., *The combination of intravenous bevacizumab and metronomic oral cyclophosphamide is an effective regimen for platinum-resistant recurrent ovarian cancer.* Journal of gynecologic oncology, 2013. **24**(3): p. 258–264.
128. Sánchez-Muñoz, A., et al., *Bevacizumab plus low-dose metronomic oral cyclophosphamide in heavily pretreated patients with recurrent ovarian cancer.* Oncology, 2010. **79**(1-2): p. 98–104.
129. Mayer, E.L., et al., *Combination antiangiogenic therapy in advanced breast cancer: a phase 1 trial of vandetanib, a VEGFR inhibitor, and metronomic chemotherapy, with correlative platelet proteomics.* Breast cancer research and treatment, 2012. **136**: p. 169–178.
130. Montagna, E., et al., *Metronomic chemotherapy combined with bevacizumab and erlotinib in patients with metastatic HER2-negative breast cancer: clinical and biological activity.* Clinical breast cancer, 2012. **12**(3): p. 207–214.
131. Saloustros, E., et al., *Metronomic vinorelbine plus bevacizumab as salvage therapy for patients with metastatic breast cancer.* Journal of BUON, 2011. **16**(2): p. 215–218.
132. Kelley, R.K., et al., *A phase 1 trial of imatinib, bevacizumab, and metronomic cyclophosphamide in advanced colorectal cancer.* British journal of cancer, 2013. **109**(7): p. 1725–1734.
133. Correale, P., et al., *Phase II trial of bevacizumab and dose/dense chemotherapy with cisplatin and metronomic daily oral etoposide in advanced non-small-cell-lung cancer patients.* Cancer biology & therapy, 2011. **12**(2): p. 112–118.
134. Chen, Y.-M., et al., *A phase II randomized trial of gefitinib alone or with tegafur/uracil treatment in patients with pulmonary adenocarcinoma who had failed previous chemotherapy.* Journal of thoracic oncology, 2011. **6**(6): p. 1110–1116.
135. Hsu, C.-H., et al., *Phase II study of combining sorafenib with metronomic tegafur/uracil for advanced hepatocellular carcinoma.* Journal of hepatology, 2010. **53**(1): p. 126–131.
136. Kummar, S., et al., *A phase I study of veliparib in combination with metronomic cyclophosphamide in adults with refractory solid tumors and lymphomas.* Clinical cancer research, 2012. **18**(6): p. 1726–1734.
137. Koumarianou, A., et al., *Combination treatment with metronomic temozolomide, bevacizumab and long-acting octreotide for malignant neuroendocrine tumours.* Endocrine-related cancer, 2012. **19**(1): p. L1–L4.

138. Turner, D.C., et al., *Combination metronomic oral topotecan and pazopanib: a pharmacokinetic study in patients with gynecological cancer.* Anticancer research, 2013. **33**(9): p. 3823–3829.
139. Dickler, M.N., et al., *A phase II trial of erlotinib in combination with bevacizumab in patients with metastatic breast cancer.* Clinical cancer research, 2008. **14**(23): p. 7878–7883.
140. Kruh, J. and C.S. Foster, *Corticosteroid-sparing agents: conventional systemic immunosuppressants.* Developments in ophthalmology, 2012. **51**: p. 29–46.
141. Prendergast, G.C. and E.M. Jaffee, *Cancer immunologists and cancer biologists: why we didn't talk then but need to now.* Cancer research, 2007. **67**(8): p. 3500–3504.
142. Kepp, O., et al., *Molecular determinants of immunogenic cell death elicited by anticancer chemotherapy.* Cancer and metastasis reviews, 2011. **30**: p. 61–69.
143. Liikanen, I., et al., *Oncolytic adenovirus with temozolomide induces autophagy and antitumor immune responses in cancer patients.* Molecular therapy, 2013. **21**(6): p. 1212–1223.
144. Cerullo, V., et al., *Immunological effects of low-dose cyclophosphamide in cancer patients treated with oncolytic adenovirus.* Molecular therapy, 2011. **19**(9): p. 1737–1746.
145. Kandalaft, L.E., et al., *Autologous lysate-pulsed dendritic cell vaccination followed by adoptive transfer of vaccine-primed ex vivo co-stimulated T cells in recurrent ovarian cancer.* Oncoimmunology, 2013. **2**(1): p. e22664.
146. Lasalvia-Prisco, E., et al., *Addition of an induction regimen of antiangiogenesis and antitumor immunity to standard chemotherapy improves survival in advanced malignancies.* Medical oncology, 2012. **29**(5): p. 3626–3633.
147. Ellebaek, E., et al., *Metastatic melanoma patients treated with dendritic cell vaccination, Interleukin-2 and metronomic cyclophosphamide: results from a phase II trial.* Cancer immunology, immunotherapy, 2012. **61**: p. 1791–1804.
148. Soriano, J.L., et al., *Metronomic cyclophosphamide and methotrexate chemotherapy combined with 1E10 anti-idiotype vaccine in metastatic breast cancer.* International journal of breast cancer, 2011. **2011**.
149. Walter, B., et al., *Pioglitazone, etoricoxib, interferon-α, and metronomic capecitabine for metastatic renal cell carcinoma: final results of a prospective phase II trial.* Medical oncology, 2012. **29**: p. 799–805.
150. Bottini, A., et al., *Randomized phase II trial of letrozole and letrozole plus low-dose metronomic oral cyclophosphamide as primary systemic treatment in elderly breast cancer patients.* Journal of clinical oncology, 2006. **24**: p. 3623–3628.
151. Mir, O., et al., *Feasibility of metronomic oral cyclophosphamide plus prednisolone in elderly patients with inoperable or metastatic soft tissue sarcoma.* European journal of cancer, 2011. **47**(4): p. 515–519.
152. Woo, H.Y., et al., *Efficacy and safety of metronomic chemotherapy for patients with advanced primary hepatocellular carcinoma with major portal vein tumor thrombosis.* The Korean journal of hepatology, 2012. **18**(1): p. 32.
153. Thakur, M.D. and D.D. Stuart, *The evolution of melanoma resistance reveals therapeutic opportunities.* Cancer research, 2013. **73**(20): p. 6106–6110.
154. Brugières, L., et al., *Single-drug vinblastine as salvage treatment for refractory or relapsed anaplastic large-cell lymphoma: a report from the French Society of Pediatric Oncology.* Journal of clinical oncology, 2009. **27**(30): p. 5056–5061.
155. Le Deley, M.-C., et al., *Vinblastine in children and adolescents with high-risk anaplastic large-cell lymphoma: results of the randomized ALCL99-vinblastine trial.* Journal of clinical oncology, 2010. **28**(25): p. 3987–3993.
156. André, N., et al., *Looking at the seemingly contradictory role of vinblastine in anaplastic large-cell lymphoma from a metronomic perspective.* Journal of clinical oncology, 2010. **29**(4): p. e90–e93.
157. Davidson, A., et al., *Phase II study of 21 day schedule oral etoposide in children.* European journal of cancer, 1997. **33**(11): p. 1816–1822.
158. Casanova, M., et al., *Vinorelbine in previously treated advanced childhood sarcomas: Evidence of activity in rhadbomyosarcoma.* Cancer, 2002. **94**(12): p. 3263–3268.
159. Minard-Colin, V., et al., *Phase II study of vinorelbine and continuous low doses cyclophosphamide in children and young adults with a relapsed or refractory malignant solid tumour: good tolerance profile and efficacy in rhabdomyosarcoma – a report from the Société Française des Cancers et leucémies de l'Enfant et de l'adolescent (SFCE).* European journal of cancer, 2012. **48**(15): p. 2409–2416.
160. Zapletalová, D., et al., *Metronomic chemotherapy with the COMBAT regimen in advanced pediatric malignancies: a multicenter experience.* Oncology, 2012. **82**(5): p. 249–260.

161. Kesari, S., et al., *Phase II study of metronomic chemotherapy for recurrent malignant gliomas in adults.* Neuro-oncology, 2007. **9**(3): p. 354–363.
162. Herrlinger, U., et al., *UKT-04 trial of continuous metronomic low-dose chemotherapy with methotrexate and cyclophosphamide for recurrent glioblastoma.* Journal of neuro-oncology, 2005. **71**: p. 295–299.
163. Doz, F., P. Marvanne, and A. Fagot-Largeault, *The person in personalised medicine.* European journal of cancer, 2013. **49**(5): p. 1159–1160.

9 Metronomic Chemotherapy for Metastatic Breast Cancer

Manzoor Ahmad Mir, Burhan Ul Haq, Mary Chatterjee, and Kirti Hooda

9.1 INTRODUCTION

The various subtypes of metastatic breast cancer (BC) are categorized as luminal A, which has low levels of the protein Ki-67, a marker of cell proliferation, and is hormone receptor–positive (ER+ and/or PR+) [1]. It often has a good prognosis and is frequently linked to a better reaction to hormone therapy. Luminal B is likewise hormone receptor–positive (ER+ and/or PR+), but it exhibits higher levels of the cell proliferation marker Ki-67 [2]. In addition to hormone therapy, luminal B tumors may exhibit more aggressive behavior and necessitate other therapeutic approaches like chemotherapy [3]. In another subtype of BC, HER2 protein is overexpressed or amplified, termed HER2-enriched. Most HER2-enriched cancers lack hormone receptors (ER–/PR–). They frequently expand more quickly and are more likely to recur. This subtype responds well to targeted therapies like trastuzumab and pertuzumab, which specifically block HER2 signaling. Triple-negative breast cancer (TNBC) and basal-like BC are frequently used interchangeably. Hormone receptors (ER–/PR–) and HER2 (HER2–) are absent in this subtype. Basal-like/TNBC cancers are frequently more aggressive, carry a higher risk of recurrence, and might be best treated with chemotherapy [4]. The normal-like subtype is defined by gene expression patterns that resemble those of normal breast tissue. Compared to other subtypes, it is linked to a better prognosis. Figure 9.1 shows various BC subtypes. The various processes through which a primary tumor changes into a secondary one at a distal place is termed "metastasis". This is a chief cancer hallmark and results in failure of treatment, which ends up causing death [5]. Thus, there is a direct connection between the patient's prognosis and metastasis. This diagnosis is regarded as the final stage in most types of cancers. The regimen of conventional cytostatic cure is based on the "maximum tolerated dose" (MTD) approach, where elevated doses of a drug are provided at 2- to 4-week intervals and target fast-dividing cells. It is a well-known fact that chemotherapy disrupts the normal cell division and does not specifically kill cancer cells. The normal cells that undergo division at a higher rate are exposed to damage as well; this leads to typical side effects, including hair loss, mucositis, and bone marrow suppression. In comparison, the lengthy breaks between therapy cycles may allow the cancer cells to recover and develop resistance, ultimately enhancing the disease progression [6]. Figure 9.2 depicts the process that can lead to BC metastasis.

9.2 METRONOMIC CHEMOTHERAPY IN PRACTICE

Metronomic chemotherapy (MCT) is the smallest possible dose of a drug or a chemotherapeutic agent that can show anticancer functions. This is coupled with a continuous dosing schedule without any major breaks in the drug doses. Right from its inception, MCT has been considered a treatment option employed in highly advanced or palliative settings, especially in cancers for which sophisticated treatments are available like BC, and mistakenly regarded as more effective [7]. One of the highly examined tumors for MCT is probably BC. This treatment is intended to reduce the symptoms of the disease as well as enhance the quality of life in subjects with metastatic disease. It is an excellent option in place of traditional chemotherapy both in terms of efficacy and fewer side effects. Figure 9.3 represents the mechanism through which MCT operates.

DOI: 10.1201/9781032632674-9

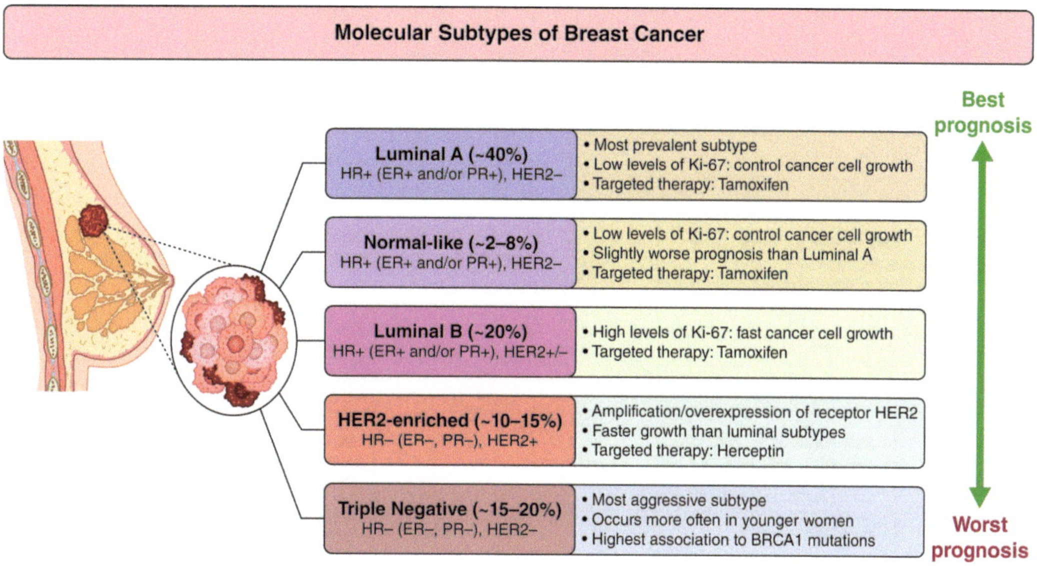

FIGURE 9.1 The subtypes of breast cancer.

MCT focuses on more time and a smaller dose to achieve minimal side effects. As per the available literature, metronomic chemotherapy has achieved significant interest in pediatric oncology and in several adult solid tumors. Specifically in BC, a lot of clinical trials examined the feasibility and efficacy of this approach. MCT is a good choice in a metastatic setting. A major setback in cancer therapy for medical oncologists is to attain cures that are non-invasive, effective, and with lesser toxicity as well as no change in the quality of life of patients, which is important especially in economically weaker nations. And MCT for sure addresses this issue.

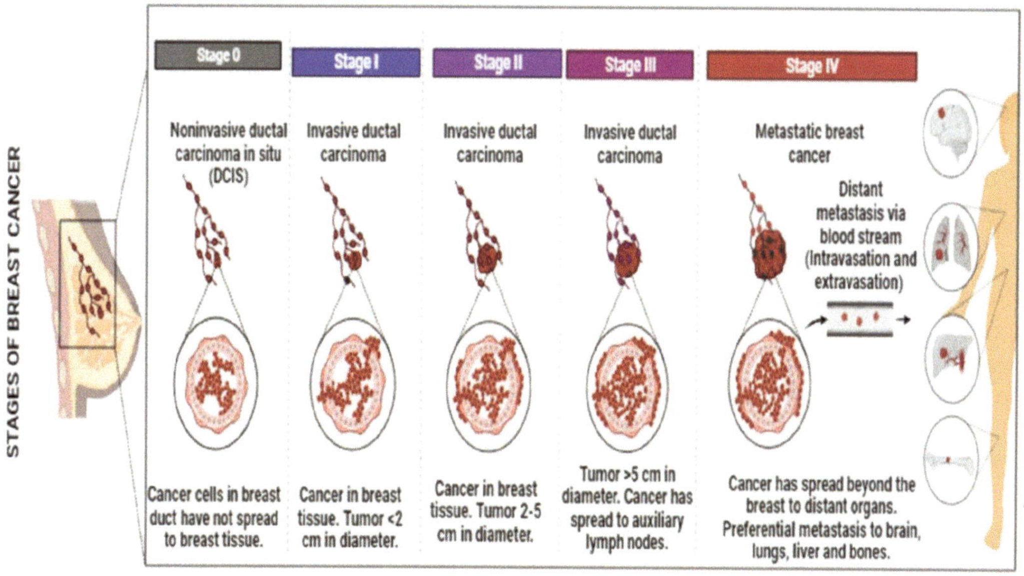

FIGURE 9.2 The stages involved in breast cancer metastasis.

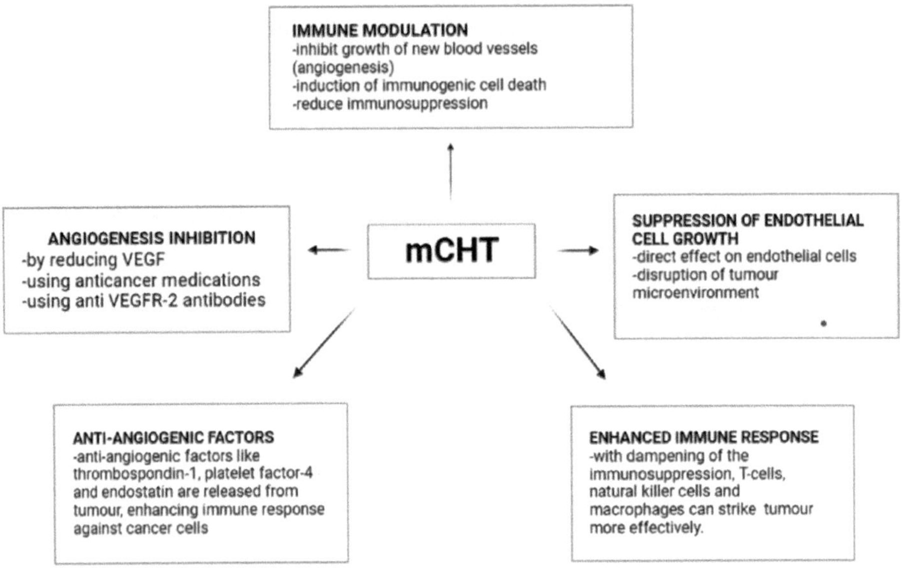

FIGURE 9.3 Mechanism of action of metronomic chemotherapy.

9.3 EPITHELIAL TO MESENCHYMAL TRANSITION IN METASTATIC BREAST CANCER

The epithelial to mesenchymal transition (EMT) represents a process in which epithelial cell polarity as well as the intercellular cohesion are no longer present, and this transition is necessary for metastasis initiation in BC [8]. The SNAIL, TWIST, and Zinc finger E-box-binding [ZEB] represent the pleiotropic transcription factors that are considered essential molecules for EMT modulation. It has been documented that the absence of expression or GATA3 dysfunctioning becomes the cause for the promotion of BC metastasis [9]. Interleukin-6 (IL-6), an inflammatory cytokine, has a significant role to play in BC metastasis [10]. The cell surface proteins also have an essential part to play in BC metastasis. The interactions between tumor cells and the cell surface carry out the cell adhesion and invasion to start metastasis [11]. An example is provided by metatherin, which shows overexpression in >40% of BCs, and it attaches to the pulmonary vasculature by the C-terminal portion of the extracellular domain [11], leading to poor clinical results in BC via metastasis. The process of EMT is represented in Figure 9.4.

In addition, the contribution to BC metastasis comes from immune cells and the tumor microenvironment. The tumor-associated macrophages (TAMs) have varied parts to play in microenvironmental signaling. In BC, the TAM M2 is frequently activated by IL-4, which is secreted by CD4+ T cells. The growth factors, chemokines, and inflammatory factors, which are the TAM-secreting cytokines, have been seen to be highly linked with metastasis by enhancing attachment to the extracellular matrix [12]. Thus, these cells and factors may act as targets for curing metastatic breast cancer (MBC). All the substances or drugs that halt the growth of cancer by disturbing the role of some particular molecules that are the cause of the survival and proliferation of tumor cells come under the targeted therapeutics for BC. For MBC, this targeted treatment is decided on by the presence or absence of hormone receptors, mainly HER2 (Figure 9.5), cancer recurrence, rate of metastasis, and the site of metastasis [13]. The personalized cure for particular subjects could amplify the therapeutic efficacy and decrease the toxicity caused by chemotherapy in BC [14]. However, the chances of recurrence and metastasis of all kinds of BC still exist. Moreover, there is still no specified cure for subjects with TNBC, the highly aggressive and recurrent form that leads to intensive metastasis. The most common sites where primary breast tumors can spread to is shown in Figure 9.6.

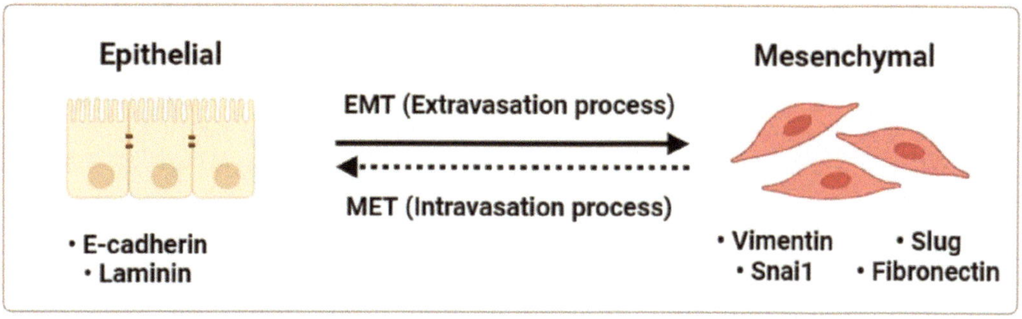

FIGURE 9.4 The process of epithelial to mesenchymal transition.

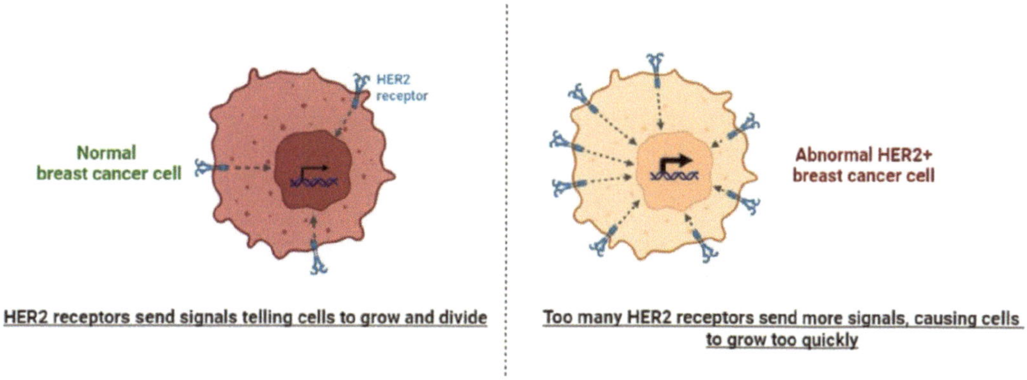

FIGURE 9.5 The normal and abnormal breast cancer cell with reference to HER2 receptor.

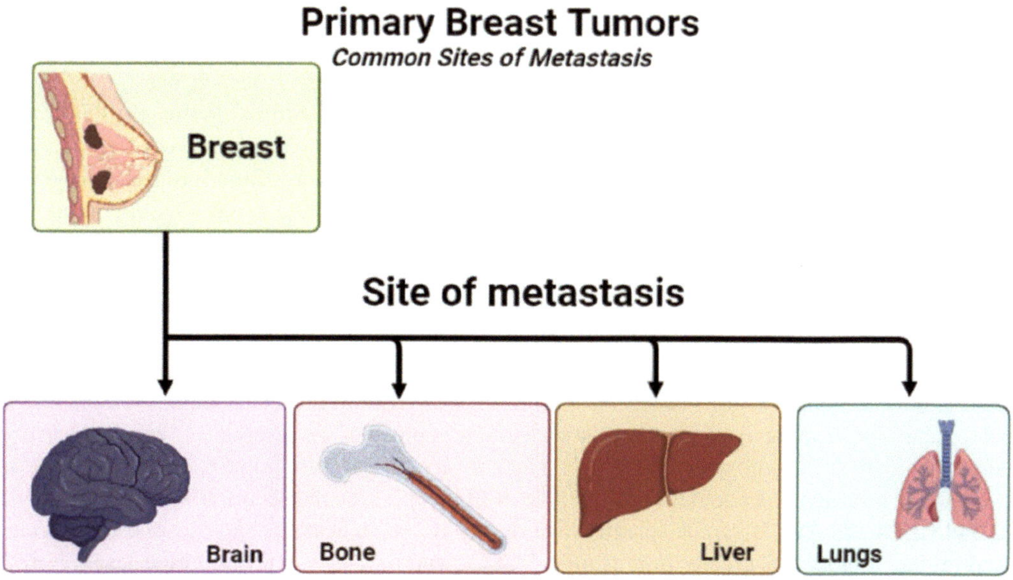

FIGURE 9.6 The most common sites to which primary breast tumors can spread.

EMT is one of the primary mechanisms responsible for metastasis, and through this process, cells having epithelial properties are converted into mesenchymal cells. Because mesenchymal cells can redifferentiate into epithelial cells or other types of cells, it is reversible [14]. EMT and the reverse process, mesenchymal to epithelial transition (MET), take place during the formation of tumors, fibrosis, and wounds. High differentiation in developing cell populations within an organism is a normal process that is induced by the precise set of regulatory alterations that permit EMT [15].

However, as tumor cells advance to malignancy during EMT, they gain the ability to move and become active [15]. Strong intercellular linkages in epidermal cells are broken down by the EMT program, permitting tumor cells to move to new tissues. Tumor cells that have traveled via lymphatic and blood vessels to other tissues go through MET, the opposite of EMT, where they are again transformed into epithelial cells with strong connections in distant regions. Furthermore, due to their stem cell characteristics, tumor cells produced from EMT are resistant to therapy. Therefore, several EMT-targeting treatments are being researched for patients with metastases [16]. By attaching to chromatin, transcription factors (TFs) regulate gene expression processes, which are crucial for the development and spread of cancer [17]. As per their capacity to either directly or indirectly block E-cadherin, EMT-TFs can be categorized. By enabling the invasion of cancer cells, the epithelial cell adhesion protein E-cadherin contributes to tumor metastasis. The increase of transforming growth factor (TGF), reactive oxygen species, and genes that affect apoptotic signaling pathways are specifically linked to the loss of E-cadherin [17]. N-cadherin and vimentin levels of mesenchymal cells increased as E-cadherin was inhibited [18]. It has been discovered that members of the SNAIL, ZEB2 (SIP1), TWIST1, SLUG, ZEB1 (TCF8/dEF1), and FOX families are transcriptional inhibitors of E-cadherin. By controlling different gene expressions through diverse combinations, these EMT-TFs are significantly linked to cancer start, progression, invasion, and metastasis in addition to resistance to treatment [19]. BC overexpression of EMT-TFs is linked with poor prognosis in MBC and promotes the growth and spread of the disease. By enlisting TWIST1 and Mi2/nucleosome remodeling and deacetylase (Mi2/NuRD) protein complexes, for instance, TWIST1 expression was strongly correlated with hypermethylation and hypoacetylation of the E-cadherin promoter in MBC [17]. Evidence demonstrating that suppression of TWIST1 plays a crucial role in the prevention of MBC [17] shows that the TWIST1/Mi2/NuRD protein complex is halted by E-cadherin expression to promote EMT and generate BC metastasis. By attaching to the E-cadherin promoter, SNAIL also reduces the expression of E-cadherin, increasing the expression of vimentin and initiating the EMT process [20]. Dong et al. discovered that SNAIL interacts with Suv39H1 (a repressor of variant 3-9 homolog 1), a significant methyl-transferase responsible for H3K9me3, and that this influences the epigenetic control of BC metastasis [21]. Similar to this, the ZEB family upregulates vimentin and N-cadherin expression by recruiting other chromatin modulators and suppressing E-cadherin expression. According to a whole genome study, tumor-promoting genes are activated to cause aggressive BC types via ZEB1-mediated activating protein-1 (AP-1) and Yes1-associated transcription regulator (YAP1) [22]. These studies offer patients with MBC a treatment alternative that targets these EMT-TFs or mediators. Recent research has demonstrated that cells possessing stem cell features, those that can self-renew and differentiate into other cell types, contribute significantly to the development, progression, and metastasis of tumors [23, 24]. Biomarkers of BC stem cells (BCSCs) have been rather thoroughly verified, although the origin of cancer stem cells is mostly unknown. For instance, a subset of CD44/CD24-+ BC cells exhibits a more aggressive phenotype, leading to treatment resistance and the spread of the disease [25, 26]. Furthermore, elevated aldehyde dehydrogenase 1 (ALDH1) activity encourages tumor initiation and self-renewal, which may be a sign of a poor clinical outcome in BC [27]. These results imply that cells displaying CD44/CD24 and ALDH1 are frequently regarded as BCSC indicators. The capacity to self-renew and initiate tumors was greatly improved in BCSCs, and they also expressed high levels of EMT markers like TWIST1, SLUG, and FOXF2 [28]. According to Storci et al., SLUG increased the stability of cytoplasmic-catenin by promoting tumor necrosis factor (TNF) and IL-8 expression.

By reducing Rad51 and ER targeting, miR-221 coupled to catenin maintained a pro-inflammatory phenotype in BCSCs [29]. BC has been demonstrated to produce TWIST1, which results in the development of cancer stem cell traits. According to one study, TWIST1 binds to the CD24 promoter and transcriptionally controls CD24 expression, leading to stem cell characteristics in BC stem cells such as self-renewal and the capacity to create mammospheres [30], indicating that EMT-TFs are primarily responsible for the phenotype in BCSCs. When tumor-derived factors and terminal stromal components interact to form a metastatic niche, CSC-TFs are thought to be key regulators in the process. One of the key modulators of ESC self-renewal that corresponds with clinical tumor grade and the final stage of BC metastasis is the RNA-binding protein Lin28 [27]. EMT is a desirable target for cancer therapies because it identifies the most deadly characteristics and chemical resistance of metastasis growth. Pirfenidone, an anti-fibrotic and anti-inflammatory drug, inhibits the TGF signaling pathway, a crucial EMT regulatory signal, which in turn decreases fibroblast proliferation [31, 32]. Pirfenidone reduced tumor burden in a mouse BC model by targeting TGF-downstream targets like SMAD3, p38, and AKT [33]. A significant additional benefit of pirfenidone during the treatment of BC was that it interfered with the hedgehog pathway signaling [34]. In addition to the surrounding tumor cells, tumors are diverse disorders that contain a variety of different cells. Proliferation and metastasis of cancer are caused by associations with these cells, which compose the adjacent tumors [35]. The term "TME" refers to the collection of these cellular and extracellular signals that include both non-cellular elements like extracellular matrix (ECM) and cytokines and cellular ones like cancer-associated fibroblasts (CAFs), adipocytes, endothelial cells, and immunological cells. These many TME constituents play roles in the development, spread, metastasis, and resistance to the treatment of cancer [36]. In addition to controlling the self-renewal of healthy mammary stem cells, the Notch pathway is crucial for encouraging the onset of BC, the growth of tumors, cell maintenance, tumor cell destiny, proliferation, survival, and motility [37]. Notch can self-renew and may function as a modulator of tumor-initiating cells (TICs), which replenish tumor cells, in BC [38–40]. Through interactions between non-malignant and malignant parts, non-malignant cells in the TME of BC, such as vascular endothelial cells, immune cells, and CAFs, enhance the activation of the Notch pathway [41]. TME is crucially regarded as a druggable target since stromal cells in the TME of BC have added to BC stemness, treatment resistance, and metastasis. Immune cells known as TAMs are crucial to the growth and spread of BC tumors. A poor prognosis results from the invasion and differentiation of TAMs, which secrete a variety of cytokines and chemokines that control the anti-tumor immune responses [42–44]. BC metastasis is highly correlated with epigenetic alterations. Heritable alterations in gene expression or stability of chromosomes are the formal definitions of epigenetics. In humans, tissue-specific gene expression must be properly maintained and developed by epigenetic mechanisms [45]. Cancer, however, leads to a deregulation of this mechanism. Normal cells have "well-balanced" epigenetic states, but cancer cells can change them in a number of different ways. BC epigenetic alterations can change a cell's status from normal to malignant, resulting in the genesis, development, and spread of the tumor [46]. Methylation of DNA, acetylation, and histone alterations of target genes are significant epigenetic modifications. BC is a varied condition for which there are numerous therapeutic options (Figure 9.1). Trastuzumab and pertuzumab are two examples of targeted therapy that are frequently used in conjunction with other anticancer medications. Different treatments that target HER2 have been developed for the HER2-positive subtype. In fact, docetaxel plus pertuzumab or trastuzumab were effective treatments for metastatic BC [47]. Additionally, tucatinib in conjunction with trastuzumab and capecitabine showed promising benefits for subjects with HER2+ MBC [48]. TNBC with a mutation in *BRCA1* particularly reacts to poly-ADP ribose polymerase (PARP) medications [49]. Nevertheless, all forms of BC still carry the risk of recurrence and metastasis. The most common reason for death in people with BC, notwithstanding recent treatment improvements, is metastasis. Research has revealed several pathways that contribute to the spread of BC (Figure 9.2), and medications have been created to block these mechanisms. But it still can't stop patients from dying from metastases. This is because there isn't a single

TABLE 9.1
Exploiting Tumor Dormancy as a Window of Therapeutic Opportunity to Target MRD

S. No.	Potential Strategy	Actionable Mechanism(s)	Proviso
1.	Stop dormancy	• Early diagnosis and removal of primary tumors • To stop tumor cell dissemination toward secondary sites • Differentiation therapies	• Early dissemination of DTCs (parallel model of metastasis) • Intrinsic or acquired resistance to treatments
2.	Reverse dormancy	• Angiogenic factors + traditional anti-proliferative treatments	• Intrinsic or acquired resistance to treatment
3.	Extend dormancy	• Immunotherapies • Antiangiogenic therapies • Dormancy-maintaining therapies (e.g., cytostatic agents)	• Immunoediting (escape) • Intrinsic or acquired resistance to treatments
4.	Eradicate dormancy	• Immunotherapies • Differentiation therapies • To halt tumor cells • Achilles heels (e.g., induces programmed cell death, halts survival, or changes metabolism pathways)	• Immunoediting (escape) • Intrinsic or acquired resistance to treatment being applied

Abbreviations: DTC, Disseminated tumor cell; MRD, minimal residual disease.

mechanism that causes BC metastasis; rather, a variety of mechanisms work together to create it. Recent research has demonstrated the considerable anticancer effect of anti–PD-L1 therapeutic immunoglobulins, such as atezolizumab or pembrolizumab, in MBC [50–52], indicating that immune checkpoints dramatically change the metastatic cascade of BC. Finally, it could be possible to use tumor dormancy in few cases as a chance to combat minimal residual disease (MRD) and/or halt metastatic relapse (Table 9.1).

9.4 TOXICITY AND SIDE EFFECTS OF METRONOMIC CHEMOTHERAPY FOR METASTATIC BREAST CANCER

MCT can nevertheless have an impact on blood cell production even though it tries to reduce bone marrow damage. This may result in a reduction in platelets, neutrophils, and red blood cells (anemia, thrombocytopenia) [53], which raises the risk of fatigue, infections, and bleeding. Gastrointestinal adverse effects, such as nausea, vomiting, diarrhea, or constipation, could occur in some patients [54]. Compared to high-dose chemotherapy, these effects are typically less severe, but they are still possible. The prevalent side effect of chemotherapy, including MCT, is fatigue. It may affect daily activities and quality of life, and its intensity can vary. MCT reduces the likelihood of peripheral neuropathy, but some medications, like vinorelbine, can still damage the nerves, causing symptoms like numbness, tingling, or pain in the extremities. Some chemotherapy drugs used in MCT can cause skin reactions, including rash, dryness, or changes in pigmentation [55]. When used with MCT, endocrine medications like tamoxifen or aromatase inhibitors might lead to hormonal alterations and their associated adverse effects, such as hot flashes, vaginal dryness, or mood swings [56]. In addition, numerous investigations have demonstrated that no one medication therapy can completely eradicate the tumor. Therefore, immune checkpoint inhibition can be a promising treatment approach in addition to currently available medications for MBC, and clinical treatment trials are urgently required.

9.5 FUTURE DIRECTIONS AND CONCLUSION

Continuous delivery of low-dose medications enables longer treatment times while reducing drug-related toxicity and any associated side effects. For MBC, there is currently no established regimen for MCT. Drug selection, dosage, and therapy duration might all differ between treatment plans. This absence of standardization makes it hard to compare outcomes across studies and hinders the development of precise treatment guidelines [57]. MCT has the potential to contribute to the emergence of drug resistance, much like conventional chemotherapy. As the disease progresses, tumors may learn to reject the low-dose chemotherapy medications over time. Research is being done on methods to prevent or delay the development of medication resistance, such as combination therapy and drug rotation. BC that has spread to other organs is a complex illness with numerous subtypes and molecular variations. Depending on the tumor biology, MCT may not always be as successful. There are still problems with treating patients according to their unique traits and finding biomarkers that indicate whether they will respond to MCT. The availability of payment for MCT may be restricted or nonexistent in various healthcare systems [58]. MCT's affordability and accessibility, particularly the availability of medications, can be difficult, which may prevent it from being widely used. The constant and protracted nature of MCT may affect patient compliance and treatment adherence. Frequent clinic visits and the requirement to provide oral drugs daily can be logistically challenging and may also reduce patient compliance with the recommended course of therapy.

For further reading, some textbooks mentioned here also give a further understanding of MCT:

- *Progress in Anti-cancer Chemotherapy* by Gabriel N. Hortobagyi and David Khayat.
- *Triple-Negative Breast Cancer: Biomarkers, Emerging Therapeutic Strategies and Clinical Challenges* by Marion Montgomery.
- *Current Cancer Treatment: Novel Beyond Conventional Approaches* by Oner Ozdemir.
- *Breast Cancer: Current and Alternative Therapeutic Modalities* by Esra Gunduz and Mehmet Gunduz.

The following links for video lectures may also be helpful:

https://youtu.be/TAqO0Mq19JQ
https://youtu.be/6wEnYfvoBqk
https://youtu.be/A9fRRCUlMms
https://youtu.be/zIWvmCsKr34

For more insight about the topic, we would suggest detailed findings from the books of (Mir MA, 2022) https://doi.org/10.1016/C2021-0-02565-7, https://doi.org/10.1016/C2022-0-00074-X, (Mir MA, 2021) https://doi.org/10.52305/WXJL6770, and (Mir MA, 2015) https://doi.org/10.1016/C2014-0-02898-5 and from the cancer.net website on the following mentioned links:

https://www.cancer.net/cancer-types/breast-cancer/types-treatment
https://discovery.ucl.ac.uk/id/eprint/1472740/
https://www.jmedsciences.com/doi/JMEDS/pdf/10.5005/jp-journals-10045-00138

For diagrammatic illustrations and descriptive tables (Lazzeroni, 2012): http://www.eurekaselect.com/article/49928

See video links on the overall status of cancer, its various types, and current and new possible treatment options available: https://www.sciencedirect.com/science/article/pii/S205970292032278X

REFERENCES

1. Gao, J.J. and S.M. Swain, *Luminal a breast cancer and molecular assays: a review*. 2018. The oncologist, **23**(5): p. 556–565.
2. Ades, F., et al., *Luminal B breast cancer: molecular characterization, clinical management, and future perspectives*. Journal of clinical oncology, 2014. **32**(25): p. 2794–2803.
3. Cazzaniga, M., et al., *Metronomic chemotherapy for advanced breast cancer patients in the real world practice: final results of the VICTOR-6 study*. The Breast, 2019. **48**: p. 7–16.
4. Cazzaniga, M.E., et al., *Metronomic chemotherapy for metastatic breast cancer treatment: clinical and preclinical data between lights and shadows*. Journal of clinical medicine, 2022. **11**(16): p. 4710.
5. Van Denderen, B.J. and E.W. Thompson, *The to and fro of tumour spread*. Nature, 2013. **493**(7433): p. 487–488.
6. Romiti, A., et al., *Metronomic chemotherapy for cancer treatment: a decade of clinical studies*. Cancer chemotherapy and pharmacology, 2013. **72**: p. 13–33.
7. Cazzaniga, M.E., et al., *Metronomic chemotherapy*. Cancers, 2021. **13**(9): p. 2236.
8. Brabletz, T., et al., *EMT in cancer*. Nature reviews. Cancer, 2018. **18**(2): p. 128–134.
9. Si, W., et al., *Dysfunction of the reciprocal feedback loop between GATA3-and ZEB2-nucleated repression programs contributes to breast cancer metastasis*. Cancer cell, 2015. **27**(6): p. 822–836.
10. Siersbæk, R., et al., *IL6/STAT3 signaling hijacks estrogen receptor α enhancers to drive breast cancer metastasis*. Cancer cell, 2020. **38**(3): p. 412–423. e9.
11. Brown, D.M. and E. Ruoslahti, *Metadherin, a cell surface protein in breast tumors that mediates lung metastasis*. Cancer cell, 2004. **5**(4): p. 365–374.
12. Chen, J., et al., *CCL18 from tumor-associated macrophages promotes breast cancer metastasis via PITPNM3*. Cancer cell, 2011. **19**(4): p. 541–555.
13. Caswell-Jin, J.L., et al., *Change in survival in metastatic breast cancer with treatment advances: meta-analysis and systematic review*. JNCI cancer spectrum, 2018. **2**(4): p. pky062.
14. Marino, N., et al., *Breast cancer metastasis: issues for the personalization of its prevention and treatment*. The American journal of pathology, 2013. **183**(4): p. 1084–1095.
15. Heerboth, S., et al., *EMT and tumor metastasis*. Clinical and translational medicine, 2015. **4**(1): p. 1–13.
16. Tanabe, S., et al., *Interplay of EMT and CSC in cancer and the potential therapeutic strategies*. Frontiers in pharmacology, 2020. **11**: p. 904.
17. Mendonsa, A.M., T.-Y. Na, and B.M. Gumbiner, *E-cadherin in contact inhibition and cancer*. Oncogene, 2018. **37**(35): p. 4769–4780.
18. Imani, S., et al., *Prognostic value of EMT-inducing transcription factors (EMT-TFs) in metastatic breast cancer: a systematic review and meta-analysis*. Scientific reports, 2016. **6**(1): p. 28587.
19. Nieto, M.A., et al., *EMT: 2016*. Cell, 2016. **166**(1): p. 21–45.
20. Liu, F., et al., *Biomarkers for EMT and MET in breast cancer: an update*. Oncology letters, 2016. **12**(6): p. 4869–4876.
21. Dong, C., et al., *Interaction with Suv39H1 is critical for Snail-mediated E-cadherin repression in breast cancer*. Oncogene, 2013. **32**(11): p. 1351–1362.
22. Feldker, N., et al., *Genome-wide cooperation of EMT transcription factor ZEB 1 with YAP and AP-1 in breast cancer*. The EMBO journal, 2020. **39**(17): p. e103209.
23. Su, C., et al., *The key roles of cancer stem cell-derived extracellular vesicles*. Signal transduction and targeted therapy, 2021. **6**(1): p. 109.
24. Song, K., M. Farzaneh, *Signaling pathways governing breast cancer stem cells behavior*. Stem cell research & therapy, 2021. **12**(1): p. 1–11.
25. Badve, S. and H. Nakshatri, *Breast-cancer stem cells—beyond semantics*. The Lancet. Oncology, 2012. **13**(1): p. e43–e48.
26. Al-Hajj, M., et al., *Prospective identification of tumorigenic breast cancer cells*. Proceedings of the National Academy of Sciences of the United States of America, 2003. **100**(7): p. 3983–3988.
27. Ginestier, C., et al., *ALDH1 is a marker of normal and malignant human mammary stem cells and a predictor of poor clinical outcome*. Cell stem cell, 2007. **1**(5): p. 555–567.
28. Da Silveira, W., et al., *Transcription factor networks derived from breast cancer stem cells control the immune response in the basal subtype*. Scientific reports, 2017. **7**(1): p. 2851.
29. Storci, G., et al., *Slug/β-catenin–dependent proinflammatory phenotype in hypoxic breast cancer stem cells*. The American journal of pathology, 2013. **183**(5): p. 1688–1697.
30. Vesuna, F., et al., *Twist modulates breast cancer stem cells by transcriptional regulation of CD24 expression*. Neoplasia, 2009. **11**(12): p. 1318–1328.

31. Schaefer, C., et al., *Antifibrotic activities of pirfenidone in animal models.* European respiratory review, 2011. **20**(120): p. 85–97.
32. Ballester, B., J. Milara, and J. Cortijo, *Pirfenidone anti-fibrotic effects are partially mediated by the inhibition of MUC1 bioactivation.* Oncotarget, 2020. **11**(15): p. 1306–1320.
33. Polydorou, C., et al., *Pirfenidone normalizes the tumor microenvironment to improve chemotherapy.* Oncotarget, 2017. **8**(15): p. 24506–24517.
34. Jeng, K.-S., et al., *Glioma-associated oncogene homolog inhibitors have the potential of suppressing cancer stem cells of breast cancer.* International journal of molecular sciences, 2018. **19**(5): p. 1375.
35. Maman, S. and I.P. Witz, *A history of exploring cancer in context.* Nature reviews. Cancer, 2018. **18**(6): p. 359–376.
36. Place, A.E., S. Jin Huh, and K. Polyak, *The microenvironment in breast cancer progression: biology and implications for treatment.* Breast cancer research, 2011. **13**: p. 1–11.
37. Politi, K., N. Feirt, and J. Kitajewski. Notch in mammary gland development and breast cancer. in Seminars in cancer biology. 2004. Elsevier.
38. Shen, Q., M. Reedijk, *Notch signaling and the breast cancer microenvironment.* Advances in experimental medicine and biology, 2021: p. 183–200.
39. Venkatesh, V., et al., *Targeting Notch signalling pathway of cancer stem cells.* Stem cell investigation, 2018. **5**: 5.
40. Choi, S., et al., *BMP-4 enhances epithelial mesenchymal transition and cancer stem cell properties of breast cancer cells via Notch signaling.* Scientific reports, 2019. **9**(1): p. 11724.
41. Zhang, Y., et al., *Notch and breast cancer metastasis: current knowledge, new sights and targeted therapy.* Oncology letters, 2019. **18**(3): p. 2743–2755.
42. Noy, R. and J.W. Pollard, *Tumor-associated macrophages: from mechanisms to therapy.* Immunity, 2014. **41**(1): p. 49–61.
43. Cassetta, L. and T. Kitamura, *Macrophage targeting: opening new possibilities for cancer immunotherapy.* Immunology, 2018. **155**(3): p. 285–293.
44. Bense, R.D., et al., *Relevance of tumor-infiltrating immune cell composition and functionality for disease outcome in breast cancer.* Journal of the National Cancer Institute, 2017. **109**(1): p. djw192.
45. Park, M., et al., *Breast cancer metastasis: mechanisms and therapeutic implications.* International journal of molecular sciences, 2022. **23**(12): p. 6806.
46. Garcia-Martinez, L., et al., *Epigenetic mechanisms in breast cancer therapy and resistance.* Nature communications, 2021. **12**(1): p. 1786.
47. Wurcel, A.G. *Drug-associated infective endocarditis trends: What's all the buzz about?* Annals of internal medicine, 2019. **170**(1): p. 68–69. doi: 10.7326/M18-3026
48. Murthy, R.K., et al., *Tucatinib, trastuzumab, and capecitabine for HER2-positive metastatic breast cancer.* The New England journal of medicine, 2020. **382**(7): p. 597–609.
49. Geenen, J.J., et al., *PARP inhibitors in the treatment of triple-negative breast cancer.* Clinical pharmacokinetics, 2018. **57**: p. 427–437.
50. Nanda, R., et al., *Pembrolizumab in patients with advanced triple-negative breast cancer: phase Ib KEYNOTE-012 study.* Journal of clinical oncology, 2016. **34**(21): p. 2460–2467.
51. Adams, S., et al., *Pembrolizumab monotherapy for previously treated metastatic triple-negative breast cancer: cohort A of the phase II KEYNOTE-086 study.* Annals of oncology, 2019. **30**(3): p. 397–404.
52. Emens, L.A., et al., *Long-term clinical outcomes and biomarker analyses of atezolizumab therapy for patients with metastatic triple-negative breast cancer: a phase 1 study.* JAMA oncology, 2019. **5**(1): p. 74–82.
53. Souto, M., et al., *Low-dose metronomic chemotherapy in metastatic breast cancer: a retrospective analysis of 40 patients.* Journal of cancer therapy, 2016. **07**: p. 480–486.
54. Liu, Y., et al., *The efficacy and toxicity profile of metronomic chemotherapy for metastatic breast cancer: A meta-analysis.* PLOS ONE, 2017. **12**(3): p. e0173693.
55. Panthi, V.K., et al., *Nanoformulations-based metronomic chemotherapy: mechanism, challenges, recent advances, and future perspectives.* Pharmaceutics, 2023. **15**(4): 1192.
56. Caparica, R., et al., *Metronomic chemotherapy combined with endocrine therapy: are we challenging some dogmas?* Expert review of anticancer therapy, 2020. **20**(7): p. 563–573.
57. Cazzaniga, M.E., et al., *Metronomic chemotherapy for advanced breast cancer patients in the real world practice: Final results of the VICTOR-6 study.* Breast, 2019. **48**: p. 7–16.
58. Romiti, A., et al., *Current achievements and future perspectives of metronomic chemotherapy.* Investigational new drugs, 2017. **35**(3): p. 359–374.

10 Recent Advances, Challenges, and the Future of Metronomic Chemotherapy

Manzoor Ahmad Mir, Sania Bashir, Shristi Handa, and Mary Chatterjee

10.1 INTRODUCTION

Metronomic chemotherapy (MCT) is a therapeutic approach in cancer treatment that involves the frequent administration of low-dose chemotherapeutic agents. Unlike traditional chemotherapy regimens that employ high-dose intermittent administration, MCT delivers continuous, lower doses of chemotherapy drugs. This treatment strategy has gained significant attention in recent years due to its potential advantages over maximum tolerated dose regimens [1, 2].

Traditional chemotherapy regimens often face challenges related to the development of drug resistance in cancer cells over time. However, MCT has shown promise in addressing this issue. By delivering continuous low doses of chemotherapeutic agents, MCT can target tumor cells more consistently and disrupt their ability to develop resistance mechanisms [3]. This sustained treatment approach can limit the emergence of drug-resistant cancer cell populations and potentially improve long-term treatment outcomes. Moreover, MCT is being investigated for its potential in combination with other treatment modalities, such as immunotherapies and targeted therapies, to further enhance its anti-tumor effects and overcome resistance mechanisms. The ability of MCT to target drug-resistant tumors represents a significant advancement in cancer treatment strategies and opens up new avenues for combating treatment resistance in various cancer types [4].

While MCT shows great promise, there are still challenges and considerations to be addressed. Optimal dosing schedules, drug combinations, and patient selection criteria need to be determined to maximize the effectiveness of this treatment approach. Additionally, further research is needed to better understand the mechanisms underlying the anti-tumor effects and resistance-reversal properties of MCT.

As a result, MCT represents a significant breakthrough in cancer treatment strategies. Its unique ability to overcome drug resistance, suppress angiogenesis, modulate the immune system, and reduce treatment-related toxicity has revolutionized the field of oncology. The synergistic potential of combining MCT with other treatment modalities further expands its therapeutic possibilities. The remarkable advancements in MCT offer renewed hope for patients and healthcare professionals, bringing us closer to more effective and personalized cancer treatments. With ongoing research and clinical trials, MCT holds great promise for the future of cancer care, providing new avenues to combat treatment resistance and improve patient outcomes.

10.2 AN INTRICATE MECHANISTIC PERSPECTIVE OF METRONOMIC CHEMOTHERAPY IN CANCER TREATMENT

MCT elicits its therapeutic effects through multifaceted mechanisms, encompassing anti-angiogenic properties, immune system modulation, and additional contributing factors. A comprehensive understanding of these mechanisms is imperative for optimizing treatment strategies and augmenting the efficacy of MCT, as demonstrated in Figure 10.1.

DOI: 10.1201/9781032632674-10

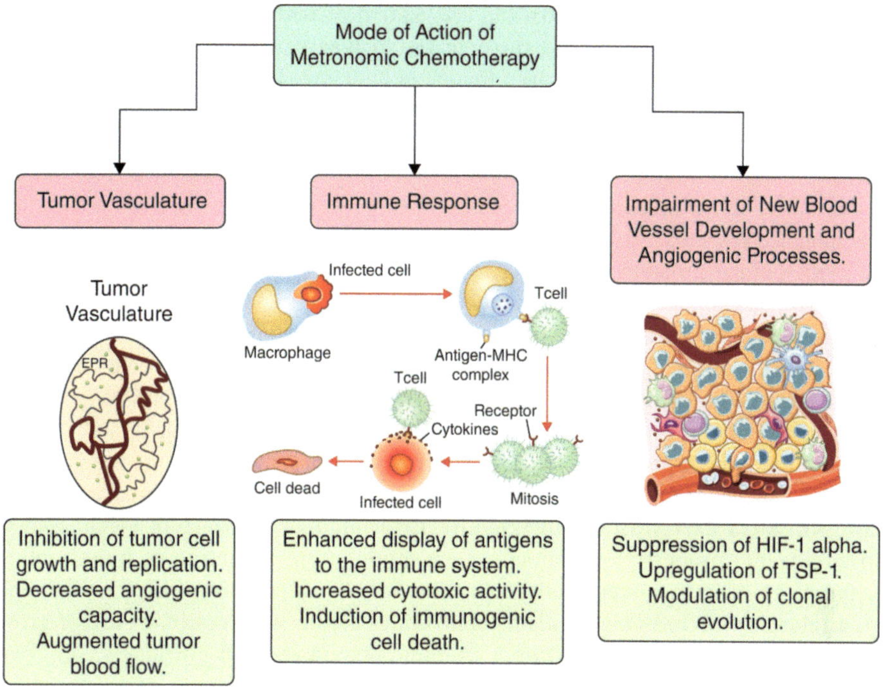

FIGURE 10.1 Schematic representation of different modes of action for metronomic chemotherapy.

Anti-angiogenic effects serve as a well-established mechanism of MCT. Tumorigenic processes necessitate a robust blood supply to sustain their growth and metastasis. MCT intervenes in this phenomenon by impeding neoangiogenesis, which involves the formation of new blood vessels [5]. Unlike conventional chemotherapy, which primarily targets rapidly proliferating neoplastic cells, MCT selectively acts upon the tumor vasculature. It impedes endothelial cell proliferation, migration, and tubulogenesis, ultimately culminating in reduced tumor vessel density and compromised tumor blood perfusion. This anti-angiogenic effect engenders tumor starvation as cancerous cells are deprived of the essential supply of nutrients and oxygen, ultimately leading to tumor growth suppression [6]. Circulating endothelial cells (CECs) play a pivotal role in tumor growth by instigating neovascularization, thereby facilitating the formation of new blood vessels to sustain tumor progression. Remarkably, MCT interventions have been shown to diminish the abundance of CECs in murine tumor models, while conventional chemotherapy treatments exhibit a contrasting effect, stimulating CEC mobilization post-treatment cycles.

Hypoxia-inducible factor-1α (HIF-1α) and vascular endothelial growth factor (VEGF) are widely acknowledged as crucial pro-angiogenic factors. In a hypoxic microenvironment, HIF-1α orchestrates the upregulation of VEGF and its receptors (VEGFR 1–2) in both normal and neoplastic cells by enhancing the transcription of HIF-1α. Significantly, MCT has been associated with decreased HIF-1α expression in diverse tumor types, with analogous outcomes reported in ovarian cell lines following MCT topotecan treatment.

In addition to its anti-angiogenic effects, MCT has exhibited potential immunomodulatory properties indicative of an intricate interaction between the treatment and the immune system. Extant research suggests that MCT can modulate immune responses and potentiate anti-tumor immunity. Prolonged exposure to low-dose chemotherapy agents can exert an influence on various immune cell subsets, such as T cells, natural killer (NK) cells, and dendritic cells, fostering immune activation and augmenting anti-tumor immune responses [7]. Notably, MCT has been demonstrated to bolster the infiltration of effector T cells into tumor sites, facilitate the release of pro-inflammatory cytokines, and enhance the recognition and eradication of malignant cells by immune effectors. These immunomodulatory effects have the potential to synergize with the direct cytotoxic actions of chemotherapy, culminating in improved treatment outcomes [8].

Furthermore, additional mechanisms contribute to the therapeutic efficacy of MCT. MCT has been observed to elicit tumor cell differentiation and senescence [9]. By promoting cellular differentiation, MCT impedes the proliferative and invasive capacities of cancerous cells. Additionally, MCT is capable of targeting cancer stem cells, a resilient subpopulation of cells within tumors that exhibit heightened resistance to conventional therapies and play a pivotal role in tumor recurrence. The elimination of cancer stem cells by MCT may avert tumor regrowth and ameliorate long-term treatment outcomes [10].

Moreover, the disruption of the tumor microenvironment constitutes another mechanism contributing to the efficacy of MCT. Tumors orchestrate a supportive microenvironment conducive to their growth and metastasis. MCT can modulate this microenvironment by diminishing the production of pro-angiogenic factors, altering the composition of the extracellular matrix, and suppressing the recruitment of immunosuppressive cells. These alterations within the tumor microenvironment create a less hospitable niche for tumor cells, thereby further impairing their growth and progression [11].

As a result, MCT exerts its therapeutic effects through a repertoire of mechanisms, including anti-angiogenic properties, immune system modulation, induction of tumor cell differentiation, targeting of cancer stem cells, and modulation of the tumor microenvironment. The concerted interplay of these mechanisms contributes to the suppression of tumor growth, curtailed angiogenesis, amplified anti-tumor immune responses, and ultimately, improved treatment outcomes.

10.3 EXPLORING THE RECENT ADVANCES IN METRONOMIC CHEMOTHERAPY'S CLINICAL APPLICATION AND EFFICACY IN ONCOLOGY

MCT has garnered considerable attention in clinical research, with numerous trials and studies evaluating its efficacy and exploring its clinical applications. These investigations have demonstrated promising results across various cancer types, including breast, ovarian, colorectal, and others. Clinical trials have assessed the effectiveness of MCT both as a standalone treatment and in combination with other therapeutic modalities.

In breast cancer, MCT has shown encouraging outcomes. Studies have revealed improved progression-free survival rates and increased overall response rates in patients receiving metronomic regimens [12]. Moreover, MCT has been found to enhance disease control and reduce the risk of relapse, particularly in patients with advanced or recurrent breast cancer. The lower toxicity profile of MCT compared to traditional chemotherapy regimens has also made it an attractive option, especially in patients who may be less tolerant of higher doses [13, 14].

In ovarian cancer, MCT has emerged as a potential strategy for improving treatment outcomes. Clinical trials have reported prolonged progression-free survival, increased response rates, and enhanced disease control in patients receiving MCT. Additionally, studies have demonstrated its efficacy in managing recurrent ovarian cancer, either as a maintenance therapy or in combination with other agents [15]. The anti-angiogenic effects of MCT have been particularly beneficial in ovarian cancer, given its propensity for tumor angiogenesis and reliance on an adequate blood supply [16].

Colorectal cancer has also been a focus of investigation in MCT studies. Research has indicated that MCT can exert notable benefits in this cancer type. It has shown efficacy in delaying disease progression, reducing tumor burden, and improving overall survival in patients with metastatic colorectal cancer. Moreover, MCT has been explored in combination with targeted therapies, such as anti-angiogenic agents and immune checkpoint inhibitors, showing potential for enhanced treatment responses and improved patient outcomes.

Comparative studies between MCT and traditional chemotherapy or targeted therapies have provided valuable insights. These studies have examined the efficacy, safety, and tolerability profiles of MCT in comparison to standard chemotherapy regimens. Results have demonstrated similar or superior efficacy of MCT, along with reduced toxicity and improved quality of life for patients. Additionally, MCT has been evaluated in combination with targeted therapies, revealing potential synergistic effects and enhanced treatment responses.

The optimal selection of chemotherapy agents, dosing schedules, and treatment duration remains an area of ongoing research and exploration. Furthermore, the identification of predictive biomarkers and patient selection criteria will be crucial for effectively applying MCT in clinical practice.

10.4 METRONOMIC CHEMOTHERAPY: CHALLENGES ON THE FRONTLINE OF CANCER TREATMENT

One of the primary challenges of MCT is drug resistance. Continuous exposure to low-dose chemotherapy drugs may lead to the development of resistance mechanisms within tumor cells. These mechanisms can arise from various sources, such as genetic mutations, enhanced DNA repair mechanisms, or the selection of drug-resistant cell populations. Over time, these resistant tumor cells can proliferate, limiting the effectiveness of MCT [17].

Another challenge lies in determining the optimal dosing schedules for MCT. Unlike traditional chemotherapy, which often involves administering high-dose drugs in cycles with rest periods, MCT lacks a standardized dosing regimen. The choice of drug, dosage, and schedule can vary depending on the specific tumor type, patient characteristics, and treatment goals [18]. Finding the right balance between efficacy and toxicity is crucial. Additionally, the continuous nature of MCT requires close monitoring and careful adjustment of drug dosages to avoid excessive toxicity while maintaining therapeutic benefits.

MCT may cause unique toxicities compared to traditional chemotherapy. Although the doses used are lower, continuous exposure can still result in adverse effects such as myelosuppression, gastrointestinal toxicity, and peripheral neuropathy [1]. These toxicities can impact the patient's quality of life and may necessitate dose adjustments or treatment interruptions. Identifying and managing these toxicities promptly is crucial for optimizing patient outcomes and minimizing treatment-related complications [14, 19].

In addition to tumor cell resistance, the tumor microenvironment poses another challenge for MCT. The microenvironment consists of various cell types, including immune cells, blood vessels, and extracellular matrix components, which play a crucial role in tumor growth and metastasis [20]. The continuous low-dose chemotherapy approach may modulate the tumor microenvironment differently than traditional chemotherapy, potentially affecting the tumor's response to treatment. Understanding the complex interplay between MCT and the tumor microenvironment is essential for developing effective combination therapies and improving treatment outcomes [11, 21].

The lack of robust clinical data and standardized protocols for MCT hinders its widespread adoption. While preclinical studies and early-phase clinical trials have shown promising results, larger, well-designed clinical trials are needed to establish the efficacy and safety of this treatment approach across different tumor types [22]. Developing consensus guidelines and treatment algorithms based on evidence-based practices would facilitate the integration of MCT into routine clinical practice.

10.5 FUTURE HORIZONS OF METRONOMIC CHEMOTHERAPY IN ONCOLOGY: PIONEERING A PARADIGM SHIFT IN CANCER MANAGEMENT

MCT represents a compelling prospect in oncology, offering the potential for improved treatment outcomes, reduced toxicities, and enhanced cost-effectiveness. As research continues to unveil the underlying mechanisms and identify optimal treatment combinations, MCT may become an integral component of personalized cancer treatment strategies, benefiting patients across a wide spectrum of cancer types and clinical settings. In light of these prospects, several key areas hold promise for MCT's future applications in oncology.

10.5.1 Personalized Medicine

MCT and advancements in cancer biology and genetics enable tailoring this therapeutic approach to individual patients based on their tumor characteristics, genetic profiles, and treatment responses. Through a deeper understanding of the molecular and genetic makeup of tumors, clinicians can identify specific biomarkers that determine a patient's likelihood of responding favorably to MCT [23]. This knowledge allows for the selection of the most appropriate chemotherapeutic agents and dosing schedules, resulting in optimized treatment plans that effectively target the unique vulnerabilities of each patient's cancer.

By adopting a personalized approach to MCT, several benefits emerge. First, treatment outcomes are potentially improved, as patients receive therapies that are precisely matched to their specific cancer subtype, leading to better response rates and increased overall survival [24]. Second, personalized medicine in MCT aims to minimize adverse effects by avoiding unnecessary exposure to chemotherapy agents that might not be effective for a particular patient's tumor. The reduced toxicity profile enhances patients' quality of life during treatment and minimizes the risk of treatment-related complications [25].

Moreover, this tailored approach paves the way for the development of innovative clinical trials focused on patient subgroups that are more likely to benefit from MCT. By enrolling patients with specific genetic mutations or biomarkers in these trials, researchers can gather more targeted and meaningful data, potentially accelerating drug development and approval processes.

10.5.2 Combination Therapies

The integration of MCT with other cancer treatment modalities presents an exciting avenue in oncology to enhance therapeutic efficacy through synergistic effects. Combining MCT with immunotherapy or targeted therapies has the potential to improve treatment outcomes by leveraging the distinct mechanisms of action of each modality.

10.5.2.1 Combination with Immunotherapy

Immunotherapy has revolutionized cancer treatment by harnessing the patient's immune system to recognize and attack cancer cells. However, not all patients respond equally to immunotherapy, and some tumors may develop resistance to immune checkpoint inhibitors. By combining immunotherapy with MCT, there is a chance to overcome these limitations and enhance the immune response against cancer cells [7, 26]. MCT, with its continuous low-dose regimen, could help create a more favorable tumor microenvironment, promoting the recruitment of immune cells and making cancer cells more susceptible to immune attack. Additionally, certain chemotherapeutic agents used in MCT might have immunomodulatory effects, further boosting the anti-cancer immune response [27].

10.5.2.2 Combination with Targeted Therapies

Targeted therapies specifically inhibit the signaling pathways or molecules that drive tumor growth, making them more selective than traditional chemotherapy. However, resistance to targeted therapies can develop over time. When combined with MCT, targeted therapies might achieve a dual effect [28]. MCT could reduce the tumor's overall burden and mitigate the development of resistance to targeted agents. Furthermore, the anti-angiogenic effects of MCT may complement the actions of targeted therapies by inhibiting the formation of new blood vessels that supply nutrients to tumors, thus impeding tumor growth [29].

10.6 OVERCOMING DRUG RESISTANCE

MCT's continuous dosing schedule has the potential to delay or circumvent drug resistance development, a major challenge in cancer therapy. By targeting tumor cells through different mechanisms, MCT could contribute to overcoming drug resistance and improving long-term treatment outcomes.

One mechanism by which MCT may counter drug resistance is by targeting tumor cells through multiple pathways simultaneously [30]. Unlike traditional chemotherapy, which often relies on a single cytotoxic mechanism, MCT employs multiple modes of action. For example, it can disrupt tumor vasculature, hamper tumor angiogenesis, and suppress cancer cell proliferation concurrently. By attacking cancer cells from multiple angles, MCT reduces the likelihood of tumor cells developing resistance to all the targeted pathways simultaneously [31].

Furthermore, the continuous exposure of tumor cells to MCT can inhibit the regrowth of resistant tumor cell clones. In traditional chemotherapy, drug-free intervals can allow for the recovery and expansion of drug-resistant tumor cells during the off-treatment periods. In contrast, MCT's uninterrupted dosing minimizes these opportunities for tumor cells to recover and reestablish resistance.

10.6.1 Adjuvant and Maintenance Therapy

The low toxicity profile of MCT makes it a promising candidate for adjuvant therapy (given after primary treatment to reduce the risk of cancer recurrence) or as maintenance therapy to keep cancer in remission and prevent relapse [32].

10.6.2 Low-Resource Settings

The cost-effectiveness of MCT could make it a viable option for cancer treatment in low-resource settings where access to expensive treatments is limited. This could significantly impact cancer care and outcomes in resource-constrained regions due to positive aspects such as lower drug costs, potential for oral administration, and versatility in treating diverse cancers.

10.6.2.1 Lower Drug Costs

MCT involves the administration of lower, more frequent doses of chemotherapy drugs, which can significantly reduce overall drug expenses. In comparison, traditional chemotherapy often requires higher doses, leading to increased drug costs. The cost savings associated with MCT can be substantial, making it more affordable and accessible for patients in low-resource settings.

Traditional chemotherapy regimens often necessitate frequent hospital visits and intensive supportive care measures to manage treatment-related side effects. In contrast, MCT's lower toxicity profile may lead to decreased reliance on expensive supportive care interventions, such as hospitalizations for severe adverse effects. This reduction in supportive care costs further contributes to the cost-effectiveness of the treatment [33].

Moreover, MCT's outpatient-based treatment approach requires fewer healthcare resources and facilities compared to the more complex and resource-intensive settings needed for traditional chemotherapy. This aspect is particularly advantageous in areas where healthcare infrastructure and medical facilities are limited.

10.6.2.2 Potential for Oral Administration

Some MCT regimens utilize oral medications, which can be administered at home by patients themselves or their caregivers. This feature further enhances treatment accessibility and convenience, particularly in areas where travel to healthcare facilities can be challenging [34]. For instance, Moes and colleagues [35] demonstrated the development of an oral solid dispersion formulation to facilitate low-dose MCT using paclitaxel, a potent anticancer drug. The conventional administration of paclitaxel often involves high doses, leading to severe side effects. To overcome this limitation, the researchers aimed to create a formulation that allows for frequent, lower doses of paclitaxel while maintaining its therapeutic efficacy. The authors conducted a comprehensive investigation to optimize the oral solid dispersion formulation of paclitaxel, considering various excipients and processing methods to enhance drug solubility and bioavailability. The dispersion formulation is designed to enhance drug dissolution, absorption, and distribution, which are crucial for achieving the desired therapeutic effect at lower doses.

The findings of the research demonstrated promising results, indicating improved drug solubility and bioavailability of paclitaxel through the oral solid dispersion formulation. This advancement in drug delivery may enable the implementation of MCT, which involves frequent administration of low doses, leading to reduced toxicity and enhanced patient compliance.

10.6.2.3 Versatility in Treating Diverse Cancers

MCT has shown promise in various types of cancers, offering versatility in its applications. This broadens its potential impact in treating different malignancies prevalent in low-resource settings, where a variety of cancer types may be encountered [36].

10.7 CONCLUSION

MCT has emerged as a promising and novel approach to cancer treatment, witnessing significant advances in recent years. This innovative strategy involves administering low doses of chemotherapy agents continuously, aiming to target the tumor microenvironment, suppress angiogenesis, and mitigate drug resistance, while simultaneously reducing toxic side effects associated with conventional high-dose chemotherapy. The research conducted so far highlights the potential benefits of MCT as an effective and well-tolerated treatment option for various types of cancer. Its ability to disrupt angiogenesis, which is crucial for tumor growth, holds promise in starving tumors of essential nutrients and inhibiting their metastatic potential. Furthermore, the continuous administration of chemotherapy agents may deter the development of drug resistance, prolonging treatment efficacy over time.

However, MCT is not without challenges; its effectiveness in aggressive cancer types and specific patient populations remains a topic of investigation, and optimizing dosing schedules and identifying predictive biomarkers are ongoing areas of research. Looking ahead, the future of MCT appears promising with the potential for personalized treatment regimens, combining it with immunotherapy and targeted therapies to enhance synergistic effects. To fully realize the potential of MCT, collaboration among scientists, clinicians, and pharmaceutical companies is crucial. Robust clinical trials, preclinical studies, and translational research will refine treatment protocols, identify biomarkers, and explore novel drug combinations.

In conclusion, MCT represents an exciting frontier in cancer therapy, showing promise as a well-tolerated and effective treatment strategy. By addressing challenges and embracing innovative approaches, MCT has the potential to significantly improve the lives of cancer patients, bringing us closer to managing cancer as a chronic condition rather than an insurmountable disease.

For further reading, some textbooks mentioned here also give a further understanding of MCT:

- *Progress in Anti-cancer Chemotherapy* by Gabriel N. Hortobagyi and David Khayat.
- *Triple-Negative Breast Cancer: Biomarkers, Emerging Therapeutic Strategies and Clinical Challenges* by Marion Montgomery.

- *Current Cancer Treatment: Novel Beyond Conventional Approaches* by Oner Ozdemir.
- *Breast Cancer: Current and Alternative Therapeutic Modalities* by Esra Gunduz and Mehmet Gunduz.

The following links for video lectures may also be helpful:

https://youtu.be/TAqO0Mq19JQ
https://youtu.be/6wEnYfvoBqk
https://youtu.be/A9fRRCUlMms
https://youtu.be/zIWvmCsKr34

For more insight about the topic, we would suggest detailed findings from the books of (Mir MA, 2022) https://doi.org/10.1016/C2021-0-02565-7, https://doi.org/10.1016/C2022-0-00074-X, (Mir MA, 2021) https://doi.org/10.52305/WXJL6770, and (Mir MA, 2015) https://doi.org/10.1016/C2014-0-02898-5 and from the cancer.net website on the following mentioned links:

https://www.cancer.net/cancer-types/breast-cancer/types-treatment
https://discovery.ucl.ac.uk/id/eprint/1472740/
https://www.jmedsciences.com/doi/JMEDS/pdf/10.5005/jp-journals-10045-00138

For diagrammatic illustrations and descriptive tables (Lazzeroni, 2012): http://www.eurekaselect.com/article/49928

See video links on the overall status of cancer, its various types, and current and new possible treatment options available: https://www.sciencedirect.com/science/article/pii/S205970292032278X

REFERENCES

1. C. Simsek, E. Esin, and S. Yalcin, "Metronomic chemotherapy: A systematic review of the literature and clinical experience," *J Oncol*, vol. 2019, p. 5483791, Mar. 2019, doi: 10.1155/2019/5483791.
2. R. Maiti, "Metronomic chemotherapy," *J Pharmacol Pharmacother*, vol. 5, no. 3, pp. 186–192, 2014, doi: 10.4103/0976-500X.136098.
3. M. E. Cazzaniga, N. Cordani, S. Capici, V. Cogliati, F. Riva, and M. G. Cerrito, "Metronomic chemotherapy," *Cancers*, vol. 13, no. 9, Art. no. 9, Jan. 2021, doi: 10.3390/cancers13092236.
4. Y.-L. Chen, M.-C. Chang, and W.-F. Cheng, "Metronomic chemotherapy and immunotherapy in cancer treatment," *Cancer Lett*, vol. 400, pp. 282–292, Aug. 2017, doi: 10.1016/j.canlet.2017.01.040.
5. A. Pircher, N. Steiner, and E. Gunsilius, "Cytotoxics and Anti-angiogenics: Metronomic Therapies," in *Tumor Angiogenesis: A Key Target for Cancer Therapy*, D. Marmé, Ed., Cham: Springer International Publishing, 2019, pp. 327–347. doi: 10.1007/978-3-319-33673-2_12.
6. J. Y. Kim and Y.-M. Kim, "Tumor endothelial cells as a potential target of metronomic chemotherapy," *Arch Pharm Res*, vol. 42, no. 1, pp. 1–13, Jan. 2019, doi: 10.1007/s12272-018-01102-z.
7. M. S. Nars and R. Kaneno, "Immunomodulatory effects of low dose chemotherapy and perspectives of its combination with immunotherapy," *Int J Cancer*, vol. 132, no. 11, pp. 2471–2478, Jun. 2013, doi: 10.1002/ijc.27801.
8. L. Galluzzi, J. Humeau, A. Buqué, L. Zitvogel, and G. Kroemer, "Immunostimulation with chemotherapy in the era of immune checkpoint inhibitors," *Nat Rev Clin Oncol*, vol. 17, no. 12, Art. no. 12, Dec. 2020, doi: 10.1038/s41571-020-0413-z.
9. E. Muraro *et al.*, "Metronomic Chemotherapy: Anti-tumor pathways and combination with immune checkpoint inhibitors," *Cancers*, vol. 15, no. 9, Art. no. 9, Jan. 2023, doi: 10.3390/cancers15092471.
10. T.-S. Chan *et al.*, "Metronomic chemotherapy prevents therapy-induced stromal activation and induction of tumor-initiating cells," *J Exp Med*, vol. 213, no. 13, pp. 2967–2988, Nov. 2016, doi: 10.1084/jem.20151665.

11. R. Muñoz, A. Girotti, D. Hileeto, and F. J. Arias, "Metronomic anti-cancer therapy: A multimodal therapy governed by the tumor microenvironment," *Cancers (Basel)*, vol. 13, no. 21, p. 5414, Oct. 2021, doi: 10.3390/cancers13215414.
12. J. Liu, M. He, Z. Wang, Q. Li, and B. Xu, "Current research status of metronomic chemotherapy in combination treatment of breast cancer," *Oncol Res Treat*, vol. 45, no. 11, pp. 681–692, Oct. 2022, doi: 10.1159/000526481.
13. B. Jain, "The role of adjuvant metronomic chemotherapy in increasing overall response rate in patients with breast cancer," *Int J Cancer Clin Res*, vol. 9, no. 2, p. 176, Sep. 2022, doi: 10.23937/2378-3419/1410176.
14. E. Munzone and M. Colleoni, "Clinical overview of metronomic chemotherapy in breast cancer," *Nat Rev Clin Oncol*, vol. 12, no. 11, Art. no. 11, Nov. 2015, doi: 10.1038/nrclinonc.2015.131.
15. A. Pandey *et al.*, "Outcomes of advanced epithelial ovarian cancer with integration of metronomic chemotherapy: An Indian rural cancer centre experience," *South Asian J Cancer*, vol. 05, no. 2, pp. 59–62, Apr. 2016, doi: 10.4103/2278-330X.181627.
16. M. Vives *et al.*, "Metronomic chemotherapy following the maximum tolerated dose is an effective antitumour therapy affecting angiogenesis, tumour dissemination and cancer stem cells," *Int J Cancer*, vol. 133, no. 10, pp. 2464–2472, 2013, doi: 10.1002/ijc.28259.
17. M. Nikolaou, A. Pavlopoulou, A. G. Georgakilas, and E. Kyrodimos, "The challenge of drug resistance in cancer treatment: a current overview," *Clin Exp Metastasis*, vol. 35, no. 4, pp. 309–318, Apr. 2018, doi: 10.1007/s10585-018-9903-0.
18. B. Laquente, F. Viñals, and J. R. Germà, "Metronomic chemotherapy: an antiangiogenic scheduling," *Clin Transl Oncol*, vol. 9, no. 2, pp. 93–98, Feb. 2007, doi: 10.1007/s12094-007-0018-3.
19. K. Lien, S. Georgsdottir, L. Sivanathan, K. Chan, and U. Emmenegger, "Low-dose metronomic chemotherapy: A systematic literature analysis," *Eur J Cancer*, vol. 49, no. 16, pp. 3387–3395, Nov. 2013, doi: 10.1016/j.ejca.2013.06.038.
20. F. Mpekris, C. Voutouri, M. Panagi, J. W. Baish, R. K. Jain, and T. Stylianopoulos, "Normalizing tumor microenvironment with nanomedicine and metronomic therapy to improve immunotherapy," *J Control Release*, vol. 345, pp. 190–199, May 2022, doi: 10.1016/j.jconrel.2022.03.008.
21. R. S. Kerbel, "A decade of experience in developing preclinical models of advanced- or early-stage spontaneous metastasis to study antiangiogenic drugs, metronomic chemotherapy, and the tumor microenvironment," *Cancer J*, vol. 21, no. 4, p. 274, Aug. 2015, doi: 10.1097/PPO.0000000000000134.
22. M. Riesco-Martinez, K. Parra, R. Saluja, G. Francia, and U. Emmenegger, "Resistance to metronomic chemotherapy and ways to overcome it," *Cancer Lett*, vol. 400, pp. 311–318, Aug. 2017, doi: 10.1016/j.canlet.2017.02.027.
23. S. Benzekry *et al.*, "Metronomic reloaded: Theoretical models bringing chemotherapy into the era of precision medicine," *Semin Cancer Biol*, vol. 35, pp. 53–61, Dec. 2015, doi: 10.1016/j.semcancer.2015.09.002.
24. I. I. Dedov, A. N. Tiul'pakov, V. P. Chekhonin, V. P. Baklushev, A. I. Archakov, and S. A. Moshkovskiĭ, "[Personalized medicine: state-of-the-art and prospects]," *Vestn Ross Akad Med Nauk*, no. 12, pp. 4–12, 2012.
25. E. V. Shlyakhto, "Scientific basics of personalized medicine: Realities and opportunities," *Her Russ Acad Sci*, vol. 92, no. 6, pp. 671–682, Dec. 2022, doi: 10.1134/S1019331622060041.
26. H. Sheng Sow and S. R. Mattarollo, "Combining low-dose or metronomic chemotherapy with anticancer vaccines," *Oncoimmunology*, vol. 2, no. 12, p. e27058, Dec. 2013, doi: 10.4161/onci.27058.
27. N. André, K. Tsai, M. Carré, and E. Pasquier, "Metronomic chemotherapy: Direct targeting of cancer cells after all?," *Trends Cancer*, vol. 3, no. 5, pp. 319–325, May 2017, doi: 10.1016/j.trecan.2017.03.011.
28. S. Banavali, E. Pasquier, and N. Andre, "Targeted therapy with propranolol and metronomic chemotherapy combination: sustained complete response of a relapsing metastatic angiosarcoma," *Ecancermedicalscience*, vol. 9, p. 499, Jan. 2015, doi: 10.3332/ecancer.2015.499.
29. E. Biziota, L. Mavroeidis, E. Hatzimichael, and P. Pappas, "Metronomic chemotherapy: A potent macerator of cancer by inducing angiogenesis suppression and antitumor immune activation," *Cancer Lett*, vol. 400, pp. 243–251, Aug. 2017, doi: 10.1016/j.canlet.2016.12.018.
30. A. Gnoni *et al.*, "Metronomic chemotherapy from rationale to clinical studies: A dream or reality?," *Crit Rev Oncol Hematol*, vol. 95, no. 1, pp. 46–61, Jul. 2015, doi: 10.1016/j.critrevonc.2015.01.008.
31. E. Pasquier, M. Kavallaris, and N. André, "Metronomic chemotherapy: new rationale for new directions," *Nat Rev Clin Oncol*, vol. 7, no. 8, Art. no. 8, Aug. 2010, doi: 10.1038/nrclinonc.2010.82.
32. Y. Liu *et al.*, "The efficacy and toxicity profile of metronomic chemotherapy for metastatic breast cancer: A meta-analysis," *PLoS One*, vol. 12, no. 3, p. e0173693, Mar. 2017, doi: 10.1371/journal.pone.0173693.

33. P. M. Parikh, S. S. Hingmire, and C. D. Deshmukh, "Selected current data on metronomic therapy (and its promise) from India," *South Asian J Cancer*, vol. 5, no. 2, pp. 37–47, 2016, doi: 10.4103/2278-330X.181623.
34. S. Kweon *et al.*, "Metronomic dose-finding approach in oral chemotherapy by experimentally-driven integrative mathematical modeling," *Biomaterials*, vol. 286, p. 121584, Jul. 2022, doi: 10.1016/j.biomaterials.2022.121584.
35. J. Moes, S. Koolen, A. Huitema, J. Schellens, J. Beijnen, and B. Nuijen, "Development of an oral solid dispersion formulation for use in low-dose metronomic chemotherapy of paclitaxel," *Eur J Pharm Biopharm*, vol. 83, no. 1, pp. 87–94, Jan. 2013, doi: 10.1016/j.ejpb.2012.09.016.
36. P. Muthusamy, K. V. Chary, and G. Nalini, "Metronomic chemotherapy: Seems prowess to battle against cancer in current scenario," *J Clin Diagn Res*, vol. 10, no. 11, pp. FC09–FC13, Nov. 2016, doi: 10.7860/JCDR/2016/23782.8802.

11 Targeted Therapies in Combination with Metronomic Chemotherapy

Manzoor Ahmad Mir, Fatima Manhas, Kirti Hooda, and Mary Chatterjee

11.1 INTRODUCTION

Cancer is the second-leading cause of mortality worldwide, killing 10 million people each year, with those from low- to middle-income countries bearing the brunt of the burden. Paul Ehrlich was the one who coined the word "chemotherapy" to describe the use of chemicals to treat illnesses. These methods were convenient but bore large toxic stress on the patient's body. So fresh research was conducted in which we can combine numerous therapy strategies for curing cancer, leading to metronomic chemotherapy development (3).

Metronomic chemotherapy treatment (MCT) refers to the chronic administration of modest doses of chemotherapy that can sustain prolonged and active plasma levels of medicines, creating favorable tolerability. MCT can be viewed as a multitarget therapy in and of itself because it affects tumor cells directly in addition to the microenvironment of the cell by suppressing tumor angiogenesis or enhancing immune response (4). In a traditional chemotherapy regimen, patients receive cytotoxic medications that are timed to be provided at intervals that are different from the maximum tolerable dose. The fundamental principle of metronomic chemotherapy is to target angiogenesis while giving the drug for a long time at a low dose. It lessens blood toxicity, which develops as a result of cancer treatment and typically causes several other illnesses. This notion was developed as a result of ground-breaking research by Klement et al. and Browder et al., which showed that mice with subcutaneous tumors could react to repeated low-dose chemotherapy treatments, even when they had developed acquired drug resistance to the same medicines (5).

So, the key characteristics of metronomic chemotherapy include chemotherapy administered regularly without pauses, replacing the maximum tolerated dosage (MTD) with a biological optimum dose, no hematopoietic growth factor application, preferring oral medications, low frequency of side effects after treatment, and potential for delayed resistance development (6).

Chemotherapeutic drugs used in directed cancer therapy target a particular genetic biomarker discovered in a particular malignancy, either directly or indirectly. Monoclonal antibodies, small molecule inhibitors, antibody-drug conjugates, and immunotherapy are all examples of targeted oncology. For instance, HER2-positive breast cancer is treated with the monoclonal antibodies pertuzumab and trastuzumab, which target the human epidermal growth factor receptor 2 (HER2) (7). Therefore, the synergistic interactions between metronomic chemotherapy and targeted therapy in the treatment of cancer offer several possible benefits by focusing on particular molecules or processes involved in the growth and progression of cancer. These targeted therapies can improve the efficacy of metronomic chemotherapy, and combining metronomic chemotherapy with targeted therapy can further lower the toxicity of the treatment because it is meant to minimize toxicity and side effects. It has also been demonstrated that the union of both can increase response rates in some cancer types, including breast cancer. Targeted medicines enable a more individualized approach to treatment by being adapted to

DOI: 10.1201/9781032632674-11

the unique genetic profile of the tumor (8). By focusing on particular molecules or pathways that are implicated in the resistance mechanisms, targeted medicines can help patients who have developed a resistance to chemotherapy (9). So, metronomic chemotherapy and targeted therapy can work together synergistically to create a more potent treatment than either one would alone.

11.2 RECENT ADVANCES IN METRONOMIC CHEMOTHERAPY

A better option than the traditional method of chemotherapy is metronomic chemotherapy. In a study by Muraro et al., a synergistic therapeutic role was seen following the concurrent administration of metronomic chemotherapy and immune checkpoint inhibitors (ICIs). ICIs are a type of immunotherapy that increases the activity of immune cells against cancer cells by blocking inhibitory pathways on those cells, such as PD-1/PD-L1 and CTLA-4. ICIs may be more effective in a setting that is immune-modulating from metronomic chemotherapy. The action of ICIs can be amplified by the chemotherapy-induced decrease in immunosuppressive cells and substances, resulting in an increase in immune cell activity against malignancies. Immune cells that have been stimulated by ICIs can target tumor antigens from metronomic chemotherapy–induced immunogenic cell death. This may result in a stronger and longer-lasting immunological response. Combining these two treatments can further regulate the tumor microenvironment, promoting immune cell infiltration and activation. Some of the immunosuppressive elements in the microenvironment may be countered by this. While metronomic chemotherapy can target remaining cancer cells and stop the spread of the disease following immune attacks, ICIs can overcome some types of tumor immune evasion. So, the combination reduces the likelihood of immunological escape. In conclusion, metronomic chemotherapy and ICIs work together to maximize the advantages of each strategy. Metronomic chemotherapy improves the effects of ICIs and produces a favorable tumor microenvironment for immune attack, while ICIs assist the immune system in fighting cancer cells to the fullest extent possible (10).

In another study, MCT based on nanoformulations showed exceptional anticancer efficacy. In this method, nanoparticles are filled with cytotoxic or antiangiogenic metronomic chemotherapeutic medicines. Due to the tumor blood vessels' tendency to leak, the nanoparticles can collect inside the tumor and shield the medications from degradation while also enabling regulated release. Targeted drug delivery minimizes systemic toxicity and negative effects by reducing exposure to healthy tissues. Combination therapies that target several aspects of tumor growth and progression are possible because of the incorporation of many medications or therapeutic agents into a single nanoparticle formulation. Medication pharmacokinetics can be altered by nanoformulations, which also improve medication absorption and circulation. In contrast to monoclonal antibodies, a mixture of antiangiogenic tyrosine kinase inhibitors (TKIs) was administered. Accordingly, pazopanib, an oral multikinase inhibitor showing encouraging effects in juvenile solid tumors and gynecological cancer, could be advantageously paired with MCT, various laboratory and preliminary clinical research has shown. MCT can be used in this situation to bridge the gap between anticancer medications and biologicals. The potential relevance of long-term usage of MCT and targeted angiogenic inhibitors may be useful in maintenance. The goal of MCT goes beyond palliative care; a neoadjuvant setting is a useful area of investigation (5).

Recent prospective clinical studies have investigated metronomic cyclophosphamide (CTX) in combination with active immunotherapy, mostly enrolling patients with breast and prostate cancer. Most of the studies included participants with advanced or metastatic disease, even in individuals who had received a lot of prior treatment. Results were generally encouraging: Combined therapies were efficient, safe, and well tolerated, and they were linked to a decline in T reg. A cell line from chronic myelogenous leukemia was used to create the allogenic tumor cell (K-562) vaccine. K-562 cells produce many antigens that are identical to those seen on the surface of many cancer cells (melanoma, sarcoma, or those originated from the lung, esophagus, and thymus), and they can activate the immune system. Celecoxib and metronomic CTX have been combined to increase the effectiveness of this vaccination (11).

In an intriguing study, Bocci et al. evaluated the pharmacoeconomic effects of metronomic cyclophosphamide-methotrexate and several other novel phase II regimens for the palliative treatment of metastatic breast cancer and concluded that the MCT scheme could lower healthcare costs, particularly those related to the combined use of the new, extremely expensive, molecularly targeted therapies (12).

When it comes to breast cancer, triple-negative breast cancer (TNBC) is an appealing candidate for immunotherapy despite the lack of specific targeted therapies for metastatic TNBC, compared to other breast cancer subtypes. This is because TNBC has a higher concentration of immune checkpoint molecules and tumor-infiltrating lymphocytes (TILs). Although programmed cell death-1 or programmed cell death ligand-1 (PD-1/PD-L1) inhibition has shown long-lasting responses in metastatic TNBC, the advantages of immunotherapy are only available to a small number of patients because tumor immune microenvironments are diverse and heterogeneous. The addition of short-term or metronomic chemotherapy in the Treatment of Nonalcoholic Fatty Liver Disorders in Children (TONIC) trial showed the ability to improve the immune milieu, hence increasing the responsiveness to PD-1 blocking in metastatic TNBC. Chue et al. described their use of immunotherapy and successive metronomic chemotherapy in two cases of recurrent metastatic TNBC patients. Both cases experienced extended remissions, providing a potentially effective therapy option. Additionally, in preclinical models of TNBC in mice, the combination of metronomic PTX and PD-1 monoclonal antibody (mAb) revealed a significant antitumor impact. Additionally, this study advanced the hypothesis that metronomic PTX enhanced the efficacy of immunotherapy by causing modifications to the immunological milieu. This phase II clinical trial (NCT04389073) was conducted as a single-arm study to evaluate the therapeutic efficacy of metronomic vinorelbine (VNR) with the anti–PD-1 antibody toripalimab for patients with metastatic HER2-negative breast cancer. Additionally, they are researching a strategy that combines metronomic chemotherapy with immunotherapy and antiangiogenesis, the results of which are highly anticipated (13).

So, utilizing metronomic and combination dosage methodologies made it possible to target tumor endothelium cells as well as other potential tumor components in addition to the obvious cancer cells, opening up a promising path for true success.

11.3 NOVEL DRUGS AND DRUG COMBINATIONS

Recent studies have looked at novel medications and treatment options for metronomic chemotherapy, including multidrug combos, ICI combinations, targeted antiangiogenic medications, and first-line therapy for advanced disease. These methods have demonstrated promise in enhancing results and lowering toxicity in some cancer types.

In a study, a mouse model with subcutaneous ectopic implantation of B16 melanoma cells was used to assess the counterbalancing effect of a novel metronomic chemotherapy protocol on such an immunosuppressive milieu. A unique medication cocktail, including taxanes and alkylating agents, was used as the chemotherapy, which was given every day in a metronomic manner. It was established that the newly developed technique was secure, well-tolerated, and highly effective. Comparing treated mice to the control group, treated animals showed a striking inhibition of tumor growth and increased survival. The decrease in CD4+ T cells and the increase in CD8+ T cells were both directly connected to this effect. Furthermore, both the spleens and the tumor lesions showed a significant decline in the percentage of regulatory T cells that were CD25+FoxP3+ and CD25+CD127 low. The metronomic treatment also produced an intrinsic CD8+ T-cell response that was directed against Trp2 TAA, which is naturally expressed in B16 (14).

This method involves administering low dosages of several antineoplastic medications, including methotrexate, etoposide, vinblastine, and PTX, repeatedly and often. While sparing nonendothelial cells and leukocytes, these levels have lethal effects on both circulating endothelial cells and circulating endothelial progenitor cells (CEPs). In comparison to tumor cells, endothelium cells are less susceptible to the development of treatment resistance due to their genetic stability (2).

In a mouse lung cancer model, intratumoral injection of a dendritic cell (DC) vaccination followed by low-dose PTX has been shown to be much more successful for therapy than either the DC vaccine or PTX alone. This combined strategy has shown the ability to inhibit tumor growth, cause CD4+ and CD8+ T-cell infiltration into the tumor site, and start an immune response specific to the tumor in the adjacent lymph nodes. The bone marrow cells showed no toxicity toward the chosen dose of PTX. Additionally, it demonstrated the capacity to promote DC maturation and function while reversing the inhibitory effects of tumor cells on DC maturation and mobility. The subcutaneous spread of mouse lung cancer is effectively stopped by this combination strategy, which also promotes increased infiltration of CD4+ and CD8+ lymphocytes (15). In addition, many chemotherapy drugs, including CTX, demonstrated the ability to cause uric acid synthesis in response to the death of tumor cells. This phenomenon can stimulate DCs, hence encouraging tumor rejection (16). Additionally, vinblastine-treated mature DCs outperform untreated DCs in terms of inducing CD8+ T-cell responses (17).

As a novel metric for MCT, the optimal biological dosage (OBD), which is the drug's lowest dose with the maximum efficacy and the absence of or lowest toxicity, may be employed. Shaked et al. (18) examined the OBD of four metronomic chemotherapeutic regimens (cyclophosphamide [CTX], vinblastine, VNR, and cisplatin) in tumor-bearing mice models. Each regimen's OBD was calculated, and 1 week after treatment, levels of viable CEPs in blood samples were assessed and compared to those in untreated control animals. According to the study, viable CEP levels significantly decreased in a dose-dependent manner, indicating that the amount of circulating CEPs may be a useful biomarker of antiangiogenic action and may be used in clinical trials (18).

A fully immune-competent mouse was implanted with a brain tumor, and it was treated with CTX on a 6-day repeating metronomic schedule. This work by Wu et al. showed that this treatment elicited a strong antitumor innate and adaptive immunity that facilitated tumor ablation. On the other hand, the immunological response was not elicited by increasing the frequency of metronomic administration from every 6 days to every 3 days or daily (19).

According to Chen et al., an intermittent 6-day schedule of 140 mg/kg of CTX caused a powerful activation of innate immune cells (macrophages and natural killer [NK] cells and DCs) that resulted in tumor regression. Interestingly, the same dose of CTX dramatically reduced the number of invading NK cells and tumor regrowth after 24 days when given on an extensive intermittent schedule (every 9 or 12 days). Therefore, metronomic medication therapy must be administered at a high enough dose and at regular intervals to induce prolonged antitumor immune cell recruitment (10).

11.4 MECHANISM OF ACTION OF METRONOMIC CHEMOTHERAPY

Both in pre-clinical and clinical settings, various modes of action have been studied. The medications used in metronomic therapy mostly impede tumor angiogenesis – the physiological process by which new blood vessels develop from pre-existing vessels – and regulate the immune system's reaction to malignancies, rather than actively destroying the tumor cells.

MCT largely shows its anticancer effects by impeding tumor angiogenesis. CTX and taxanes are two antineoplastic medicines with considerable in vitro potentiality against angiogenesis that have been shown to benefit from metronomic drug administration. In vivo, studies have also shown that MCT possesses an antiangiogenic mechanism that selectively inhibits the proliferation of activated endothelial cells and inhibits endothelial cell migration. Additional linked processes include increased thrombospondin-1 (the innermost barrier of angiogenesis) expression and a protracted drop in the number and viability of bone marrow–derived endothelial progenitor cells (20).

Both the innate and adaptive immune systems are crucial to the emergence and management of cancer. Neutropenia (a problem in which there are fewer neutrophils than normal) and lymphopenia (a disorder in which there are few lymphocytes) are well-known side effects of chemotherapy that have contributed to a patient's cancer recurring in their lifetime. Studies have demonstrated that a low dose of cyclophosphamide (CTX) on regulatory T cells reduced their quantity while also assisting in enhancing lymphocyte proliferation and memory T cells, making their impact on this situation highly relevant (10).

Cell cycle disruption can cause tumor dormancy, or it can result from a powerful balanced scenario where the start of apoptosis balances cell growth. Both during the outset of cancer formation and following the successful completion of anticancer therapy during the remission stage, tumors might resume their progression from lingering residual illness. Three key mechanisms by which MCT promotes tumor dormancy are immune surveillance, suppression of angiogenesis, and cancer cell programmed death (21).

Shaked and colleagues have investigated cellular pharmacodynamic biomarkers capable of identifying the OBD for various metronomic regimens, given that the main mechanism of metronomic chemotherapy revolves around antiangiogenesis, achieved by targeting both actively dividing endothelial cells within a tumor's developing vasculature and bone marrow–derived circulating VEGFR-2+ cells, which include CEPs. Their investigations have covered some crucial areas, including:

1. Prior observations show that extended daily low-dose metronomic chemotherapy results in significant and long-lasting decreases in circulating VEGFR-2+ CEPs.
2. Preclinical confirmation of using the concentrations of these cells as a blood-based indicator of angiogenesis and the effectiveness of targeted antiangiogenic drugs, including the best biological dose.
3. The possibility of using CEPs as a clinical marker to assess the efficacy of specific antiangiogenic medications.

Notably, their research showed a relationship between each OBD and the greatest drop in active circulating VEGFR-2+ CEPs in peripheral blood. These findings suggest that CEPs may serve as a substitute biomarker for establishing the ideal low dose in metronomic chemotherapy regimens, hence lowering the amount of treatment planning guesswork (18). These potential biomarkers have been investigated in particular to show how the medication has antiangiogenic effects. Blood biomarkers include vascular endothelial growth factor (VEGF), angiopoietin, thrombospondin 1/2 (TSP-1/2), and CECs among them (22). In addition, single nucleotide polymorphisms and immunohistochemical markers, such as VEGF and TSP-2, are being studied (23).

More recently, it was found that CDK-12 overexpression could predict the response to methotrexate-based metronomic therapy using patient-derived xenografts (PDXs) from human breast tumors with CDK-12HIGH and CDK-12LOW expression levels. Additionally, CDK-12 expression was confirmed to be a useful biomarker in breast cancer patients in a retrospective analysis involving lymph node–positive patients who were randomized to receive metronomic cyclophosphamide plus methotrexate (CM) chemotherapy or no chemotherapy after receiving standard adjuvant treatments (24).

11.5 ADVANTAGES AND LIMITATIONS OF METRONOMIC CHEMOTHERAPY

Low toxicity: MCT is well tolerated and cost-effective treatment option as demonstrated by various clinical studies conducted. The toxic effects were either rare or not found in case of MCT. The most common toxic effects were mild nausea or vomiting,mild to moderate anemia, neutropenia, leucopenia, and lymphopenia as well as low grade fatigue. Overall MCT offers a significant clinical improvement including improvement in QOL with minimal toxicity.

Cost-effective: Metronomic chemotherapy uses lower medication doses, which may cut overall treatment costs as well as hospital and supportive care expenditures.

Flexibility in administration: Metronomic chemotherapy schedules are frequently more adaptable, facilitating outpatient treatment and lowering the requirement for inpatient stays.

Improved immune response: According to some studies, metronomic chemotherapy may activate the immune system, making it more able to identify and combat tumor cells. This might work in conjunction with immunotherapy techniques.

While metronomic chemotherapy offers potential advantages such as reduced toxicity and enhanced antitumor effects, it also comes with its own set of challenges and hurdles:

Drug resistance: The emergence of drug resistance is a significant issue in metronomic chemotherapy, just as it is with conventional chemotherapy. Long-term low-dose chemotherapy exposure can select tumor cells that are resistant to the medications, reducing the efficacy of treatment.

Tumor heterogeneity: Tumors can consist of a heterogeneous collection of cell types with different chemotherapeutic drug sensitivity levels. Metronomic chemotherapy might not eliminate all tumor cell populations, which might allow for the survival of more resistant cells and the failure of the treatment.

Optimal dosing: Determining the optimal dose and schedule of metronomic chemotherapy can be challenging. Too high a dose might lead to increased toxicity, while too low a dose might not effectively control tumor growth. Finding the right balance is crucial for achieving therapeutic benefits.

Patient compliance can be affected by metronomic chemotherapy, which frequently necessitates more frequent drug administration. For treatment to be effective, patients must be dedicated to following the prescribed regimen.

11.6 CLINICAL APPLICATIONS AND EFFICACY

A metronomic chemotherapy regimen is usually a combination of drugs like CTX, methotrexate, vinblastine, etc., in short doses for a long time. Many investigators have used various metronomic chemotherapy regimens for patients with advanced and recurrent ovarian carcinoma, advanced multiple myeloma, hormone-resistant prostate cancer, and non-Hodgkin lymphoma.

According to the American Cancer Society, breast cancer is the second most common type of cancer that affects females after skin cancer, while for males it is less than 1% of the population. From various studies, it was found that faulty pre-cancerous genes *BRCA1* and *BRCA2* are one of the responsible factors, and progeny who inherit it have a 20% higher chance of getting breast cancer later in life. Available conventional methods can treat cancer but cannot control the recurrence and toxicity. So, metronomic chemotherapy has now become a method of central importance that has shown high efficiency for patients' well-being, as they showed little or no side effects from treatment. The most repeatedly examined drugs in trials are CTX, methotrexate, VNR, and capecitabine. In 2010, Taguchi et al. evaluated the efficiency of capecitabine by administering it in low doses to 33 metastatic breast cancer patients by following a routine of giving the drug twice daily for a month, which showed that patients were able to tolerate the dose, and the median progression-free survival (the length of time during and after the treatment of a disease) has also improved (26).

Non–small cell lung carcinoma is known to constitute 20% of cancer mortality. Even though there are many advanced treatments for lung cancer, the treatment can still cause toxicity inside the patient, which is now studied to be overcome by metronomic chemotherapy. The most studied antineoplastic agent in non–small cell lung carcinoma is VNR, which was administered to the patient three times a week and was specifically considered unsuitable for platinum-based treatment. The patient showed a great result in terms of response rate and overall disease control rate, demonstrating a stabilization of the disease for a longer period.

Colorectal cancer – affecting more than 40% of the cancer population according to the Global Burden of Disease Study 2017 – is the second most common reason for cancer-related mortality. Recurrence of this type of cancer is common. Fluorouracil with oxacillin (OXA) is one of the basic drugs given with CTX, a type of drug from the fluorouracil class, in combination with other drugs. The doses given were able to stop cancer due to the antiangiogenic property of CTX, but recurrence was fast and toxicity due to drug administration was also high. So, a metronomic regimen was designed and studied on 56 patients, which was a low dose of CTX as a maintenance therapy, i.e., the drug was given to them as a means of trying to control the adverse effect of the drug in their body while trying to tackle tumors inside for a long time to make them stabilize with treatment, as no other option works well on them (2).

Prostate cancer is one of the most frequent types of cancer. Many prostate cancers grow slowly and are restricted to the gland, where they may not cause major harm, but with further growth, they can pose a serious hazard because they spread aggressively, limiting treatment options. Several metronomic chemotherapies were evaluated in advanced-stage prostate cancer throughout the last three decades, but they were confined to treating cancer with hormonal choices, such as providing the patient drugs such as apalutamide, abiraterone acetate, darolutamide, or cabazitaxel. However, employing a metronomic chemotherapy regimen directly on this form of cancer is not recommended because the patient's body can fight the medicine; hence, it is utilized as a treatment option in conjunction with traditional chemotherapy (2).

11.7 TARGETED THERAPIES IN COMBINATION WITH METRONOMIC CHEMOTHERAPY IN VARIOUS TYPES OF CANCERS

Two unique methods are employed to treat various types of cancers: Metronomic chemotherapy and targeted medicines. Combining these strategies has the potential to enhance patient outcomes, decrease side effects, and boost treatment effectiveness. Targeted therapies try to specifically affect cancer cells while preserving healthy cells, in contrast to conventional chemotherapy, which can harm both healthy and diseased cells. Compared to conventional chemotherapy, this selectivity frequently has fewer adverse effects. While metronomic chemotherapy can offer a continuous low-dose treatment that addresses various elements of cancer growth and progression, targeted therapies can focus on the molecular drivers of cancer cells with specificity (Figure 11.1).

Targeted therapies in combination with metronomic chemotherapy in various types of cancers are discussed in the following sections.

11.7.1 Non-Hodgkin Lymphomas

With regard to non-Hodgkin lymphoma (NHL), all-oral MCT may help to build more effective, controllable, and cost-efficient treatments in older patients with aggressive NHL. MCT has been characterized in several ways over the years. A sensible description of this specific schedule was published in 2015 by Bocci and Kerbel, which encompasses both pharmacodynamic and pharmacokinetic components of the medication (27).

Since NHL is extremely chemosensitive and a variety of physiologically targeted medicines have become accessible, the next natural approach was to combine traditional chemo with new

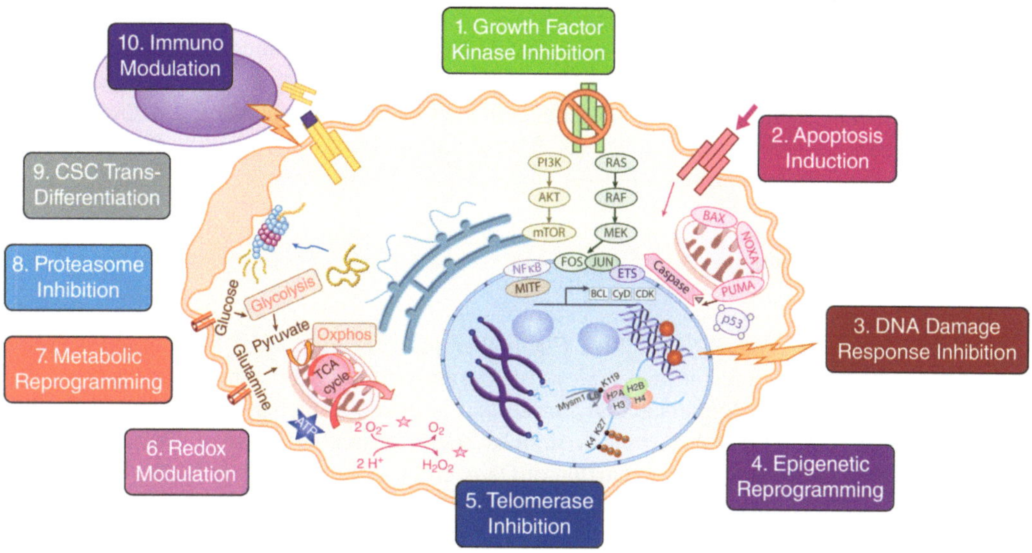

FIGURE 11.1 The principles of targeted cancer treatment. Small molecules and other drugs (1–9) have the ability to target certain physiological processes that are interrupted in a cancer cell. Furthermore, additional cells in the tumor microenvironment and immune cells can be targeted (10).

medications. While coupling MTD-chemotherapy with monoclonal antibodies resulted in ground-breaking advances, partnering MTD-chemotherapy with lenalidomide, ibrutinib, or venetoclax resulted in little improvement. This was largely owing to the higher toxicity of the innovative regimens, which necessitated therapy termination, particularly in the elderly.

In contrast, the combination of MCT with new medications has often been a success story in solid tumors, acute myeloid leukemia, and myeloma. In this section, we will present a few interesting instances that should motivate us to "recycle" existing cytotoxic medications into new "smart" NHL regimens. A phase II randomized trial combining thalidomide, PEP-C (prednisone, etoposide, procarbazine, and cyclophosphamide), and rituximab (RT-PEP) in rrMCL (relapsed/refractory mantle cell lymphoma) who had previously received a median of two lines of MTD-chemotherapy was conducted in 2010. The median overall response rate (ORR) was 73% (32% CR (complete response)/Cru (complete response unconfirmed), 41% PR (partial response), n = 22), and progression-free survival (PFS) was 10 months after a median follow-up of 38 months. Interestingly, correlative investigations revealed that RT-PEP suppressed angiogenic indicators. It is worth mentioning that the results of the pivotal phase II rituximab-lenalidomide (R2) study are quite comparable to the results of the RT-PEP. Thus, it is quite tempting to predict that MCT might further enhance both R2 and the tafasitmab + lenalidomide combos. Indeed, in myeloma, the triplet metronomic cyclophosphamide plus lenalidomide-dex or with pom-dex has been found to virtually completely eradicate the disease (28).

Azacytidine (AZA) and decitabine are two ancient cytotoxic medicines that have found fresh life in myeloid diseases and also in other areas. They were shown to exhibit strong epigenetic actions that can affect DNA transcription at low or metronomic concentrations. Indeed, AZA, either alone or in combination with romidepsin, resulted in clinical remission in T-follicular-helper type lymphomas (tFHT-PTCL). Recently, low-dose/metronomic AZA in conjunction with MTD-chemotherapy has been investigated in NHL (ClinicalTrials.gov). The justification for this is the possibility of enhancing MTD-chemotherapy activity by bio-modulating DNA transcription. Indeed, early clinical data seems to already confirm this hypothesis in PTCL. In fact, during the last American Society of Hematology meeting, it was shown that AZA-CHOP allowed a higher-than-expected remission rate in rrPTCL, especially in the tFHT subtype. This active combination will be further evaluated in the ALLIANCE/Intergroup randomized study A051902, while more trials are evaluating "MTD-chemotherapy + Aza" in front-line or in the salvage setting in B-cell NHL (ClinicalTrials.gov). This is indeed good news, but MCT + AZA may be an even more intriguing combination to assess. It would allow an all-oral schedule and the continuous interaction of different drugs (27).

11.7.2 Colorectal Cancer

Colorectal cancer is the second biggest cause of cancer-related mortality globally, according to the Global Burden of Disease Study 2017. Chemotherapy with OXA and 5-fluorouracil is used as first-line treatment for colorectal cancer. However, fluoropyrimidine resistance is frequent in a variety of malignancies, including colorectal cancer, and is linked to increased thymidylate synthase expression. Pemetrexed (PMX), a strong thymidylate synthase inhibitor, may be a better option than fluoropyrimidine. Although the combination of OXA and PMX, known as PEMOX, has exhibited exceptional clinical results in multiple clinical studies in patients with advanced colorectal cancer, it has also shown dose-related toxicities such as neutropenia and thrombocytopenia. The most serious side effects of OXA therapy include neurotoxicity, nephrotoxicity, and thrombocytopenia (29).

11.7.3 Angiosarcomas

In a small number of patients, metronomic chemotherapy was shown to provide long-term management of recurrent angiosarcoma. Thus, a multidrug metronomic combination of low-dose Trofosfamide, the peroxisome proliferator-activated receptor gamma agonist pioglitazone, and the selective cyclooxygenase-2 inhibitor rofecoxib were used in patients with advanced malignant vascular tumors. As a result, MCT has most certainly contributed to the patient's long-term tumor control (30).

11.7.4 Breast Cancer

Up to this point, various MCT combinations have been investigated, including those with celecoxib, sorafenib, veliparib, and some others. Perroud et al. hypothesized that the combination of metronomic CTX and the anti-COX2 medication celecoxib might be appropriate given the high expression of the prostaglandin synthesis enzyme cyclooxygenase-2 (COX2) in breast cancer. Along with quality of life, these scientists also examined a wide range of angiogenesis-related biomarkers, including VEGF, TSP-1, and others. The authors found no significant changes in TSP-1 over the study period, nor did the percentage of CEPs and endothelial cells (ECs) change over time, along with an excellent safety profile, specifically the absence of G3 adverse events, when looking at angiogenesis biomarkers. Combining MCT with PARP inhibitors (PARPi) was one of the most intriguing combos that was tested. Since these drugs are taken orally, they could be combined with MCT in an all-oral regimen. The justification for this combination is based on the activity of PARPi in patients with mutant *BRCA* genes, which can enhance the DNA damage brought on by chemotherapy (31).

11.7.5 Prostate Cancer

In patients who were resistant to docetaxel and had received a great deal of pre-treatment, the combination of metronomic CTX and prednisolone was feasible and showed meaningful clinical activity, leading to a confirmed prostate specific antigen (PSA) decrease of 50% or more in 26% of patients (95% confidence interval [CI]: 11–48%), with a median duration of 12.5 months in these responders. With a median response time of 10 months, 48% of patients (with a PSA drop of 30%; 95% CI: 27–69%) experienced this. It's interesting to note that metronomic CTX-prednisolone also appears to have a palliative effect as measured by a visual analog scale, since half of the patients reported a pain response. These findings are extremely significant since palliation and quality of life are still important objectives in the context of hormone refractory prostate cancer (HRPC), particularly in patients who have had a lot of pre-treatment. Metronomic CTX and prednisolone appear to be well tolerated in this situation. There were no cardiovascular events and only very infrequent cases of grade 3 and grade 4 poisoning. Furthermore, we validated previously reported findings showing that this schedule is free of hemorrhagic cystitis or secondary cancers, even in the subset of patients receiving continuous care for less than 6 months.

So, even in individuals who have received a lot of previous treatment, the combination of CTX and prednisolone seems to be able to improve pain and change PSA kinetics positively without causing much toxicity. After docetaxel-based chemotherapy fails, the response rates seen in this retrospective study are encouraging for the use of a metronomic CTX-prednisolone combination as a second-line treatment. This metronomic combination is therefore a strong candidate for combination with other immunotherapies or other antiangiogenic agents (32).

11.7.6 Lung Cancer

CEP activity and number were both decreased by CTX MCT, which also resulted in long-term tumor inhibition. It has been suggested that low-dose MCT may work well as a therapy to stop CEPs from mobilizing. MCT also causes the blood arteries in tumors to function normally, which enhances tumor perfusion. As a result, MCT slows the growth of primary and secondary tumors by cutting off the flow of vital nutrients and clearing out metabolites. In a xenograft model of human lung cancer, VNR MCT and an angiogenesis inhibitor (Endostar) dramatically improved antiangiogenic responses by preventing tumor growth by lowering the expression of cluster of differentiation (CD), VEGF, hypoxia-inducible factor 1, and CEPs (33).

11.7.7 Ovarian Cancer

Metronomic chemotherapy in combination with other antiangiogenetic drugs has been a notable field of research for first-line, maintenance, and salvage therapy, since angiogenesis plays a significant role in the pathogenesis of ovarian cancer. Compared favorably to previously published studies on metronomic chemotherapy in ovarian cancer, the clinical efficacy of the CyTo (topotecan and cyclophosphamide) regimen shows relatively higher rates of patients achieving at least disease stabilization (86.3%) and lower rates of patients who are completely resistant to this treatment (13.6%) (2).

For ovarian cancer, bevacizumab was evaluated as an antiangiogenic drug and its interaction with metronomic treatment. In a phase III research study by Burger et al., bevacizumab was added to first-line conventional therapy, strengthening its position by contrasting patients receiving carboplatin and paclitaxel in three groups. The first group did not contain bevacizumab, but the second and third groups contained initial and continuous bevacizumab, respectively. According to reports, bevacizumab increased the median PFS by about 4 months, from 10.3 months for control groups to 14.1 months on average (34).

11.7.8 Renal Cell Cancer

For renal cancer, targeted therapy is now considered the norm. The survival rate of these patients has increased as a result of focusing on angiogenesis, a crucial process in the oncogenesis of malignant cancers. Pazopanib and metronomic topotecan were combined for testing against human renal cell cancer cell lines. In metastatic foci, the combination caused and sustained dormancy. It was also demonstrated that pazopanib raised intracellular topotecan levels. Bellmunt et al. conducted an early trial looking at six cycles of combination therapy with MTD gemcitabine, metronomic capecitabine, and sorafenib for six cycles, then metronomic sorafenib. For these patients, the median PFS was 11.1 months.

Another trial employed capecitabine twice daily orally for 4 days every 3 weeks, low-dose interferon three times a week, and etoricoxib with pioglitazone daily. For the entire cohort, the median overall survival (OS) and PFS were 26.9 and 7.2 months, respectively. A reported 48.8% of cases of poisoning were grade 4 (25).

So, when treating different types of cancer, the use of targeted medicines in conjunction with metronomic chemotherapy can have many benefits. To increase therapy effectiveness and boost patient outcomes, this synergistic strategy makes use of the advantages of both treatments.

11.8 FUTURE ASPECTS OF METRONOMIC CHEMOTHERAPY

The findings of the experimental and clinical trials have documented metronomic chemotherapy as a revolutionary therapeutic approach to combat various types of cancer. Optimizing metronomic anticancer medication remains difficult after a decade of clinical research. Thus, cancer research should be focused on identifying the most appropriate therapeutic agents that can be used for cancer care and as a combination therapy with other treatment choices. Synergizing antiangiogenesis and immunomodulation with contemporary treatment options has opened a new door for exploring metronomic chemotherapy on tumors that were previously left untreated due to late discovery or because the patient's body was resisting treatment.

11.9 CONCLUSION

In conclusion, new developments in metronomic chemotherapy have revealed intriguing avenues for the treatment of cancer. Frequent, low-dose chemotherapy has demonstrated many benefits, including decreased toxicity, increased efficacy, and potential methods to combat resistance. The goal of research has been to find useful indicators for classifying patients and evaluating treatments, particularly those with antiangiogenic effects. A technique to lessen the problem of acquired drug resistance shown in cancer cells has been revealed by the investigation of metronomic regimens: The promise of targeting genetically stable ECs. These methods present a fresh perspective on increasing treatment efficacy by reducing the genetic instability causing resistance.

To improve the results of metronomic chemotherapy, further research is needed to better understand the complex molecular landscapes of cancer and the intricate interactions between many components. Capitalizing on the advantages of metronomic chemotherapy while overcoming challenges like identifying appropriate doses and individualized patient responses has the potential to further transform the landscape of cancer treatment as we traverse the complexity of cancer therapy.

For further reading, some textbooks mentioned here also give a further understanding of metronomic chemotherapy:

- *Progress in Anti-cancer Chemotherapy* by Gabriel N. Hortobagyi and David Khayat.
- *Triple-Negative Breast Cancer: Biomarkers, Emerging Therapeutic Strategies and Clinical Challenges* by Marion Montgomery.
- *Current Cancer Treatment: Novel Beyond Conventional Approaches* by Oner Ozdemir.
- *Breast Cancer: Current and Alternative Therapeutic Modalities* by Esra Gunduz and Mehmet Gunduz.

The following links for video lectures may also be helpful:

https://youtu.be/TAqO0Mq19JQ
https://youtu.be/6wEnYfvoBqk
https://youtu.be/A9fRRCUlMms
https://youtu.be/zIWvmCsKr34

For more insight about the topic, we would suggest detailed findings from the books of (Mir MA, 2022) https://doi.org/10.1016/C2021-0-02565-7, https://doi.org/10.1016/C2022-0-00074-X, (Mir MA, 2021) https://doi.org/10.52305/WXJL6770, and (Mir MA, 2015) https://doi.org/10.1016/C2014-0-02898-5 and from the cancer.net website on the following mentioned links:

https://www.cancer.net/cancer-types/breast-cancer/types-treatment
https://discovery.ucl.ac.uk/id/eprint/1472740/
https://www.jmedsciences.com/doi/JMEDS/pdf/10.5005/jp-journals-10045-00138

For diagrammatic illustrations and descriptive tables (Lazzeroni, 2012): http://www.eurekaselect.com/article/49928

See video links on the overall status of cancer, its various types, and current and new possible treatment options available: https://www.sciencedirect.com/science/article/pii/S205970292032278X

REFERENCES

1. Romiti A, Cox MC, Sarcina I, Di Rocco R, D'Antonio C, Barucca V, et al. Metronomic chemotherapy for cancer treatment: a decade of clinical studies. Cancer Chemotherapy and Pharmacology. 2013;72(1):13–33.
2. Wysocki PJ, Łobacz M, Potocki P, Kwinta Ł, Michałowska-Kaczmarczyk A, Słowik A, et al. Metronomic chemotherapy based on topotecan or topotecan and cyclophosphamide combination (CyTo) in advanced, pretreated ovarian cancer. Cancers. 2023;15(4):1067.
3. Su NW, Chen YJ. Metronomic therapy in oral squamous cell carcinoma. J Clin Med. 2021;10(13):2818.
4. Cazzaniga ME, Cordani N, Capici S, Cogliati V, Riva F, Cerrito MG. Metronomic chemotherapy. Cancers (Basel). 2021;13(9): 2236.
5. Panthi VK, Dua K, Singh S, Gupta G, Hansbro P, Paudel K. Nanoformulations-based metronomic chemotherapy: mechanism, challenges, recent advances, and future perspectives. Pharmaceutics. 2023;15:1192.
6. Maiti R. Metronomic chemotherapy. Journal of Pharmacology & Pharmacotherapeutics. 2014;5(3):186–92.
7. Smith CEP, Prasad V. Targeted cancer therapies. American Family Physician. 2021;103(3):155–63.
8. Kerbel RS. Improving conventional or low dose metronomic chemotherapy with targeted antiangiogenic drugs. Cancer Research and Treatment. 2007;39(4):150–9.
9. Kareva I. A combination of immune checkpoint inhibition with metronomic chemotherapy as a way of targeting therapy-resistant cancer cells. International Journal of Molecular Sciences. 2017;18(10):2134.
10. Muraro E, Vinante L, Fratta E, Bearz A, Höfler D, Steffan A, et al. Metronomic chemotherapy: antitumor pathways and combination with immune checkpoint inhibitors. Cancers. 2023;15(9): 2471.
11. Romiti A, Falcone R, Roberto M, Marchetti P. Current achievements and future perspectives of metronomic chemotherapy. Investigational New Drugs. 2017;35(3):359–74.
12. Scharovsky OG, Mainetti LE, Rozados VR. Metronomic chemotherapy: changing the paradigm that more is better. Current Oncology. 2009;16(2):7–15.
13. Jiaxuan L, He M, Wang Z, Li Q, Xu B. Current research status of metronomic chemotherapy in combination treatment of breast cancer. Oncology Research and Treatment. 2022;45:681–92.
14. Tagliamonte M, Petrizzo A, Napolitano M, Luciano A, Rea D, Barbieri A, et al. A novel multi-drug metronomic chemotherapy significantly delays tumor growth in mice. Journal of Translational Medicine. 2016;14(1):58.

15. Moschella F, Proietti E, Capone I, Belardelli F. Combination strategies for enhancing the efficacy of immunotherapy in cancer patients. Annals of the New York Academy of Sciences. 2010;1194:169–78.
16. Melief CJ. Cancer immunotherapy by dendritic cells. Immunity. 2008;29(3):372–83.
17. Tanaka H, Matsushima H, Nishibu A, Clausen BE, Takashima A. Dual therapeutic efficacy of vinblastine as a unique chemotherapeutic agent capable of inducing dendritic cell maturation. Cancer Research. 2009;69(17):6987–94.
18. Shaked Y, Emmenegger U, Man S, Cervi D, Bertolini F, Ben-David Y, et al. Optimal biologic dose of metronomic chemotherapy regimens is associated with maximum antiangiogenic activity. Blood. 2005;106(9):3058–61.
19. Wu J, Waxman DJ. Immunogenic chemotherapy: dose and schedule dependence and combination with immunotherapy. Cancer Letters. 2018;419:210–21.
20. Bocci G, Nicolaou KC, Kerbel RS. Protracted low-dose effects on human endothelial cell proliferation and survival in vitro reveal a selective antiangiogenic window for various chemotherapeutic drugs. Cancer Research. 2002;62(23):6938–43.
21. Bergers G, Benjamin LE. Tumorigenesis and the angiogenic switch. Nature Reviews Cancer. 2003;3(6):401–10.
22. Natale G, Bocci G. Does metronomic chemotherapy induce tumor angiogenic dormancy? A review of available preclinical and clinical data. Cancer Letters. 2018;432:28–37.
23. Cramarossa G, Lee EK, Sivanathan L, Georgsdottir S, Lien K, Santos KD, et al. A systematic literature analysis of correlative studies in low-dose metronomic chemotherapy trials. Biomarkers in Medicine. 2014;8(6):893–911.
24. Filippone MG, Gaglio D, Bonfanti R, Tucci FA, Ceccacci E, Pennisi R, et al. CDK12 promotes tumorigenesis but induces vulnerability to therapies inhibiting folate one-carbon metabolism in breast cancer. Nature Communications. 2022;13(1):2642.
25. Simsek C, Esin E, Yalcin S. Metronomic chemotherapy: a systematic review of the literature and clinical experience. Journal of Oncology. 2019;2019:5483791.
26. Cao Y. Future options of anti-angiogenic cancer therapy. Chinese Journal of Cancer. 2016;35:21.
27. Ramakrishnan R, Assudani D, Nagaraj S, Hunter T, Cho HI, Antonia S, et al. Chemotherapy enhances tumor cell susceptibility to CTL-mediated killing during cancer immunotherapy in mice. The Journal of Clinical Investigation. 2010;120(4):1111–24.
28. Cox MC, Bocci G. Metronomic chemotherapy regimens and targeted therapies in non-Hodgkin lymphoma: The best of two worlds. Cancer Letters. 2022;524:144–50.
29. Maharjan R, Choi J, Kweon S, Pangeni R, Lee NK, Park S-J, et al. A novel oral metronomic chemotherapy provokes tumor specific immunity resulting in colon cancer eradication in combination with anti-PD-1 therapy. Biomaterials. 2021;281:121334.
30. Vogt T, Hafner C, Bross K, Bataille F, Jauch K-W, Berand A, et al. Antiangiogenetic therapy with pioglitazone, rofecoxib, and metronomic trofosfamide in patients with advanced malignant vascular tumors. Cancer. 2003;98:2251–6.
31. Cazzaniga ME, Capici S, Cordani N, Cogliati V, Pepe FF, Riva F, et al. Metronomic chemotherapy for metastatic breast cancer treatment: clinical and preclinical data between lights and shadows. Journal of Clinical Medicine. 2022;11(16):4710.
32. Ladoire S, Eymard JC, Zanetta S, Mignot G, Martin E, Kermarrec I, et al. Metronomic oral cyclophosphamide prednisolone chemotherapy is an effective treatment for metastatic hormone-refractory prostate cancer after docetaxel failure. Anticancer Research. 2010;30(10):4317–23.
33. Shu Y, Weng S, Zheng S. Metronomic chemotherapy in non-small cell lung cancer. Oncology Letters. 2020;20(6):307.
34. Burger RA, Brady MF, Bookman MA, Fleming GF, Monk BJ, Huang H, et al. Incorporation of bevacizumab in the primary treatment of ovarian cancer. New England Journal of Medicine. 2011;365(26):2473–83.

12 Recent Advances in Metronomic Chemotherapy

Nanotechnology-Based Low-Dose Chemotherapy in Breast Cancer

Manzoor Ahmad Mir, Burhan Ul Haq, and Nasreen Rashid

12.1 INTRODUCTION

Metronomic chemotherapy (MCT) is the term used to describe the smallest biologically active dose of a drug that can cause anticancer activity when administered as part of a continuous dosing schedule without significant interruptions in drug administration. Instead, standard medical care chemotherapy is predicated on giving patients the maximum tolerated dosage (MTD) of a therapeutic agent or combination of drugs throughout numerous cycles with lengthy drug-free intervals in between. From its start, MCT has been thought of as a therapy that should only be utilized in very advanced or palliative situations, particularly in cases of diseases like breast cancer (BC) when more sophisticated and contemporary treatments are available and incorrectly thought to be more successful [1]. Despite the challenging conditions that MCT initially encountered, it soon became clear that it has some unusual characteristics, mostly linked to its diverse mechanisms, and that these characteristics may be used to enhance the pattern of treatment.

Either as a single delivery or in combination with radiation therapy, the highly useful and widely employed method for cancer treatment is supposedly chemotherapy [2–4]. Although traditional chemotherapy is used to a certain extent, the significant limitations of chemotherapy include the occurrence of multiple drug resistance (MDR), lack of non-specific targeting, marginal (negligible) therapeutic indices, lower bioavailability, and its high-dose needs in addition to undesirable side effects [5]. To relieve the aforementioned side effects, a relatively novel treatment strategy termed "metronomic chemotherapy" (MCT) has been introduced. This option without expanded rest durations is concerned with long-term, uniformly spaced delivery of low strengths of multiple chemotherapeutic agents. In addition to the effectiveness of this therapeutic strategy against tumors coupled with minimum toxicity, its novelty also exists in terms of the cell target switch, emphasizing tumor endothelial cells. This option (MCT) will help in introducing alterations in the therapeutic guideline designs against cancer by virtue of the proficiency and experience gained in the preclinical and clinical practice domain of MCT [6, 7]. In the previous two decades, the treatment has been significantly influenced by the emergence of nanotechnology. The nano-drug delivery systems have revealed the possibility of avoiding some of the issues associated with traditional cytotoxic drugs by ameliorating the effects of the therapeutic agents while preventing toxicity in normal cells, and this is possible due to their selectivity regarding tumors via an enhanced pervasion and aggregation effect and active uptake by cells [8, 9]. Significant promising clinical impacts have been shown by micelles, polymeric nanoparticles, and liposomes among various nanoformulation drug delivery systems. Currently, multiple nanoparticle-moderated cytotoxic agents are permitted for clinical use, and many are at different preclinical and clinical phases. The low bioavailability, lesser biodistribution, absence of biodegradation, possible toxicity, and unpredictability in the circulation increase worries over safety, especially long-term delivery, although these nanoscale drug carriers

DOI: 10.1201/9781032632674-12

provide significant uses as drug delivery systems. In response to chemoresistance, which frequently appears in cancer treatment, novel drug carriers are needed possessing better targeting ability; this will help prevent cancer, minimize the adverse effects, and help manage related pain [5, 10]. The enzyme-responsive nanoparticles (NPs) have been perceived as one of the most promising stimulus-responsive NPs. A significant promise for treating cancer is provided by bio-enzymes, which help in the reaction catalyses within the biological systems. These bio-enzymes are used in combination with NPs to increase their accumulation in tumor areas. The nanomedicines have the potential to cause tumor death when administered with harmful biological enzymes in cancer therapy. Primarily, the alterations in the expression of specific enzymes like glycosidases, proteases, and phosphatases may be witnessed in inflammatory portions, which could be controlled to get a particular class of drugs at the location of preference via enzyme-derived drug release [11–13]. There is also an economic advantage in the case of MCT, as most of the drugs employed in the metronomic plan have generic alternatives and are relatively less costly as well as available in oral form. This leads to inexpensive hospital stays and IV injections.

This treatment can be used widely, especially in developing countries, due to the minimum expenditure in applying it as a treatment option. It creates a problem for the onset of a huge number of trials supported by industry [14, 15]. Metronomics can be a good option to augment cancer cures in low- and middle-income nations (LMINs). This is more encouraging in a maintenance setting, where the application is needed for an extended period and is more useful because of low expenses and low toxicity profile [16, 17]. Thus, the MCT method is composed of nanoscale drug delivery systems to cure multiple malignancies in both preclinical and clinical domains.

12.2 TRADITIONAL CHEMOTHERAPY AND THE ASSOCIATED CONS

Traditional chemotherapy employs compounds either solo or in combination to cure diseases like cancer until the highest cytotoxicity is gained by providing the MTD. It may also be employed with surgery and radiotherapy [18]. Surgery, radiotherapy, X-rays, and chemotherapy approaches are prevalent among the traditional treatments to cure cancer. During the initial phase of cancer progression, surgery is maximally efficacious. However, the organs, tissues, and healthy cells can be impaired using radiotherapy. To eradicate transformed cells along with bearable toxicity and complications, intravenous administration of cytotoxic agents is given at an MTD [18–20]. There have been fewer morbidities and deaths witnessed by employing chemotherapy, but almost all the antitumor agents kill normal cells, especially the ones that are speedily dividing and proliferating. Thus, it is very important to work on ways to increase efficacy and lower the side effects to ameliorate the capability of cancer therapies [21]. The cancer stem cells (CSCs) represent the tumor-initiating cells, which display self-renewal and cause a small percentage of heterogeneous tumors. Tumor resistance is often associated with these CSCs [22]. The other cons associated with traditional chemotherapy include the following (see Figure 12.1):

1. Hindrance in dosage selection.
2. Poor accuracy.
3. Quick drug metabolism.
4. Undesirable side effects (like infertility [23], bone thinning [24], heart damage [25], and leukemia [26]).

The process that modulates the availability of drugs, epithelial-mesenchymal transition (EMT), as well as the oncogenic signaling routes includes several ways that permit transformed cells to fight the fatal effects of drugs [27]. The quantity of drug that reaches transformed cells is determined by the alterations in the activity or constitution of proteins that are engaged in transporting drugs, and thus they could expose the cancer cells to lower concentrations of drugs and develop resistance

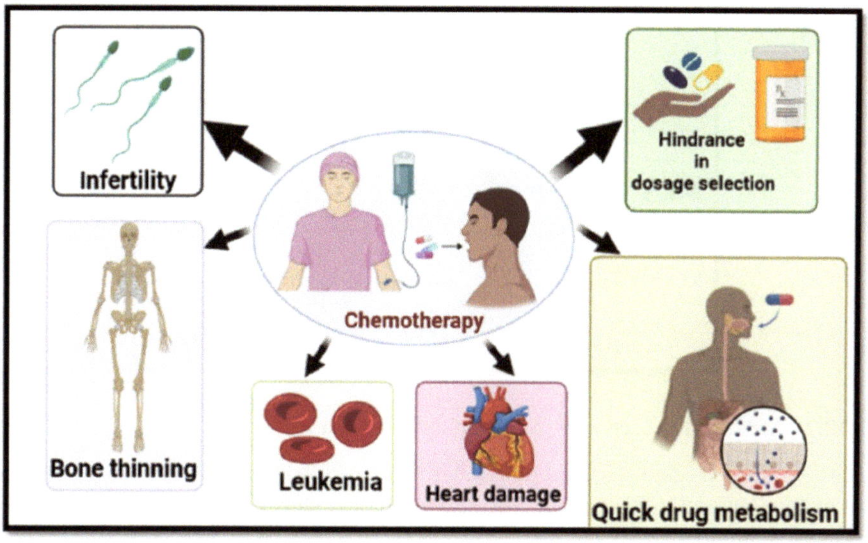

FIGURE 12.1 Undesirable effects and cons associated with traditional chemotherapy.

[28]. In response to this, a new therapeutic method was developed, termed MCT, in which one-tenth to one-third of the MTD is usually administered in the absence of lengthy drug-free intervals [29].

12.3 METRONOMIC CHEMOTHERAPY

MCT is concerned with the frequent anticancer drug administration and is recurrently given in small doses, with no long time for the drug-free condition [30]. Recently, alternative ways have been observed uncovering MCT as part of a multitargeted cure plan in addition to MCT targeting angiogenesis to control cancer progression. The term "metronomic" was introduced by Hanahan et al. As mentioned, MCT is the recurrent delivery of anticancer agents at low concentrations without lengthy drug-free intervals. Therefore, it acquires a sustained and minimal systemic drug percentage with no significant toxic side effects. There are a few important characteristics associated with MCT like repeated administration of drugs at minimal strengths with no breaks, employing a biologically optimized dose over the MCT with no application of hematopoietic growth factors (GFs), and the drug preference through the oral route [30, 31].

12.3.1 The Mechanism of Action

Both straight and incidental impacts of MCT have been witnessed on tumor cells and the tumor microenvironment. In addition, it can stop tumor angiogenesis, trigger an immune response against tumors, and increase tumor dormancy (Figure 12.2) [30].

12.3.2 Preventing Angiogenesis

MCT principally exhibits its antitumor properties by impeding blood vessel formation in tumors. Cyclophosphamide (CTX) and taxanes are two antineoplastic medicines with considerable in vitro potentiality against angiogenesis that have been shown to benefit from metronomic drug administration. In vivo studies have also shown that MCT possesses an antiangiogenic mechanism that selectively inhibits the proliferation of activated endothelial cells and inhibits endothelial cell migration. Additional linked processes include increased thrombospondin-1 (the innermost angiogenesis

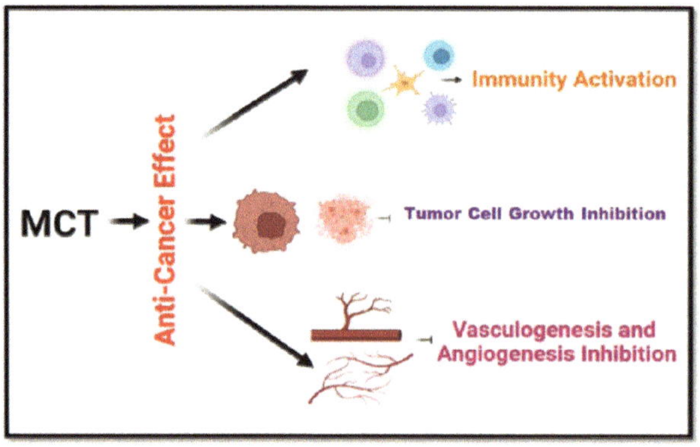

FIGURE 12.2 Mechanism of action of metronomic chemotherapy.

obstacle) expression and a protracted drop in the number and viability of bone marrow–derived endothelial progenitor cells (EPCs) [32].

12.3.3 Triggering Immunity

Both intrinsic and acquired immune systems have a significant impact on the development and control of cancer. Neutropenia and lymphopenia are the most frequent side effects of chemotherapy when it comes to the immune system. However, extensive research has demonstrated and evaluated that only a few anticancer medications, including taxanes, CTX, and anthracyclines, exhibit immune-activating properties [33]. That is because metronomic therapies are somewhat compatible with the effect on modulatory T cells. Additionally, T cells are CD+ CD25+ Foxp3+ lymphocytes that could suppress an immune reaction to an antigen in two separate ways, including through cytokine and cell contact. Inhibiting the activity of both tumor-specific (CD8+ cytotoxic T lymphocytes and CD4+ T helper cells) and tumor-unspecific effector cells such as natural killer (NK) and NK T cells allows T cells to delay the immune response to the tumor. T cells have also been seen in higher quantities in relation to the formation of tumors and a poor response to treatment in many human cancers [34]. To boost the immune response to antigens linked with the tumor, it is therefore necessary to reduce or specifically halt the function of T cells. Additionally, the effect of low-dose CTX on T cells has been demonstrated in multiple experimental and clinical trials. It diminishes the T-cell proportion, impairs their functionality, and speeds up lymphocyte and memory T-cell growth [30, 35].

12.3.4 Initiation of Tumor Dormancy

Cell cycle disruption can cause tumor dormancy, or it can result from a powerful balanced scenario where the start of apoptosis balances cell growth. Both during the onset of cancer formation and throughout the remission stage following the completion of anticancer therapy, tumors might recommence their progression from lingering residual illness [36]. Three key mechanisms through which MCT enhances tumor dormancy are immune surveillance, angiogenesis suppression, and programmed cell death of malignant cells [37].

12.4 CHEMOTHERAPEUTIC DIFFERENCE: CONVENTIONAL AND METRONOMIC

In cycles close to or at the MTD, cytotoxic medicines are administered in accordance with conventional chemotherapy regimens. To allow the patient to recuperate from severe drug responses, the

TABLE 12.1
Comparison between Traditional Chemotherapy and Metronomic Chemotherapy

S. No.	Parameters	Traditional Chemotherapy	Metronomic Chemotherapy	References
1.	Dosage	Intended to deliver at MTD	Lower than MTD	[37]
2.	Frequency of dosage	Defined gaps (weekly, fortnightly, three-weekly)	Per day, on alternate days, weekly	[30]
3.	Concentration in the plasma	Altered	Maintained	[30, 42]
4.	Cells aimed at	Multiplying tumor cells	Endothelial cells	[32]
5.	Purpose of administration	Killing cancer by inhibiting or killing quickly multiplying cancer cells directly	To attain cancer control by concentrating on blood vessel formation	[30]
6.	Toxicity	Toxicity is a central hurdle because of drugs employed as MTD	No marked undesirable side effects due to the attainment of a steady low drug concentration in the blood	[37]
7.	Efficacy	Less success in metastasis than the primary tumor	Advanced cancer	[30]

Abbreviations: MCT, Metronomic chemotherapy; MTD, Maximally tolerable dose.

dose is replaced with a sizable drug-free break. This method effectively slows the progression of disease in a large number of individuals, including both pediatric and adult patients, although it has significant drawbacks. Table 12.1 draws a comparison between MCT and traditional chemotherapy for better understanding. A common problem with metastatic and dangerous malignancies is recurrence, even if the improvement is typically seen in the early stages [30, 38]. As an alternative to conventional chemotherapy, MCT refers to the repetitive and uninterrupted use of traditional chemotherapeutics at reduced doses [39, 40]. In addition, its proper toxicity profile, reduced price, and ease of application make it superior to conventional treatments [41]. Figure 12.3 shows the problems associated with traditional chemotherapy and the ability of MCT to prevent tumor recurrence.

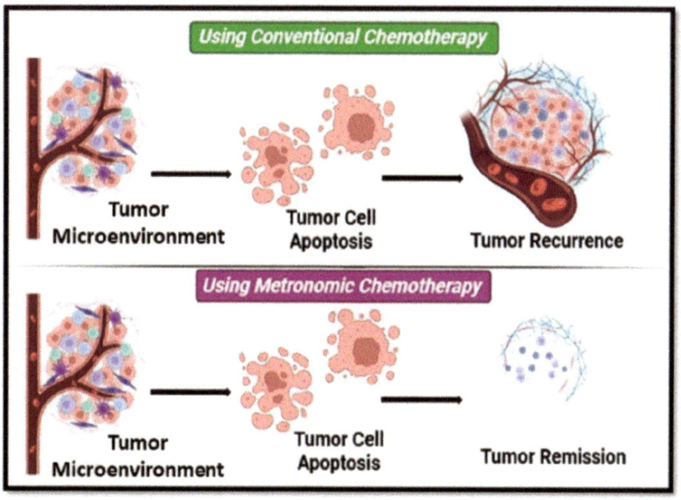

FIGURE 12.3 MCT does not result in tumor recurrence as compared to traditional chemotherapy.

12.5 ADVANCING NANOTECHNOLOGY IN BREAST CANCER TREATMENT

Nano-missiles are well-equipped with arsenals to carry out the job of eliminating the majority of cancer cells because NPs might target particular multifunctional features. The effectiveness of the chosen targeted medicine delivery is enhanced. In addition, these are adaptable in terms of dimensions [43], biocompatibility [44], materials employed, production techniques, and biodegradability. By conjugating different ligands to different NPs, high multiple abilities could encourage the combined targeted delivery of medication at the exact place to specifically kill cancer cells. The capacity to load a medicine increases the biological half-life of the antitumor treatment while also maintaining its viability. This finally results in a reduced dose being given to patients [45]. Due to improved permeability and retention, encapsulation boosted the ability and targeted the delivery of the medicine to the cancer locus, limiting the negative effects on other healthy, non-tumor cells. Theranostics are bifunctional NPSs that monitor (by imaging) and treat (via medication) cancer at the same time [37]. Due to their high surface area to volume ratio, NPs also enhance the bioavailability of substances. Favorable in vivo delivery has been demonstrated in long-lasting, bioavailable, and non-aggregating polyethylene glycol (PEG) NPs [46].

Nowadays, a gradual shift toward more sophisticated therapies and away from traditional drug formulations is witnessed. Examples include designing local drug delivery via encapsulation or surface attachment, coupling specialized gene therapies with GFs to NPs, and more [47–50]. To advance the use of NPs in the treatment of BC, numerous scientific advancements are still needed, particularly for cancers with a poor prognosis and enhanced recurrence rates, like HER-2+ and triple-negative breast cancer (TNBC). Treatments for HER-2+ BC are vastly improved by HER-2 targeting nanotherapies based on HER-2 targeting ligands like trastuzumab. In order to speed up the killing of cancer cells, trastuzumab activates immune effector cells for antibody-dependent cell-mediated cytotoxicity [51]. As a result, it is frequently used either alone or in conjunction with chemotherapy to lower the rate of recurrence and enhance the subject's prognosis. [52, 53]

NPs typically take one of two different paths: Either passive or active. The bulk of nanomedicine takes the passive method, also termed enhanced permeability and retention (EPR), depending on the kind and stage of the tumor. In the EPR effect, Chauhan et al. claim that NPs traverse leaky tumor tissue capillaries and aggregate inside cancer cells. The tumor's location, vascularity, and rate of growth all had an impact on how leaky it was [54].

The use of nanotechnology in reconstructive and cancer surgery is extremely promising. The area of BC recurrence via implants may experience significant advancements because of modified nanomedicine [54]. In vitro drug-eluting medical implants are extra prevalent, particularly for the treatment of cancer. The two highly important cell kinds in capsular contracture, macrophages and breast-derived fibroblasts, were cultured in silicone to mimic the tightly packed sphere structure of breast adipose tissue, and they demonstrated in vitro downregulation of pro-inflammatory cytokines interleukin (IL)-B1, tumor necrosis factor (TNF)-a, and IL-6 [55]. It has been proposed that breast implants with nanofiber coatings could yield novel anticancer agents [54].

By encouraging region-specific tumor bed chemotherapy and reducing some of the negative side effects of traditional systemic chemotherapy regimens, this technology can revolutionize the way cancer is treated. Similarly to this, altering the breast implant surface using antifibrotic medications might reduce the possibility of capsular contracture. It is challenging to avert the recurrence of BC by interfering with the critical connection during treatment since the primary origin of the disease's cells is yet unknown. Immunotherapy may help prevent the return of cancer since the immune system has a lasting immunological memory impact [56, 57]. The Ca^{2+} ion is considered crucial for the activity of T cells and adaptive immunity [58, 59]. A tumor acidity activatable and Ca^{2+}-aided immuno-nano agent has been created by Li et al. [60] for complementarily enhancing the activation of T cells to improve immunotherapy. As a result of the long-term memory effect, this nanodrug greatly reduced tumor growth and recurrence in mice. This immunotherapy approach may open up new avenues for medical uses like the treatment of

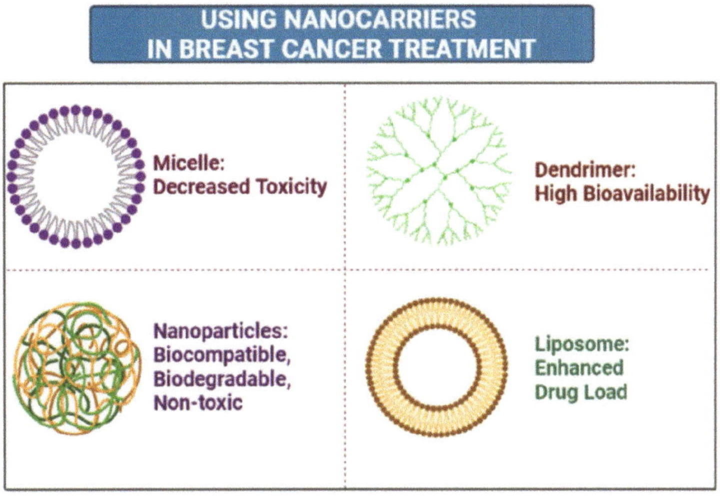

FIGURE 12.4 Properties associated with various nanocarriers, including micelles, dendrimers, NPs, and liposomes.

cancer and the prevention of relapse. Once administered for cancer therapies, plasma protein-coated nanocarriers are easily accessible for eradication by the mononuclear phagocyte system (MPS) [61]. Small interfering RNA (siRNA) delivery systems have also been created using chitosan nanoparticles that have been PEG-modified. The siRNA-loaded NPs made sure that the 4T1 breast tumor cells didn't grow or metastasize both in vitro and in vivo [62]. Altering the surface of NPs using ligands, including immunoglobulins, transferrin, and folic acid, for specific targeted therapy could be accomplished with improved selectivity and higher precision compared to the passive method [63]. The properties of various nanocarriers as a treatment option for BC are shown in Figure 12.4.

12.6 NANOFORMULATION-BASED METRONOMIC CHEMOTHERAPY

Employing nanotechnology to improve therapies is not novel; in fact, as the benefits become more clear, there has been a major increase in nanotechnology research. Continually expanding into radiotherapy, chemotherapy, prognosis, and imaging, this field has shown the ability to improve and evolve subject care [64]. Nanomedicines are also used as optional therapies for anticancer drugs. Using nanomedicines, specific site targeting could be achieved by virtue of the smaller particle size (nanometers) for the treatment of cancer (Figure 12.5), increasing their bioavailability, and planning minimally damaging side effects. Additionally, using a small amount of medications results in cost reductions [65, 66].

Additionally, nanotechnology is often used in the fabrication of these agents because their nanometer sizes are thought to be among the original cutoff scores for producing items in the field of nanotechnology. The liposomal formulations and drug conjugates like polymer [67], protein [68], and/or immunoglobulins [69], represent the majority of antitumor nanomedicines that have received approval and are intended to improve the pharmacokinetics and dynamics of the free medication and implement passive targeting. A few examples of cytotoxic medications that have been integrated into various nanocarriers to study their antitumor activity in an experimental context are etoposide (ETP), oxaliplatin (OXA), pemetrexed (PMX), and docetaxel (DTX) [70, 71]. Similar to experimental therapies, MCT built its foundation through a variety of experimental studies, starting with the trailblazing work at Folkman and Kerbel's laboratories. Judah Folkman was the first to propose that increasing tumor angiogenesis was essential for the progression of malignancy, and vascular

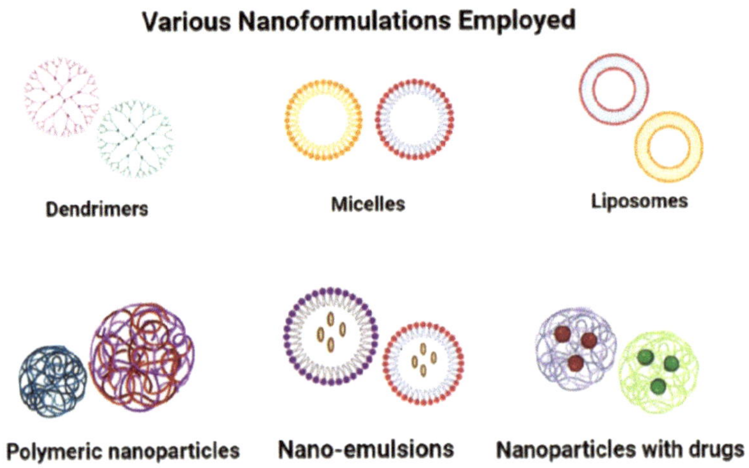

FIGURE 12.5 Different ways to deliver the drugs using NPs.

epidermal growth factor (VEGF), a subsequent angiogenesis-promoting factor, was later found to be distinct from tumors. Folkman then proposed that preventing the formation of VEGF would starve the tumor (referred to as a dormant tumor). When paired with additional cytotoxic medications like CTX, the effects of a metronomic schedule consisting of a single agent have shown a modest degree of efficacy. Excellent efficiency against tumors in tumor-possessing animals (tumors derived from breast cell lines) was revealed when CTX was administered routinely in drinking water along with an antiangiogenic compound, and it might be useful in chronic therapy [72, 73]. Figure 12.6 shows the application of nanomedicine in BC.

Numerous preclinical investigations emphasized the importance of MCT's efficacy and paved the way for its clinical evaluation. Significant research has been done in recent decades to clarify the curative efficacy of MCT and its mechanism of action. A metronomic regimen was formerly thought to only affect actively proliferating cells, more notably, the endothelial

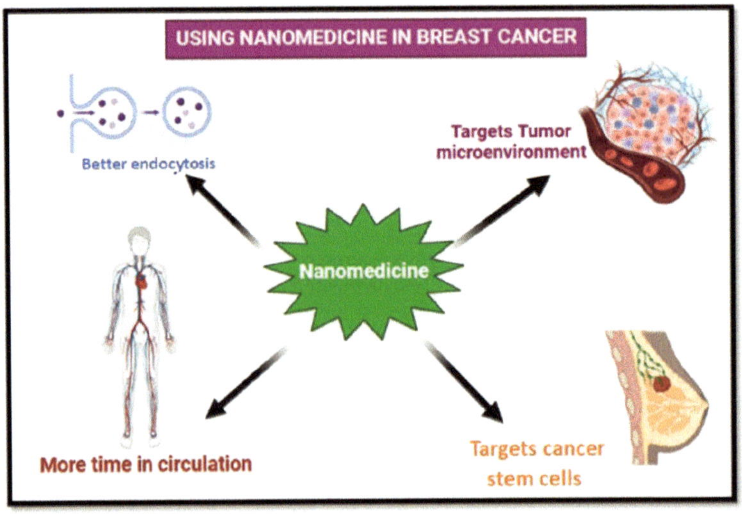

FIGURE 12.6 Usage of nanomedicine in breast cancer. It helps in better endocytosis, targets cancer stem cells, and targets the tumor microenvironment, as well as remains in circulation for more time as compared to conventional chemotherapy.

cells lining the tumor's blood vessels. The growth of new vasculature is needed for oxygen, nutrients, and proliferation in the growing tumor. In the early 20th century, Miller et al. postulated two alternative mechanisms for anticancer medications. Chemotherapeutic drugs can either directly affect cancer cells that are actively growing, or they can indirectly affect newly created blood vessels [73]. Considering this suggestion, Browder and his team suggested an approach to promote the antiangiogenic effects of chemotherapy: Continuous administration of anticancer drugs at levels below MTD with minimal intervals between phases. That a typical metronomic regimen of CTX at a low dose is more effective than the typical schedule was demonstrated in cultured breast cells that have developed drug resistance [74]. Additionally, research conducted by Klement and colleagues has demonstrated that extended administration of small-dose vinblastine combined with anti–vascular endothelial growth factor-2 (VEGFR-2) antibody combinations resulted in tumor vanquishing [75]. Using a mouse model of colon cancer, Maharjan and colleagues have assessed the immune-regulating effectiveness of a new oral MCT composed of OXA and PMX. In this investigation, MTD was linked to systemic immunosuppression, while MCT did not exhibit significant lymphotoxicity. Additionally, MCT increased the tumor's percentage of functional T cells. Combining MCT with anti–programmed cell death protein-1 led to a 95% complete response by suppressing the tumor volume by 97.85 ± 84.88% of the control group [76]. The same author and team additionally increased the antitumor immune response in mice with cancer by administering one more metronomic treatment of oral PMX-integrated colloidal dispersions. As well as increasing the percentage of infiltrating lymphocytes and impairing T-cell capabilities, metronomic oral treatment of PMX in this investigation demonstrated greater immunity against tumors [77]. Furthermore, using metronomic scheduling in tumor-bearing mice, Choi et al.'s research investigated the tumor-modulating immunity of OXA-loaded nanoemulsion (OXA-NE). The tumor volume in this study was reduced by 78.3 ± 9.67% in comparison to the control group. Additionally, OXA-NE demonstrated immunomodulatory effectiveness, increasing tumor antigen absorption and enhancing immune effector cell activity in tumor tissue [78]. Additionally, Jha and colleagues synthesized DTX-loaded NE and used metronomic dosing to test the anticancer effectiveness in tumor-bearing mice. In this investigation, the treatment group showed a noticeably greater tumor-impeding rate with the optimized formulation than the control group. Additionally, oral absorption was improved, which boosted DTX's chemotherapeutic action via MCT [79]. Moreover, research on rats and monkeys by Pangeni et al. investigated the effectiveness of oral metronomic scheduling on colorectal cancer, as well as oral absorption and antitumor activity employing OXA-loaded solid formulation. The optimized powder formulation's particle size (133 1.47 nm) fell into the nanoscale range. The results showed a significant improvement in oral absorption in both animal models, in addition to a significant obstruction of tumor growth with a metronomic dose [80].

Furthermore, using metronomic scheduling in a mouse model, Cai and colleagues' study evaluated the antiangiogenic and anticancer efficacy of new zoledronic acid–loaded liposomes. MCT has been shown to significantly slow the growth of tumors. Additionally, this treatment reduced CD206 antibodies in tumor tissues, which decreased tumor-associated macrophages [81]. Jyoti and colleagues also tested liposomal topotecan-activating MCT in tumor-endothelial spheroids in vitro in combination with radiation therapy. According to the research, liposomal topotecan inclusion increases the outcome by protecting it from systemic clearance, allowing for greater absorption, and improving tissue exposure in tumors [71]. A second investigation by Cai and colleagues coupled metronomic administration with target delivery to assess triptolide's antiangiogenic and anticancer effects in nude mice. When compared to liposome administration at MTD, the triptolide-loaded liposome in this study demonstrated a superior antitumor effect, as shown by a striking decrease in the volume of circulating EPCs in serum, the tumor volume, and the vessel density [82]. Figure 12.7 depicts one of the many ways to deliver NPs to the site of a tumor (breast cancer).

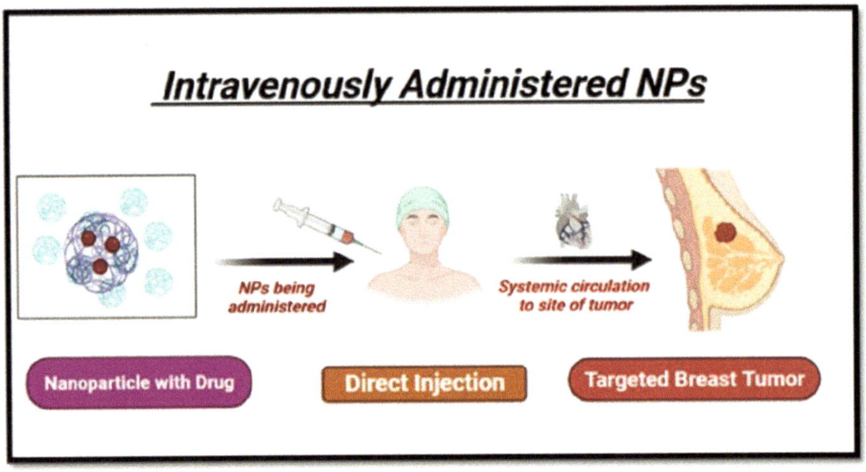

FIGURE 12.7 NPs being administered using an injection with the aim to reach the specific tumor site within the breast cancer patient.

12.7 MCT IN CLINICAL PRACTICE

Numerous investigations have so far shown that the majority of MCT-based clinical trials were well tolerated. Additionally, as evaluated in multiple clinical trials, this therapeutic approach is regarded as a cost-benefit treatment choice. The severe toxic consequences are either hardly ever observed or extremely uncommon. In general, side effects included nausea, minor-grade fatigue, mild to severe lymphopenia, leukopenia, neutropenia, and anemia. Additionally, MCT offers a remarkable clinical improvement, such as an improvement in quality of life (QOL) with minimal toxicity. But to increase the usage of MCT in clinical settings, a well-conducted phase III trial demonstrating the significant value of metronomic dosing is required. Furthermore, well-reviewed randomized phase II studies will also produce strong proof to support the use of low-dose chemotherapy. The sample size should be appropriate and should include patients who are representative of the general population [83]. Finding clinical studies that are statistically meaningful and have a randomized design has proven to be challenging thus far. In this review, we focused on the clinical research that MCT has undergone recently, namely on cancers of the breast, lungs, colon, prostate, gastroesophageal system, and pancreas.

12.8 MCT STUDIES IN BREAST CANCER

One of the tumors for MCT that has been most thoroughly researched is likely BC. Patients with metastatic disorders are typically the focus of the therapy, which aims to improve QOL and lessen disease symptoms. MCT is the most effective and least harmful alternative to conventional chemotherapy. Capecitabine (CAPE), methotrexate (MTX), vinorelbine (VRB), and CTX were the cytotoxic medications that were tested the most frequently in metronomic studies. CAPE and VRB have undergone the most extensive testing as monotherapies against cancer [84]. In 2010, the effectiveness of CAPE as a first-line treatment in 33 patients with metastatic BC was examined by Taguchi and coworkers. The following regimen was used to routinely deliver this medication: For 21 of the 28 days, 825 mg/m^2 twice every day. The median progression-free survival (PFS) and overall survival (OS) were 24.8 and 6.9 months, respectively. This trial showed that CAPE was more effective and well tolerated in this regimen [85]. Fedele and colleagues gave 60 patients with metastatic BC who had received prior treatment of 1500 mg once daily to test the impact of frequent CAPE. The median time to progression (TTP) was 7, the median OS was 17, and the clinical benefit rate (CBR) was 62% according to this study. In addition, 34 older patients had their VRB evaluated at 70 mg/m^2 three times each week, followed by a 1-week hiatus. The median FPS and OS for this MCT program were 15.9 and 7.7 months, respectively, with a grade 3 neutropenia rate of 6% [86]. Evaluation of VRB in 32 individuals was carried out by De Iuliis et al. at a dose of 30 mg administered orally every other day. This investigation demonstrated that the CBR was 50% and the safety profile was ideal (no grade 3 or 4 occurrences). Similar to this, 63 patients received a modest dose of combination oral CTX and MTX treatment, with results showing an overall CBR of 32% and an overall objective response rate (ORR) of 19% [87].

12.9 MCT AS TARGETED THERAPY

MCT is a multitargeted therapy that has both direct and indirect effects on cancer cells and their microenvironment. It works by reducing tumor angiogenesis, inducing an immune response to target tumor cells, and promoting tumor dormancy [26]. Antineoplastic drugs are used in targeted therapy, focused on specific proteins and genes that support the division, growth, and tumor cell migration. The combination of MCT with targeted therapy could eventually increase the efficacy and accuracy of maintenance, as demonstrated in several experimental and clinical settings. It is important to remember that the combined efficacy of targeted treatments and MCT has been erratic and largely failed. In addition, monoclonal antibodies were given in combination with antiangiogenic tyrosine kinase inhibitors (TKIs). Accordingly, pazopanib, an oral multikinase inhibitor showing an encouraging effect in pediatric solid tumors and gynecological cancer, could be advantageously paired with MCT, as diverse laboratory and preliminary clinical research has shown. MCT can be used in this situation to bridge the gap between anticancer medications and biologicals. The potential relevance of long-term usage of MCT and targeted angiogenic inhibitors may be useful in maintenance. The goal of MCT goes beyond palliative care; a neoadjuvant setting is a useful area of investigation. The benefits of CTX combined with liposomal doxorubicin (DXB) as a preventative measure in patients with locally advanced BC were shown in a study conducted by Dellapasqua and colleagues. A reported 44.8% of surgeries were breast-conserving procedures. Patients also displayed 62.1% partial response (PR) without exhibiting grade 4 toxicity [88]. MTX, CTX, and an immunogenic vaccination were also used to measure an immunogenic response. The accomplished ORR was 23.8%. In addition, the CAPE with DTX once weekly to a metronomic dose of taxane therapy was assessed by Young and his team to increase the expression of thymidine phosphorylase with a once-daily celecoxib regimen. The median time for disease progression (TTP) was 42%, and the observed CBR was 3.6 months in this investigation [89]. Microtubule inhibitors' antiangiogenic properties are useful in metronomic therapy. In geriatric individuals with metastatic BC, the oral delivery of VRB has been studied; the ORR in these patients was 38% [90]. In non–small cell lung cancer (NSCLC) patients receiving a substitute dose of 100 mg oral ETP, stable disease and PR were calculated at 34% and 28%, respectively. The comparable figures for OS and TTP were 9 months and 6 months, respectively [91]. Seventy-one individuals with small cell lung cancer (SCLC) who had previously received treatment received an oral metronomic regimen of ETP with lomustine and CTX. Hematological toxicity was uncommon, although the ORR reported in this trial was 38% and severe [92].

12.10 RECENT ADVANCEMENTS AND OBSTACLES

A metronomic schedule is used for continuously providing chemotherapy at low dosages. The lowest doses display fewer side effects and could encourage a shift in cancer treatment toward the management of chronic diseases [93]. Indeed, MCT as maintenance hasn't been thoroughly investigated in clinical practice, despite being thought of as a promising method. The majority of studies use small samples and single-arm observations. However, many of these evaluations have shown improved safety profiles and hopeful effects. Numerous ongoing trials are evaluating the metronomic effect as maintenance in many concomitant cancers, including BC. The results of these will support the metronomic application as maintenance therapy even more [94]. Furthermore, most metronomic therapy studies up to now have not shown appreciably improved OR and/or PFS. The metronomic regimen was assessed in a maintenance setting and concurrently given with other medicines at usual high dosages; however, no studies were performed to challenge this regimen and suitably encourage care in particular treatment lines. As a result, there is currently insufficient data to support the use of MCT as a single treatment regimen for improving outcomes. Furthermore, additional research must be done to confirm this strategy's efficacy in a clinical context [93]. According to certain studies on the immune effects of anticancer drugs, MCT is linked to immunological activity and activity against angiogenesis. In particular, checkpoint inhibition raises questions concerning how to encourage the death of an immunogenic cell. The hypoglycemic drug metformin has also received substantial study attention in the context of repurposed drugs, with retrospective analyses suggesting survival importance in diabetic patients with cancer. There is evidence of an antiangiogenic reaction with enhanced apoptosis in BC cell lines, among other anticancer mechanisms. According to preliminary findings, the use of metformin reduces particular numbers of immunosuppressive cells and might have antagonistic effects when combined with the use of a checkpoint obstacle. The unpublished outcome, on the other hand, reveals that metronomic therapy might work in concert with checkpoint inhibition, opening the door to innovative combination therapies. In conclusion, the MCT field is advancing and cutting edge, despite these sporadic issues. Additionally, clinical trials are growing to include treatments for all illnesses, not only palliative ones. The viability of combining MCT delivery with non-cancer drugs that have been repurposed is quite fascinating. Co-delivery with immunotherapeutic and targeted medicines has become more and more popular [83, 84]. On the other hand, two significant issues associated with the extended use of metronomic therapy as maintenance are cost-effectiveness and QOL. The financial impact of ongoing care for advanced malignancies has been assessed in a number of cost-effective studies. In addition, a second-line cure for disease progression might also hasten the global worsening of QOL, primarily due to illness-connected symptoms, which could ultimately jeopardize the possibility of second-line therapy. Other crucial issues include tolerance and compliance with prolonged therapy. No clinical investigation has yet shown that maintenance treatment significantly improves the QOL. The tolerable toxicity profile of extended delivery of a single anticancer medication or biological agent, however, has been revealed by various studies demonstrating that overall QOL is not compromised [83, 95]. Without question, the cost of treating cancer is a significant financial burden that needs to be resolved immediately. Cancer treatment is expensive because the medications are more expensive [96]. Bocci and the team's study contrasted metronomic doses with new treatment modalities for treating BC, comparing the outcomes and costs related to healthcare. This work, which is the only pharmacoeconomic analysis of metronomic therapy, successfully shows the potential for MCT to be more cost-effective. MCT is preferable to MTD since it requires less infrastructure because it requires smaller dosages, fewer parenteral administrations, and less complex proportions. In cancer treatment, it may be possible to integrate the medications given with the dosing schedule; metronomic regimens are more appropriate for this than MTD.

The MCT for this customization course needs to be overseen by a response and surrogate marker. The specific prognostic biomarker has not yet received recognition in the literature, although a few reports have been made in this regard. Metronomic treatment has this drawback because its treatment activities do not apply to the direction [36, 97]. A well-conducted phase III trial that identifies the scientifically notable relevance of the metronomic regimen is necessary to increase the application of MCT in clinical practice. Compelling evidence to support the use of low-dose cancer therapy will be provided by the properly executed randomized phase II studies. The research population should also consist of patients who are representative of the general community and be homogeneous. In contrast, to evaluate the impact of a metronomic regimen with a thorough inquiry, it is crucial to correlate a solo metronomic medicine with either current high-dose chemotherapy or the best symptomatic cure in a randomized scenario. As a result, we still require additional randomized trials with analytical stability. Although MCT and medication repurposing face challenges of an economic barrier in high-income nations, the low cost of both treatments is a benefit in LMINs. Medical professionals in high-income countries may learn from these settings as new, beneficial research emerges [83, 93].

12.11 CONCLUSION

These days, researchers are encouraged to deliver medicine to the specified location of the tumor by several nanoscale drug delivery methods because the MCT modality is widely used as a targeted therapy in both settings (preclinical and clinical). As a result, various problems brought on by traditional chemotherapy were resolved. The advantages of MCT over traditional treatment include a lower cost, easier application, and better toxicity profile. Growing evidence supports the use of MCT in personalized medicine, starting with palliative care and elderly patients who are physically unfit. As we highlighted in this study, several drug-incorporated nanoformulations showed noticeably higher anticancer efficacy in comparison to free drugs in various stages of clinical and preclinical practice utilizing the MCT technique. As seen in numerous experimental models and the clinic, combining MCT with a specific targeted treatment may eventually boost the efficacy and specificity of maintenance. However, there are still some questions regarding the selection of the ideal metronomic regimen for a certain illness, optimum combinational administration and programming, predictive biomarkers, and timing of organization. According to the perspective of worldwide oncology and the accomplishment of recent encouraging observations, MCT may be a hopeful therapy option for improving cancer management in LMINs [98]. We must promptly delay introducing the crucial function of MCT in treating fragile patients, since comparisons with the right supportive treatment are insufficient. Before including metronomic regimens as an optional maintenance cure, or rather than switching to remedial care in clinical settings, further clinically based comparative research studies must be conducted.

12.12 FUTURE PERSPECTIVE

The results from various experimental and clinical studies have established MCT as a new therapeutic approach for preventing many cancer types. Even after the clinical evaluation was done (>10 years), optimizing metronomic anticancer therapy remains difficult. Therefore, in the immediate future, cancer research should concentrate on identifying the maximum effective agents for managing the tumors before deciding whether to use each medicine alone or in combination, as well as how to interpret the timing of drug administration. In addition, innovative medical oncology strategies are being developed to combine MCT with conventional treatment for cancer, targeted therapy, and/or radiation therapy. Such techniques could increase a wide range of possibilities for potent pairings. At the same time, a comprehensive study of pharmaco-proteomic and pharmaco-genetics on tumor endothelial cells is now needed for analysis of their susceptibility to MCT and for estimating the maximum advantageous combination of drugs to be employed in clinical use [26]. Contrary to this, the ever-increasing cost of cancer treatment is a huge issue even in high-income countries. In LMINs, where maximum cancer incidences and mortality are witnessed, the option for several of the latest treatment plans is far from imagination. Due to their expense, fatality, and the complex framework and technology needed, the standards of care in industrialized countries are beyond what LMINs can afford. According to earlier research that has been published, metronomics may be an additional positive method for enhancing the management of cancers in LMINs. Due to its bearable toxicity profile and reasonable cost, this therapeutic method can be more advantageous even in a maintenance scenario when continued delivery is required [83]. More trials should be conducted using MCT and nano-formulations to optimize a better cure for BC patients.

For further reading, some textbooks mentioned here also give further understanding of metronomic chemotherapy:

- *Progress in Anti-cancer Chemotherapy* by Gabriel N. Hortobagyi and David Khayat.
- *Triple-Negative Breast Cancer: Biomarkers, Emerging Therapeutic Strategies and Clinical Challenges* by Marion Montgomery.
- *Current Cancer Treatment: Novel Beyond Conventional Approaches* by Oner Ozdemir.
- *Breast Cancer: Current and Alternative Therapeutic Modalities* by Esra Gunduz and Mehmet Gunduz.

The following links for video lectures may also be helpful:

https://youtu.be/TAqO0Mq19JQ
https://youtu.be/6wEnYfvoBqk
https://youtu.be/A9fRRCUIMms
https://youtu.be/zIWvmCsKr34

For more insight about the topic, we would suggest detailed findings from the books of (Mir MA, 2022) https://doi.org/10.1016/C2021-0-02565-7, https://doi.org/10.1016/C2022-0-00074-X, (Mir MA, 2021) https://doi.org/10.52305/WXJL6770, and (Mir MA, 2015) https://doi.org/10.1016/C2014-0-02898-5 and from the cancer.net website on the following mentioned links:

https://www.cancer.net/cancer-types/breast-cancer/types-treatment
https://discovery.ucl.ac.uk/id/eprint/1472740/
https://www.jmedsciences.com/doi/JMEDS/pdf/10.5005/jp-journals-10045-00138

For diagrammatic illustrations and descriptive tables (Lazzeroni, 2012): http://www.eurekaselect.com/article/49928
See video links on the overall status of cancer, its various types, and current and new possible treatment options available: https://www.sciencedirect.com/science/article/pii/S205970292032278X

REFERENCES

1. Cazzaniga, M.E., et al., *Metronomic chemotherapy.* Cancers, 2021. **13**(9): p. 2236.
2. Pangeni, R., et al., *Enhanced oral absorption of pemetrexed by ion-pairing complex formation with deoxycholic acid derivative and multiple nanoemulsion formulations: preparation, characterization, and in vivo oral bioavailability and anticancer effect.* International journal of nanomedicine, 2018. **13**: p. 3329–3351. oi: 10.2147/IJN.S167958
3. Wysocki, P.J., et al., *Metronomic chemotherapy based on topotecan or topotecan and cyclophosphamide combination (CyTo) in advanced, pretreated ovarian cancer.* Cancers, 2023. **15**(4): p. 1067.
4. Buda-Nowak, A., et al., *Metronomic Chemo-Endocrine Therapy (FulVEC) as a Salvage Treatment for Patients with Advanced, Treatment-Refractory ER+/HER2-Breast Cancer—A Retrospective Analysis of Consecutive Patients Data.* Journal of clinical medicine, 2023. **12**(4): p. 1350.
5. Senapati, S., et al., *Controlled drug delivery vehicles for cancer treatment and their performance.* Signal transduction and targeted therapy, 2018. **3**(1): p. 7.
6. Scharovsky, O.G., L.E. Mainetti, and V.R. Rozados, *Metronomic chemotherapy: changing the paradigm that more is better.* Current oncology, 2009. **16**(2): p. 7–15.
7. Liu, J., et al., *Current research status of metronomic chemotherapy in combination treatment of breast cancer.* Oncology research and treatment, 2022. **45**(11): p. 681–692.
8. Maeda, H., et al., *Tumor vascular permeability and the EPR effect in macromolecular therapeutics: a review.* Journal of controlled release, 2000. **65**(1-2): p. 271–284.

9. Koo, H., et al., *In vivo targeted delivery of nanoparticles for theranosis*. Accounts of chemical research, 2011. **44**(10): p. 1018–1028.
10. Li, Y., et al., *Biodegradable polymer nanogels for drug/nucleic acid delivery*. Chemical reviews, 2015. **115**(16): p. 8564–8608.
11. Li, M., et al., *Enzyme-responsive nanoparticles for anti-tumor drug delivery*. Frontiers in chemistry, 2020. **8**: p. 647.
12. Ding, M., et al., *Bioenzyme-based nanomedicines for enhanced cancer therapy*. Nano convergence, 2022. **9**(1): p. 7.
13. Chang, T.M.S., *ARTIFICIAL CELL evolves into nanomedicine, biotherapeutics, blood substitutes, drug delivery, enzyme/gene therapy, cancer therapy, cell/stem cell therapy, nanoparticles, liposomes, bioencapsulation, replicating synthetic cells, cell encapsulation/scaffold, biosorbent/immunosorbent haemoperfusion/plasmapheresis, regenerative medicine, encapsulated microbe, nanobiotechnology, nanotechnology*. Artificial cells, nanomedicine, and biotechnology, 2019. **47**(1): p. 997–1013.
14. Munzone, E. and M. Colleoni, *Clinical overview of metronomic chemotherapy in breast cancer*. Nature reviews. Clinical oncology, 2015. **12**(11): p. 631–644.
15. Banys-Paluchowski, M., et al., *Metronomic chemotherapy for primary non-metastatic breast cancer–a systematic review of the literature*. Geburtshilfe und Frauenheilkunde, 2017. **77**(02): p. 142–148.
16. Malik, P.S., V. Raina, and N. André, *Metronomics as maintenance treatment in oncology: time for chemo-switch*. Frontiers in oncology, 2014. **4**: p. 76.
17. She, L., et al., *Cost-effectiveness analysis of metronomic capecitabine as adjuvant chemotherapy in locoregionally advanced nasopharyngeal carcinoma*. Frontiers in oncology, 2022. **12**: p. 904372.
18. Frei Iii, E. and G.P. Canellos, *Dose: a critical factor in cancer chemotherapy*. The American journal of medicine, 1980. **69**(4): p. 585–594.
19. Hanna, N. and L.H. Einhorn, *Testicular cancer: a reflection on 50 years of discovery*. Journal of clinical oncology, 2014. **32**(28): p. 3085–3092.
20. May, T., D.P. Goldstein, and R.S. Berkowitz, *Current chemotherapeutic management of patients with gestational trophoblastic neoplasia*. Chemotherapy research and practice, 2011. **2011**: 806256.
21. Pangeni, R., et al., *Preparation, characterization, and in vivo evaluation of an oral multiple nanoemulsive system for co-delivery of pemetrexed and quercetin*. Pharmaceutics, 2018. **10**(3): p. 158.
22. Wang, X., H. Zhang, and X. Chen, *Drug resistance and combating drug resistance in cancer*. Cancer Drug Resistance, 2019. **2**(2): p. 141.
23. Poorvu, P.D., et al., *Cancer treatment-related infertility: a critical review of the evidence*. JNCI cancer spectrum, 2019. **3**(1): p. pkz008.
24. Nisha, Y., et al., *Effect of cytotoxic chemotherapy on bone health among breast cancer patients. Does it require intervention?* Supportive care in cancer, 2021. **29**: p. 6957–6972.
25. Lanza, O., et al., *New insights on the toxicity on heart and vessels of breast cancer therapies*. Medical sciences, 2022. **10**(2): p. 27.
26. Tao, J.J., K. Visvanathan, and A.C.J.T.B. Wolff, *Long term side effects of adjuvant chemotherapy in patients with early breast cancer*. Breast, 2015. **24**: p. S149–S153.
27. Phi, L.T.H., et al., *Cancer stem cells (CSCs) in drug resistance and their therapeutic implications in cancer treatment*. Stem cells international, 2018. **2018**: 5416923.
28. Peetla, C., S. Vijayaraghavalu, and V. Labhasetwar, *Biophysics of cell membrane lipids in cancer drug resistance: implications for drug transport and drug delivery with nanoparticles*. Advanced drug delivery reviews, 2013. **65**(13-14): p. 1686–1698.
29. Hanahan, D., G. Bergers, and E. Bergsland, *Less is more, regularly: metronomic dosing of cytotoxic drugs can target tumor angiogenesis in mice*. The journal of clinical investigation, 2000. **105**(8): p. 1045–1047.
30. Cazzaniga, M.E., et al., *Metronomic chemotherapy*. Cancers, 2021. **13**(9): p. 2236.
31. Maiti, R. *Metronomic chemotherapy*. Journal of pharmacology & pharmacotherapeutics, 2014. **5**(3): p. 186–192.
32. Bocci, G., K. Nicolaou, and R.S. Kerbel, *Protracted low-dose effects on human endothelial cell proliferation and survival in vitro reveal a selective antiangiogenic window for various chemotherapeutic drugs*. Cancer research, 2002. **62**(23): p. 6938–6943.
33. Zitvogel, L., et al., *Immunological aspects of cancer chemotherapy*. Nature reviews. Immunology, 2008. **8**(1): p. 59–73.
34. Kono, K., et al., *CD4 (+) CD25 high regulatory T cells increase with tumor stage in patients with gastric and esophageal cancers*. Cancer immunology, immunotherapy, 2006. **55**: p. 1064–1071.

35. Lutsiak, M.C., et al., *Inhibition of CD4+ 25+ T regulatory cell function implicated in enhanced immune response by low-dose cyclophosphamide.* Blood, 2005. **105**(7): p. 2862–2868.
36. Bergers, G. and L.E. Benjamin, *Tumorigenesis and the angiogenic switch.* Nature reviews cancer, 2003. **3**(6): p. 401–410.
37. Bahl, A. and S. Bakhshi, *Metronomic chemotherapy in progressive pediatric malignancies: old drugs in new package.* Indian journal of pediatrics, 2012. **79**: p. 1617–1622.
38. Weitman, S., E. Glatstein, and B. Kamen, *Back to the basics: the importance of concentration× time in oncology.* Journal of clinical oncology, 1993. **11**(5): p. 820–821.
39. Lien, K., et al., *Low-dose metronomic chemotherapy: a systematic literature analysis.* European journal of cancer, 2013. **49**(16): p. 3387–3395.
40. Pasquier, E., M. Kavallaris, and N. André, *Metronomic chemotherapy: new rationale for new directions.* Nature reviews. Clinical oncology, 2010. **7**(8): p. 455–465.
41. Simsek, C., E. Esin, and S. Yalcin, *Metronomic chemotherapy: a systematic review of the literature and clinical experience.* Journal of oncology, 2019. **2019**: 5483791.
42. Pasquier, E., N. Andre, and D. Braguer, *Targeting microtubules to inhibit angiogenesis and disrupt tumour vasculature: implications for cancer treatment.* Current cancer drug targets, 2007. **7**(6): p. 566–581.
43. Zhang, H., et al., *Effects of PEGylated paclitaxel nanocrystals on breast cancer and its lung metastasis.* Nanoscale, 2015. **7**(24): p. 10790–10800.
44. Ghasemiyeh, P. and S. Mohammadi-Samani, *Solid lipid nanoparticles and nanostructured lipid carriers as novel drug delivery systems: Applications, advantages and disadvantages.* Research in pharmaceutical sciences, 2018. **13**(4): p. 288–303.
45. Wakaskar, R.R., *General overview of lipid–polymer hybrid nanoparticles, dendrimers, micelles, liposomes, spongosomes and cubosomes.* Journal of drug targeting, 2018. **26**(4): p. 311–318.
46. Gref, R., et al., *Biodegradable long-circulating polymeric nanospheres.* Science, 1994. **263**(5153): p. 1600–1603.
47. Majidi, S., et al., *Magnetic nanoparticles: Applications in gene delivery and gene therapy.* Artificial cells, nanomedicine, and biotechnology, 2016. **44**(4): p. 1186–1193.
48. Chen, Y., et al., *Transdermal vascular endothelial growth factor delivery with surface engineered gold nanoparticles.* ACS applied materials & interfaces, 2017. **9**(6): p. 5173–5180.
49. Choi, S.W., et al., *Surface modification of functional nanoparticles for controlled drug delivery.* Journal of dispersion science and technology, 2003. **24**(3-4): p. 475–487.
50. Tang, C. and H. Li, *Application of nanoparticles in the early diagnosis and treatment of tumors: current status and progress.* Traditional medicine research, 2020. **5**(1): p. 34.
51. Park, S., et al., *The therapeutic effect of anti-HER2/neu antibody depends on both innate and adaptive immunity.* Cancer cell, 2010. **18**(2): p. 160–170.
52. Romond, E.H., et al., *Trastuzumab plus adjuvant chemotherapy for operable HER2-positive breast cancer.* The New England journal of medicine, 2005. **353**(16): p. 1673–1684.
53. Van Pelt, A.E., et al., *Neoadjuvant trastuzumab and docetaxel in patients with breast cancer: preliminary results.* Clinical breast cancer, 2003. **4**(5): p. 348–353.
54. Chauhan, V.P. and R.K. Jain, *Strategies for advancing cancer nanomedicine.* 2013. Nature materials, **12**(11): p. 958–962.
55. Barr, S., E. Hill, and A. Bayat, *Development, fabrication and evaluation of a novel biomimetic human breast tissue derived breast implant surface.* Acta biomaterialia, 2017. **49**: p. 260–271.
56. Chen, Q., et al., *Photothermal therapy with immune-adjuvant nanoparticles together with checkpoint blockade for effective cancer immunotherapy.* Nature communications, 2016. **7**(1): p. 13193.
57. Lollini, P.-L., et al., *Vaccines for tumour prevention.* Nature reviews. Cancer, 2006. **6**(3): p. 204–216.
58. Shi, X., et al., *Ca2+ regulates T-cell receptor activation by modulating the charge property of lipids.* Nature, 2013. **493**(7430): p. 111–115.
59. Vaeth, M., et al., *Store-operated Ca2+ entry controls clonal expansion of T cells through metabolic reprogramming.* Immunity, 2017. **47**(4): p. 664–679. e6.
60. Li, Y., et al., *A tumor acidity activatable and Ca 2+-assisted immuno-nanoagent enhances breast cancer therapy and suppresses cancer recurrence.* Chemical science, 2020. **11**(28): p. 7429–7437.
61. Carboni, E., et al., *Particle margination and its implications on intravenous anticancer drug delivery.* AAPS PharmSciTech, 2014. **15**: p. 762–771.
62. Sun, P., et al., *Chitosan-based nanoparticles for survivin targeted siRNA delivery in breast tumor therapy and preventing its metastasis.* International journal of nanomedicine, 2016: p. 4931–4945.

63. Bregoli, L., et al., *Nanomedicine applied to translational oncology: a future perspective on cancer treatment.* Nanomedicine, 2016. **12**(1): p. 81–103.
64. Kemp, J.A. and Y.J. Kwon, *Cancer nanotechnology: current status and perspectives.* Nano convergence, 2021. **8**(1): p. 34.
65. Rasool, M., et al., *New challenges in the use of nanomedicine in cancer therapy.* Bioengineered, 2022. **13**(1): p. 759–773.
66. Yang, F., et al., *The potential role of nanomedicine in the treatment of breast cancer to overcome the obstacles of current therapies.* Frontiers in pharmacology, 2023. **14**: p. 1143102.
67. Sartaj, A., et al., *Polymeric nanoparticles: exploring the current drug development and therapeutic insight of breast cancer treatment and recommendations.* Polymers, 2021. **13**(24): p. 4400.
68. Sorolla, A., et al., *Peptides, proteins and nanotechnology: a promising synergy for breast cancer targeting and treatment.* Expert opinion on drug delivery, 2020. **17**(11): p. 1597–1613.
69. Juan, A., et al., *Antibody conjugation of nanoparticles as therapeutics for breast cancer treatment.* International journal of molecular sciences, 2020. **21**(17): p. 6018.
70. Pangeni, R., et al., *Enhanced oral absorption of pemetrexed by ion-pairing complex formation with deoxycholic acid derivative and multiple nanoemulsion formulations: preparation, characterization, and in vivo oral bioavailability and anticancer effect.* International journal of nanomedicine, 2018. **13**: p. 3329–3351.
71. Jyoti, A., et al., *An in vitro assessment of liposomal topotecan simulating metronomic chemotherapy in combination with radiation in tumor-endothelial spheroids.* Scientific reports, 2015. **5**(1): p. 15236.
72. Kerbel, R. and J. Folkman, *Clinical translation of angiogenesis inhibitors.* Nature reviews. Cancer, 2002. **2**(10): p. 727–739.
73. Folkman, J.J.N.e.j.o.m., *Tumor angiogenesis: therapeutic implications.* The New England journal of medicine, 1971. **285**(21): p. 1182–1186.
74. Browder, T., et al., *Antiangiogenic scheduling of chemotherapy improves efficacy against experimental drug-resistant cancer.* Cancer research, 2000. **60**(7): p. 1878–1886.
75. Klement, G., et al., *Continuous low-dose therapy with vinblastine and VEGF receptor-2 antibody induces sustained tumor regression without overt toxicity.* The Journal of clinical investigation, 2000. **105**(8): p. R15–R24.
76. Maharjan, R., et al., *A novel oral metronomic chemotherapy provokes tumor specific immunity resulting in colon cancer eradication in combination with anti-PD-1 therapy.* Biomaterials, 2022. **281**: p. 121334.
77. Maharjan, R., et al., *Metronomic delivery of orally available pemetrexed-incorporated colloidal dispersions for boosting tumor-specific immunity.* Drug delivery, 2021. **28**(1): p. 2313–2328.
78. Choi, J.U., et al., *Modulating tumor immunity by metronomic dosing of oxaliplatin incorporated in multiple oral nanoemulsion.* Journal of controlled release, 2020. **322**: p. 13–30.
79. Jha, S.K., et al., *Enhanced antitumor efficacy of bile acid-lipid complex-anchored docetaxel nanoemulsion via oral metronomic scheduling.* Journal of controlled release, 2020. **328**: p. 368–394.
80. Pangeni, R., et al., *Improvements in the oral absorption and anticancer efficacy of an oxaliplatin-loaded solid formulation: Pharmacokinetic properties in rats and nonhuman primates and the effects of oral metronomic dosing on colorectal cancer.* International journal of nanomedicine, 2020: p. 7719–7743.
81. Cai, X.-J., et al., *Anti-angiogenic and anti-tumor effects of metronomic use of novel liposomal zoledronic acid depletes tumor-associated macrophages in triple negative breast cancer.* Oncotarget, 2017. **8**(48): p. 84248–84257.
82. Cai, X.-J., et al., *Combination of metronomic administration and target delivery strategies to improve the anti-angiogenic and anti-tumor effects of triptolide.* Drug delivery and translational research, 2020. **10**: p. 93–107.
83. Malik, P.S., V. Raina, and N. André, *Metronomics as maintenance treatment in oncology: time for chemo-switch.* Frontiers in oncology, 2014. **4**: p. 76.
84. Phi, L.T.H., et al., *Cancer stem cells (CSCs) in drug resistance and their therapeutic implications in cancer treatment.* Stem cells international, 2018. **2018**: 5416923.
85. Taguchi, T., et al., *Study of low-dose capecitabine monotherapy for metastatic breast cancer.* Chemotherapy, 2010. **56**(2): p. 166–170.
86. Fedele, P., et al., *Efficacy and safety of low-dose metronomic chemotherapy with capecitabine in heavily pretreated patients with metastatic breast cancer.* European journal of cancer, 2012. **48**(1): p. 24–29.
87. De Iuliis, F., et al., *On and off metronomic oral vinorelbine in elderly women with advanced breast cancer.* Tumori, 2015. **101**(1): p. 30–35.

88. Aurilio, G., et al., *Oral metronomic cyclophosphamide and methotrexate plus fulvestrant in advanced breast cancer patients: a mono-institutional case-cohort report.* The breast journal, 2012. **18**(5): p. 470–474.
89. Young, S., R. Lafrenie, and M.J. Clemons, *Phase II trial of a metronomic schedule of docetaxel and capecitabine with concurrent celecoxib in patients with prior anthracycline exposure for metastatic breast cancer.* Current oncology, 2012. **19**(2): p. 75–83.
90. Addeo, R., et al., *Low-dose metronomic oral administration of vinorelbine in the first-line treatment of elderly patients with metastatic breast cancer.* Clinical breast cancer, 2010. **10**(4): p. 301–306.
91. Kakolyris, S., et al., *Treatment of non-small-cell lung cancer with prolonged oral etoposide.* Journal of clinical oncology, 1998. **21**(5): p. 505–508.
92. Lebeau, B., et al., *Oral second-and third-line lomustine–etoposide–cyclophosphamide chemotherapy for small cell lung cancer.* Lung cancer, 2010. **67**(2): p. 188–193.
93. Wichmann, V., et al., *Recent clinical evidence on metronomic dosing in controlled clinical trials: a systematic literature review.* Acta oncologica, 2020. **59**(7): p. 775–785.
94. Scharovsky, O.G., L.E. Mainetti, and V.R. Rozados, *Metronomic chemotherapy: changing the paradigm that more is better.* Current oncology, 2009. **16**(2): p. 7–15.
95. Klein, R., et al., *Cost-effectiveness of pemetrexed as first-line maintenance therapy for advanced non-squamous non-small cell lung cancer.* Journal of thoracic oncology, 2010. **5**(8): p. 1263–1272.
96. Torre, L.A., et al., *Global cancer statistics, 2012.* CA: a cancer journal for clinicians, 2015. **65**(2): p. 87–108.
97. Bocci, G. and R.S. Kerbel, *Pharmacokinetics of metronomic chemotherapy: a neglected but crucial aspect.* Nature reviews. Clinical oncology. 2016. **13**(11): p. 659–673.
98. André, N., et al., *Has the time come for metronomics in low-income and middle-income countries?* The Lancet. Oncology, 2013. **14**(6): p. e239–e248.

Index

S

T

V

W

Z